Lecture Notes in Electrical Engineering 1132

The book series *Lecture Notes in Electrical Engineering* (LNEE) publishes the latest developments in Electrical Engineering—quickly, informally and in high quality. While original research reported in proceedings and monographs has traditionally formed the core of LNEE, we also encourage authors to submit books devoted to supporting student education and professional training in the various fields and applications areas of electrical engineering. The series cover classical and emerging topics concerning:

- Communication Engineering, Information Theory and Networks
- Electronics Engineering and Microelectronics
- Signal, Image and Speech Processing
- Wireless and Mobile Communication
- Circuits and Systems
- Energy Systems, Power Electronics and Electrical Machines
- Electro-optical Engineering
- Instrumentation Engineering
- Avionics Engineering
- Control Systems
- Internet-of-Things and Cybersecurity
- Biomedical Devices, MEMS and NEMS

For general information about this book series, comments or suggestions, please contact leontina.dicecco@springer.com.

To submit a proposal or request further information, please contact the Publishing Editor in your country:

China

Jasmine Dou, Editor (jasmine.dou@springer.com)

India, Japan, Rest of Asia

Swati Meherishi, Editorial Director (Swati.Meherishi@springer.com)

Southeast Asia, Australia, New Zealand

Ramesh Nath Premnath, Editor (ramesh.premnath@springernature.com)

USA, Canada

Michael Luby, Senior Editor (michael.luby@springer.com)

All other Countries

Leontina Di Cecco, Senior Editor (leontina.dicecco@springer.com)

**** This series is indexed by EI Compendex and Scopus databases. ****

Jason C. Hung · Neil Yen · Jia-Wei Chang
Editors

Frontier Computing on Industrial Applications Volume 2

Proceedings of Theory, Technologies and Applications (FC 2023)

Springer

Editors
Jason C. Hung
Department of Computer Science
and Information Engineering
National Taichung University of Science
and Technology
Taichung City, Taiwan

Neil Yen
School of Computer Science and Engineering
University of Aizu Tsuruga
Aizuwakamatsu, Japan

Jia-Wei Chang
Department of Computer Science
and Information Engineering
National Taichung University of Science
and Technology
Taichung City, Taiwan

ISSN 1876-1100 ISSN 1876-1119 (electronic)
Lecture Notes in Electrical Engineering
ISBN 978-981-99-9537-0 ISBN 978-981-99-9538-7 (eBook)
https://doi.org/10.1007/978-981-99-9538-7

This Springer imprint is published by the registered company Springer Nature Singapore Pte Ltd.
The registered company address is: 152 Beach Road, #21-01/04 Gateway East, Singapore 189721, Singapore

Paper in this product is recyclable.

Contents

Diagnosis and Optimization of Marketing Strategy Based on Association Rule Mining Algorithm

Ran Mei(✉)

Wuhan Business and Trade Vocational College Wuhan, Hubei 430205, China
651705492@qq.com

Abstract. In marketing, can not leave the *sup.* of modern information technology means. Through the Internet database, enterprises use big date mining (DM) technology to analyze the massive data collected, and provide scientific data *sup.* For marketing decision-making. DM will also help enterprises to achieve more accurate segmentation and positioning of customers, and accurately obtain the needs of potential customers. In this article, we will take the most classical Apriori algorithm of association rules as the research object, and the classical algorithm has also completed the parallelization implementation in the MapReduce framework. This article draws on the domestic and foreign marketing research theory, analyzes the inside and outside environment of Xiaomi smart mobile phone, starts from the marketing 4P theory, studies the current marketing strategy, and on this basis, formulates targeted optimization plan, and has been preliminarily confirmed. At the same time, in view of the current domestic market mobile phone product homogeneity phenomenon is serious, consumer demand diversification and other status quo. Combined with the Apriori algorithm under the association rules, this article deeply studies the influence of DM on the consumer behavior and personalized marketing of Xiaomi smart mobile phone customers, as well as the new marketing strategy.

Keywords: Association Rule Mining Algorithm · Apriori Algorithm · Mapreduce Framework · Marketing and Marketing · Diagnosis and Optimization

1 Introduction

The massive user data in the Internet platform contains a lot of useful information. If the rules and associations hidden in data sets can be analyzed and extracted through data technology, and applied to different industries, it will help users make development decisions that change the present and adapt to the future.

Association mining is widely used in various field, which can not only investigate the knowledge mode formed in the industry for a long time, but also find out new hidden rules. Scholar Chiang W Y [1] conducted an empirical study on the coffee house industry in Taiwan and proposed a new mining method of association rules to tap valuable markets for online CRM marketing strategies. Scholar Shokyar S [2] analyzed the reasons behind

J. C. Hung et al. (Eds.): FC 2023, LNEE 1132, pp. 1–8, 2024.
https://doi.org/10.1007/978-981-99-9538-7_1

the loss of customers in the mobile communication market through association rules, and on the basis of this article, draw up requiring marketing policy for each group of clients. Scholar Wang S C [3] conducted data analysis through association rules and determined the company's most valuable customers. Then, through the customer's past transaction records, the correlation relationship between the products purchased by the customer is found. It can be found that the research of domestic and foreign experts and scholars on marketing strategy has been very in-depth, involving various fields and having different characteristics.

Therefore, this article will analyze the marketing status of Xiaomi smart mobile phone through 4P marketing theory, and then optimize the marketing of Xiaomi smart mobile phone in view of the improved Apriori algorithm -- MR_Apriori algorithm.

2 Relevant Theoretical Concepts

2.1 Concept of Association Rule Mining Algorithm

Association rule mining algorithm is mainly used to find out the potential association relationship between items in data sets [4]. Association analysis is also called shopping basket analysis, and the information found is usually represented by association rules or frequent term lists. Association rule mining algorithms include: Apriori algorithm; Partition algorithm; FP-GA; DHP algorithm. This article mainly uses the classical Apriori Association rule mining algorithm in association rules.

The association rule mining step consists of two stages, as shown in Fig. 1:

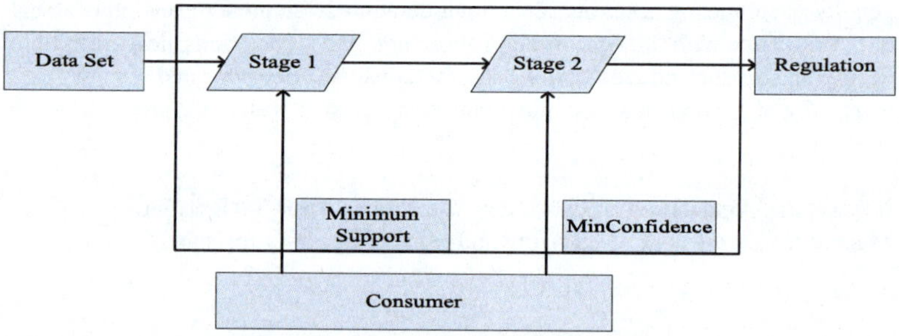

Fig. 1. Association rule mining model

Among them, *sup.* And *conf.* Are important concepts in association rules [5]:

If the $X \rightarrow Y$ association rule can be satisfied in the data set M, the formula of *sup.* is:

$$S = Support(E \rightarrow F) = P(E \cup F) \tag{1}$$

Similarly, the formula of *conf.* Can be expressed as:

$$C = Confidence(E \rightarrow F) = P(E|F) \tag{2}$$

2.2 Apriori Association Rule Mining Algorithm

Apriori algorithm is still the most widely used and classical frequent itemset mining algorithm, which uses the prior nature of relevant term lists to realize DM [6]. The Apriori algorithm uses k term lists to explore (k + 1) term lists. Firstly, Apriori algorithm is used to scan the data and find all frequent 1 term lists. Then, the frequent 1 term list is used to find all the frequent 2 term lists, the frequent 2 term list is used to find all the frequent 3 term lists, and so on.

Apriori algorithm association rules are defined as follows: Suppose a data set sum to $M, M = \{M_1, M_2, ..., M_n\}$, $M_p = \{J_1, J_2, .., J_n\}$. Where M_p is a transaction, J_n is also a transaction, and J is the list of all transaction items in the data set M. The association rule can be shown as:

$$Support(E \Rightarrow F) = \frac{count(E \cup F)}{|M|} \tag{3}$$

$$Confidence(E \Rightarrow F) = \frac{support(E \cup F)}{support(E)} \tag{4}$$

2.2.1 Defects of Apriori Algorithm

Apriori algorithm with the increase of data, the efficiency of algorithm processing will become low. The traditional algorithm has the following shortcomings [7]: ①Generate a large number of candidate term lists;.②Memory shortage due to huge amount of data.③A large number of redundant rules are generated.

2.3 Apriori Algorithm Under MapReduce Framework

MapReduce is a distributed file system that can access application data with high throughput [8, 9]. In order to improve Apriori algorithm, we introduce Mapper and Reducer functions in MapReduce framework into Apriori algorithm and name the improved algorithm after MR_Apriori [10]. The implementation process of MR_Apriori algorithm is introduced from two parts: "Generating frequent term lists" and "generating association rules". The specific process is shown in Fig. 2:

In general, MR_Apriori algorithm improves the defects of the traditional Apriori algorithm. Compared with the traditional Apriori algorithm, MR_Apriori algorithm has the following advantages: first, it makes up for the shortcomings of the traditional Apriori algorithm's single platform memory shortage and poor processing performance; Then, the Apriori algorithm needs to scan the data set for many times to improve the algorithm execution efficiency [11, 12].

2.4 4P Marketing Theory

The 4P theory is to analyze the marketing strategy of an enterprise from four aspects: product, price, place and promotion. These four factors are factors that enterprises can take measures to control, and are initiatives that enterprises can change according to market conditions and characteristics.

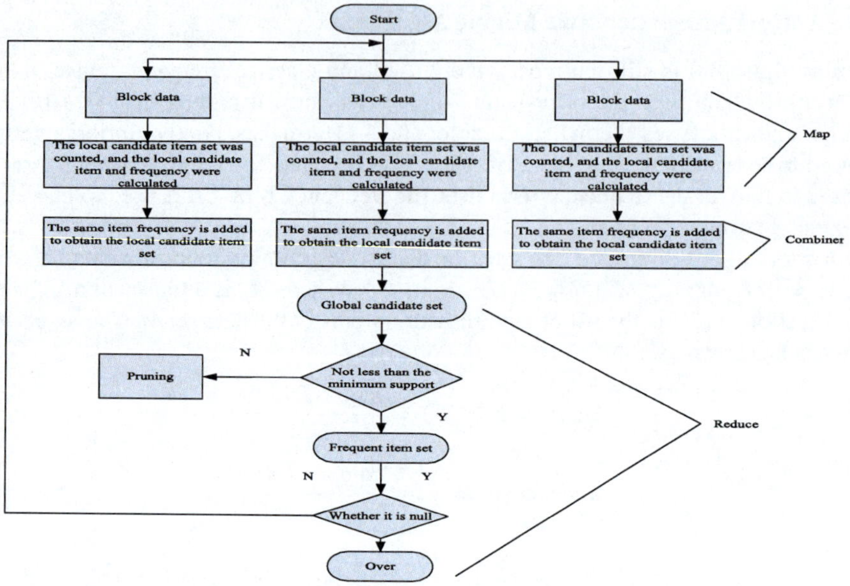

Fig. 2. Flowchart of the algorithm for producing part of frequent term lists

3 Analysis on the Marketing Status of Xiaomi's Smart Smart Mobile Phone

3.1 External Marketing Environment of Xiaomi's Smart Mobile Phone Market

As the world economy becomes more and more connected to the Internet, it also brings great opportunities for smart mobile phone marketers to locate overseas markets. Canalys released the global smart mobile phone market share data in 2022, as shown in Table 1:

Table 1. Global smart mobile phone market share data for 2022

Smart mobile phone shipments and growth in the word		
Canalys Preliminary Smart Mobile Phone Market Pulse:2022		
Brand	2021die Absatzbeteiligung	2022die Absatzbeteiligung
Samsung	20%	22%
Apple	17%	19%
Xiaomi	14%	13%
OPPO	11%	9%

(*continued*)

Table 1. *(continued)*

Smart mobile phone shipments and growth in the word		
Canalys Preliminary Smart Mobile Phone Market Pulse:2022		
vivo	10%	9%
others	28%	28%
Note:data are estimates		
Source: Canalys estimates (sell-in shipments),smart mobile phone Analysis, January 2023		

From the global smart mobile phone market share data in 2022 released by Canalys in Table 1, it can be seen that Xiaomi's global market share in 2022 reached 13%, down 1% compared with 2021, but ranking third in the world.

3.2 Internal Marketing Environment of Xiaomi's Smart Mobile Phone Market

1) In terms of product strategy: the concept of Xiaomi mobile phone was born for the sake of fever, and its customers are positioned as enthusiasts who grow up with the Internet. The strategy has bolstered Xiaomi's customer loyalty, but led to a low brand value for its smart smart mobile phone.
2) In terms of pricing strategy: Xiaomi focuses on cost performance and is cheaper than other smart mobile phone brands. This value strategy promotes the sales of Xiaomi smart mobile phone, but it makes the status of Xiaomi mobile phone brand in the eyes of consumers lower.
3) In terms of sales channels: Xiaomi sells its smart mobile phone directly through B2C network to avoid profit distribution with physical stores and distributors, avoid redundant expenses, and partially reduce the incidence of fake goods. However, this channel is too single.
4) In terms of promotion strategy, Xiaomi mobile phone uses event marketing, hunger marketing and social media marketing. These promotional strategies help Xiaomi attract many potential users. But Xiaomi's after-sales service has been repeatedly complained about by consumers.

4 Experimental Analysis

4.1 Optimize System Operation Test

The MR_Apriori algorithm is selected as the mining algorithm. The experimental platform is Hadoop cluster platform, and Ubuntu16.04.3 is used as the virtual machine system version. The min *sup.* Threshold min_sup = 15% and the mini *conf.* Threshold min_conf = 75% are set, and the data set is the purchase data of Mi 1–4 series, Redmi series, Redmi note and Mi note.

It can be known from the mining performance and operation efficiency comparison of MR-Apriori algorithm and Apriori algorithm in Table 2 and Fig. 3. With the increase of transaction volume, the difference between the efficiency and speed of MR-Apriori algorithm and Apriori algorithm in mining association rules will be larger.

Table 2. Compare the efficiency of MR-Apriori algorithm and Apriori algorithm

		MR_Apriori Algorithm	Apriori algorithm
Volume of Business	Min-Sup	Time (min)	Time (min)
10K	(0.5,0.2,0.15.0.15)	2.1	3.6
50k	(0.5,0.2,0.15.0.15)	7.7	12.5
100k	(0.5,0.2,0.15.0.15)	15.3	28.9
150K	(0.5,0.2,0.15.0.15)	22.5	34.7
200K	(0.5,0.2,0.15.0.15)	29.6	45.1

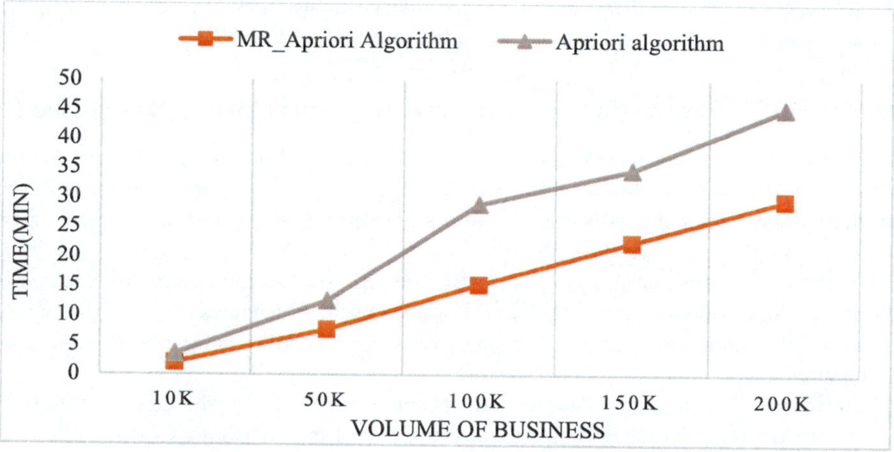

Fig. 3. MR-Apriori algorithm and Apriori algorithm mining performance analysis

4.2 Research on Marketing Optimization of Xiaomi Data on Account of MR_Apriori Association Rule Mining Algorithm

MR_Apriori algorithm is applied in the analysis of customer information of Xiaomi offline platform. By mining some useful data, such as basic personal information of users, product information, etc., it summarizes the commonalities and personalities of users who use Xiaomi smart smart mobile phone, so as to realize the precise marketing of Xiaomi smart mobile phone users.

4.2.1 Use Personalized Marketing for Xiaomi Smart Mobile Phone Users on Account of MR_Apriori Association Rule Mining Algorithm

Taking Xiaomi smart smart mobile phone and Xiaomi hardware products as the research objects, joint DM is carried out to achieve information sharing, which is beneficial to both parties and gives full play to the potential value of their stored customer data. Xiaomi Forum is equivalent to the communication between customers to find loopholes in the system, improve the superiority of the system, and drive the potential sales of Xiaomi smart mobile phone.

4.2.2 Send Customized Promotional Information

Because the needs of different people are different, different genders are also different, and they change over time. This makes enterprises need to consider more factors and requirements when designing products. Therefore, it is important to make comprehensive analysis with the help of DT to find out the main needs. At present, product production is seriously homogeneous, and in DT marketing, product marketing should be differentiated, which is an important course.

4.2.3 Cross Marketing Using MR_Apriori Association Rule Mining Algorithm

The steps of cross-marketing are divided into:i Model the consumption behavior of individuals. ii Scoring using predictive models. And on this basis, the comprehensive satisfaction of each individual is calculated. Iii is processed by the score matrix optimization. Among them, customer preference as the guidance, select a number of products and services, cross marketing. In cross-selling, one or more target markets are first determined according to customer needs, then different service combinations are selected according to the target market, and finally this series of combinations is output as the final result.

4.2.4 Evaluate Product Performance on Account of MR_Apriori Association Rule Mining Algorithm

Marketers can use DM tools to find the trend of sales, costs and profits of products, regions or markets to understand the effect of a promotion campaign or the reaction of a new product in the market, providing reliable information *sup*. For Xiaomi's R&D team to improve the decision-making of the next step.

5 Conclusion

DM is one of the rapidly developing fields in the computer industry. It is a data analysis and processing technology aiming at prediction and description. This article studies a very important technical association rule mining in DM, combining DM technology with Xiaomi smart mobile phone marketing to mine customer data. On account of the systematic review of DT, marketing and other related concepts, combined with the current situation of Xiaomi mobile phone, this article studies the relevant application of DT marketing in Xiaomi mobile phone, and mainly draws the following conclusions: (1) On the basis of MR_Apriori Association rule mining algorithm, the optimization research on millet data marketing is carried out, which can make marketing more accurate and grasp the specific needs of users more accurately. (2) MapReduce framework can complete large-scale parallel data processing. Therefore, it is an effective method to introduce association rule mining algorithm in Xiaomi Enterprise.

References

1. Chiang, W.Y.: Applying DM for online CRM marketing strategy: an empirical case of coffee shop industry in Taiwan. Brit. Food J. **120**(3), 665–675 (2018)
2. Shokyar, S., Saeidpour, P., Otarkhani, A.: Predicting customers' churn using DM technique and its effect on the development of marketing applications in value-added services in telecom industry. Int. J. Inf. Syst. Serv. Sect. **10**(4), 59–72 (2018)
3. Wang, S.C., Hsiung, W.S., Yan, K.Q., et al.: Use the relevance of product segmentation to enhance the competitiveness of enterprises. Int. J. Inf. Technol. **12**(2), 1–10 (2018)
4. Ariestya, W.W., Supriyatin, W., Astuti, I.: Marketing strategy for the determination of staple consumer products Using FP-growth and apriori algorithm. J Ilmiah Ekonomi Bisnis **24**(3), 225–235 (2019)
5. Berka, P.: Comprehensive concept description on account ofassociation rules: a meta-learning approach. Intell. Data Analysis. **22**(2), 325–344 (2018)
6. Martin D, Martinez-Ballesteros M, Garcia-Gil D, et al. MRQAR: a generic MapReduce framework to discover Quantitative Association Rules in DT problems. Knowledge-Based Systems, 2018, 153(AUG.1):176–192
7. Jabbour, S., Mazouri, F., Sais, L.: Mining negatives association rules using constraints. Procedia Comput. Sci. **127**(1), 481–488 (2018)
8. Menon, S., Ghoshal, A., Sarkar, S.: Modifying transactional databases to hide sensitive association rules. Inf. Syst. Res. **33**(1), 152–178 (2022)
9. Yang, H.J., Wu, Z.S., Cao, F., et al.: Rule of prescribing traditional Chinese medicine in the treatment of pneumoconiosis basedon association rules and k-means clustering algorithm. J. of Hainan Med Coll.: English edition **28**(10), 1–49 (2022)
10. Tong, C., Zhang, Y., Zhou, M., et al.: Online monitoring data processing method of transformer oil chromatogram on account of association rules. IEEJ Trans. Electr. Electron. Eng. **17**(3), 354–360 (2022)
11. Souza, C.R., Maia, V.A., Aguiar-Campos, N., et al.: Tree species consistent co-occurrence in seasonal tropical forests: an approach through association rules analysis. Forest. Syst. **30**(2), e006-e014 (2021)
12. Antomarioni, S., Ciarapica, F.E., Bevilacqua, M.: Association rules and social network analysis for supporting failure mode effects and criticality analysis: Framework development and insights from an onshore platform. Saf. Sci.. Sci. **150**, 105711 (2022). https://doi.org/10.1016/j.ssci.2022.105711

Short Video Creation Mode Based on Interactive Multi-objective Optimization Algorithm

Henan Xu(✉)

Harbin Guangsha College, Harbin, Heilongjiang, China
100365466@qq.com

Abstract. This paper proposes a short video creation mode based on an interactive multi-objective optimization algorithm. By adopting a multi-objective optimization algorithm, this mode can balance multiple objectives to obtain the optimal balance point. At the same time, it also uses interactive feedback, allowing users to choose and adjust optimization solutions to achieve a better user experience. Specifically, the mode includes steps such as determining optimization objectives, generating initial optimization solutions, user feedback, and repeating the optimization process. The advantage of this mode is that it fully utilizes users' personalized needs and creativity to achieve better creative results and user experience. Moreover, it can continuously improve the optimization solution through iterations to achieve the optimal solution.

Keywords: Interactive · Short Video Creation · Objective Optimization Algorithm

1 Introduction

Hort videos have quickly taken over the mobile screens of network users and established themselves as one of the most significant social platforms in the "Internet + era" as a new carrier of social culture [1]. Abstract With the advent of the era of short video, people can receive more information easy to be read and understood in a short time [2]. Short video creation resource optimization refers to the selection and optimization of multiple resources such as materials, audio, music, and editing during the short video production process to achieve the best balance between video quality and user experience. The interactive multi-objective optimization algorithm is an algorithm that can balance multiple objectives and continuously adjust the optimization solution through multiple interactive feedbacks to obtain better results. By combining the two, a short video creation resource optimization model based on interactive multi-objective optimization algorithm can be obtained.

Short video has become an important part of current life, which brings new thinking to filmandtelevision media talents [3]. As an emerging form of media, short video has become widely popular on social media platforms and has become an important way for people to quickly obtain information and entertainment. Unlike traditional videos, short videos are more innovative and diverse in their creation and content presentation,

J. C. Hung et al. (Eds.): FC 2023, LNEE 1132, pp. 9–14, 2024.
https://doi.org/10.1007/978-981-99-9538-7_2

covering various themes and types such as comedy, food, lifestyle, education, music, fashion, and more. Therefore, short video creation needs to be optimized from different perspectives to improve video quality and attract more users to watch and interact.

2 Problems Faced by Short Video Creation in the Era of Intelligent Media

Nterest in mobile short-video platforms as a new social network service tool has surged in recent years [4]. With growing popularity of mobile phones, mobile only short video sharing social media applications appeared on the market [5]. The challenges faced in short video creation include but are not limited to: a lack of high-quality material resources, requiring effective screening and editing; considering the needs and preferences of different platforms and audiences to gain better exposure and user feedback; creating high-quality and attractive video works within a limited time, requiring rapid editing and effects processing. To improve the quality and effectiveness of short videos, it is necessary to explore an effective resource optimization model and method.

3 Definition and Characteristics of Short Video

Currently, there is a limited amount of research data on the frequency of interactive and short video creation in the CNKI database in China. There are a total of five Chinese language papers from 2021 to 2023, mainly covering applied research, development research, and technical research, as shown in the figure below, As show in Table 1 and Table 2.

Table 1. Statistics of research levels

year	2021	2022	2023
number of publications	2	3	4

Table 2. Statistics of Published Articles

Research level distribution	Applied research	development research	Technical research
number of references	1	1	1

The main topics include: image building, short video services, marketing and promotion strategies, etc. Sub-topics include: communication channels, Chinese culture, etc. Research is conducted from the perspective of creative methods and content, As show in Fig. 1 and Fig. 2.

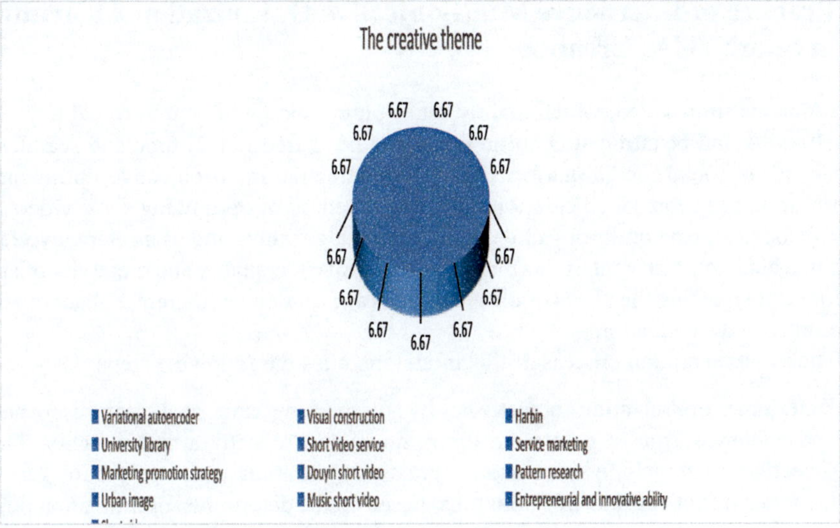

Fig. 1. Creative theme

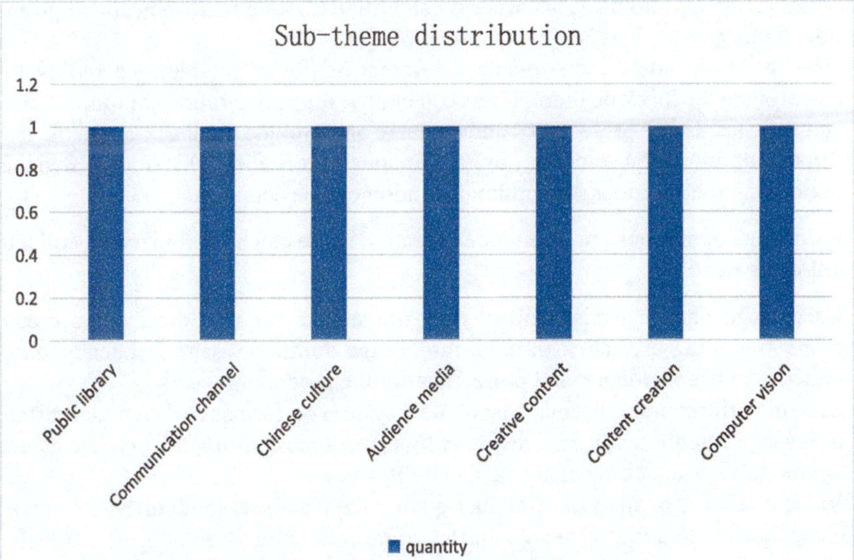

Fig. 2. Sub-theme distribution

4 Practice of Interactive Multi-objective Optimization Algorithm in Short Video Creation

Hort video is short video, which mainly relies on mobile intelligent terminal to realize fast shooting and beautification editing, and can be shared in real time and seamlessly connected on social media platform [6]. The interactive multi-objective optimization algorithm-based short video creation model is a method of optimizing short video creation resources using multi-objective optimization algorithms and user interactive feedback mechanisms. The goal of this model is to improve the quality and creativity of short videos while meeting the needs of different users and audiences, thereby enhancing user experience and satisfaction.

The implementation process of this model includes the following steps:

(1) Determine optimization objectives: by researching and analyzing user needs and audience groups, determine the optimization objectives and weights. These objectives may include video quality, creativity, duration, audience groups, etc.
(2) Generate initial optimization solutions: based on the determined optimization objectives and weights, generate initial optimization solutions. This can be achieved by using existing video materials and creative elements or by algorithm generation.
(3) Multi-objective optimization: optimize the initial solutions through multi-objective optimization algorithms. This process can utilize existing multi-objective optimization algorithms such as NSGA-II, MOEA/D, etc.
(4) User feedback: allow users to select different optimization solutions and evaluate and provide feedback on them. Users can choose the optimization solutions that best suit their needs and preferences and evaluate and provide feedback on them.
(5) Repeat optimization process: by continuously repeating the above steps and optimizing the solutions, the optimal solution can be achieved.

In concrete operations, the short video creation mode can be optimized by collecting the following data:

(1) User usage data analysis: Collect user usage data for this short video creation mode, including user click-through rate, usage duration, usage frequency, etc., to understand user feedback and preference for the mode.
(2) System performance data analysis: Collect system performance data under different user usage conditions, such as response time, resource consumption, etc., to evaluate the mode's system performance and stability.
(3) Video quality data analysis: Conduct quality data analysis for short videos created using the mode, such as video resolution, frame rate, color saturation, etc., to evaluate whether the created videos meet user expectations.
(4) Short video creation data analysis: Analyze the short videos created using the mode, such as video length, views, likes, comments, etc., to understand the videos' impact and influence among users.
(5) User satisfaction data analysis: Collect user satisfaction data for the mode through surveys or questionnaires, including user evaluations of the mode's preference, usability, efficiency, and overall feedback, to understand the user's overall evaluation and feedback for the mode.

5 The Influence of Short Videos on Teenagers

By June 2020, the number of short-form video APP users in China was 818 million, accounting for 87% of the total Internet users, that is, 87% of the online population used short-form video APP and among these users college students and other youth groups occupy a large proportion [7]. Taking the example of teenage users, their usage behavior and preferences in short video applications can be determined through factors such as their usage time and frequency, preferred video types and content, interactive behavior such as commenting, liking, sharing, as well as personal information and setting preferences such as profile pictures, usernames, and privacy settings.

For instance, statistical analysis can be conducted on the usage time and frequency of teenage users in short video applications, based on a sample size of 1000. The following conclusions can be drawn from the data analysis on teenage user behavior in short videos:

Average usage time: 45 min per day.
Longest usage time: 3 h per day.
Shortest usage time: 10 min per day.
Usage frequency: 60% use daily, 25% use weekly, and 15% use monthly.

Anging from a few seconds to a few minutes, the short video has become a popular form of learning and sharing creative skills such as drawing, photography, and crafting [8]. With the advent of the digital age, short videos have become one of the important ways for the public to obtain information. Based on the characteristics of high dissemination, wide content coverage, and concise content, this paper explores the impact of short videos on adolescents [9]. We can create related short video themes based on the statistical results, corresponding to the creation of short video content that young users prefer, based on their usage habits. In the era of self-media, the fast dissemination and strong interactivity of short videos should be fully utilized to create a good growth environment for adolescents and contribute to the healthy development of online platforms [10].

Overall, interactive multi-objective optimization algorithms for short video creation should consider multiple optimization objectives and allow users to select and provide feedback on optimization schemes. This can result in more personalized and user-oriented optimization schemes, enhancing user experience and satisfaction. The resource optimization model for short video creation based on interactive multi-objective optimization algorithms provides an effective resource optimization method for short video creators, making creation more efficient and targeted, and providing support and assurance for the sustainable development of the short video industry.

Acknowledgements. This work was Philosophy and Social Science Research Planning Project of Heilongjiang Province: Research on the Establishment of Ideological Morality and Values for Heilongjiang Youth under the New Communication Environment with Mobile Short Videos, Project Number: 22YSE451.

References

1. Li, F.: Dilemma and Solution of copyright justice system for health short video under network big data environment monitoring. J. Environ. Public Health **2022** (2022). DOI:https://doi.org/10.1155/2022/9640574
2. Liu, L., Yu, M., Feng, X.: Some characteristics and innovative development counter-measures of short weather video: a case study of the first award-winning excellent short weather video in hubei province meteorological and environmental research. Cranston **12**(3), 34–37 (2021). DOI:https://doi.org/10.19547/j.issn2152-3940.2021.03.010
3. Qiu, S., Liu, Y.: Research on the mode of short video project teaching and media talent training in colleges and universities under 5G industry chain based on multimedia technology. J. Phys. Conf. Ser. **1992**(2) (2021). DOI:https://doi.org/10.1088/1742-6596/1992/2/022057
4. Choi, Y., Wen, H., Chen, M., Yang, F.: Sustainable determinants influencing habit formation among mobile short-video platform users. Sustainability Basel **13**(6), 3216 (2021). DOI:https://doi.org/10.3390/su13063216
5. Aliaksandra, S.: Case Study of TikTok; Home social computing and social media. In: Participation, User Experience, Consumer Experience, and Applications of Social Computing Conference paper User-Generated Short Video Content in Social Media. A Case Study of TikTok ; Conference paper First Online (10 July 2020)
6. Wang, Y.: Content characteristics and limitations of original short video based on depth data. J. Phys.: Conf. Ser. **1881**, 042070 (2021)
7. Xu, X., Jia, F., Du, T.: Examining the factors influencing college students'continuance intention to use short-form video APP. Alternate title: Short-form video APP. Mou, Aslib J. Inform. Manag. Bradford **73**(6), 992–1013 (2021). DOI:https://doi.org/10.1108/AJIM-03-2021-0080
8. Zhou, Q.: Understanding User Behaviors of Creative Practice on Short Video Sharing Platforms –A Case Study of TikTok and Bilibili; search.proquest.com (2019)
9. Yao, T., Lin, Z.: Analysis on the impact of short videos on adolescents' physical exercise. Journal of Sports Goods and Technology (15 January 2023)
10. Liang, J., Li, F., Hu, C.: Seeing long-term benefits in the short-term" - the impact of short video development on adolescents' Physical and Mental Health in the Digital Economy Era. Talent and Intelligence (19 October 2022)

Research on Earthquake Detection Based on Machine Learning

Dianguang Gai[1], Tingmei Tang[1](✉), and Hui Li[2]

[1] Seismological Bureau of Shandong Province, Jinan 250014, China
770847540@qq.com

[2] Shandong Institute of Commerce and Technology, Jinan 250103, China

Abstract. Accurate and efficient detection of each earthquake is an important foundation for earthquake work. However, at present, due to the lack of observation data information, the accuracy of parameter estimation is low, especially for seismic detection of major projects, the current situation of sparse station layout makes it impossible to use multiple averages like dense network seismic detection to reduce the discreteness of parameter estimation, and the accuracy of seismic detection parameters based on a single station needs to be improved. In recent years, there has been an increasing focus on how machine learning can be used to improve seismic detection performance. In this paper, the principle of AI seismic detection and waveform mask matching is analyzed, and then a new deep learning method, TransQuake, is proposed based on the frontier sequence model Transformer for seismic wave detection. TransQuake combines STA/LTA algorithms for feature enhancement of seismic waveform data and interpretable model learning using a multi-head attention mechanism. To validate the performance of the model, this paper conducts an extensive evaluation on the 2008 Wenchuan MW7.9 earthquake aftershock dataset. The results show that TransQuake can achieve the best detection performance beyond the baseline of leading-edge algorithms.

Keywords: Machine Learning · Seismic Detection · Deep Learning · Event Identification

1 Introduction

Earthquakes are devastating wherever and whenever they are. As one of the countries prone to earthquakes, China has experienced great threats to the safety of people's lives and property caused by earthquakes [1]. Accurate determination of earthquake location, intensity and parameters is an important basic work of earthquake science research, and an important prerequisite for achieving future earthquake forecasting, disaster prevention and mitigation [2]. As the volume of seismic data has grown exponentially over the past few decades, it is becoming increasingly important to develop efficient and highly generalizable automatic identification technologies. As detection techniques have improved, more and more signals previously considered as noise have been re-determined as micro-seismic or ground tremor, a change that, in addition to reducing the completeness of the

J. C. Hung et al. (Eds.): FC 2023, LNEE 1132, pp. 15–22, 2024.
https://doi.org/10.1007/978-981-99-9538-7_3

catalogue, has revealed that the detection rate of these tiny earthquakes is highly consistent with the fault slip rate and reflects fault activity at a much higher resolution than GPS. This is undoubtedly the best solution to the problem of seismic detection with a large amount of available marker data [3]. Therefore, based on machine learning methods, the analysis and processing of the waveform data of the Chinese mainland near station before the strong earthquake in recent years to obtain the missed micro-earthquakes not only has a good supplementary effect on the earthquake catalog, but also has important indicative significance for studying the change of earthquake frequency or the anomaly extraction before the large earthquake.

2 Seismic Detection

2.1 AI Seismic Detection Principle

When dealing with highly complex and non-linear separable seismic datasets, neural network-based methods, especially deep learning, show better solving ability for complex problems and are most widely used in seismic detection research. From the original data, each layer of neural network can extract different levels of features [4].

The model of this paper is based on the attention mechanism. The attention mechanism is to imitate the visual selective attention of the human brain, the most typical example is the human eye watching the web page, the human eye will selectively observe certain areas in the web page [5].

The component of long-head attention is the zoom dot product attention mechanism. Multiple zoom dot attention mechanisms calculate input features multiple times at a time, and each zoom dot attention mechanism is calculated in the samse way. A diagram of the structure of the scaling dot product attention mechanism is shown in Fig. 1.

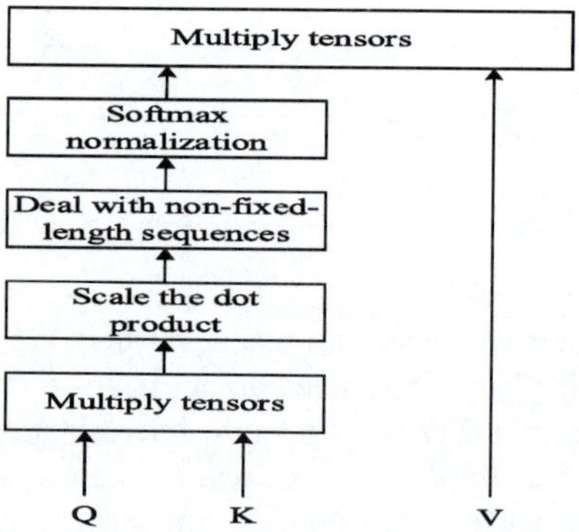

Fig. 1. Zoom dot product attention mechanism structure

Among them, K, Q, and V represent key, query, and value, respectively, and all three are matrices. The constituent vectors corresponding to K and V are a pair of key-value pairs, which are one-to-one correspondence [6]. The scaling dot product attention mechanism first performs a dot product calculation on Q and K, and adds a scaling layer(softmax) after the calculation result to avoid model underfitting caused by excessive Q and K values [7, 8]. The calculation formula for the zoom dot product attention mechanism is shown in Eq. (1).

$$Attention(Q, K, V) = softmax(\frac{QK^T}{\sqrt{d_k}})V \tag{1}$$

Figure 2 shows the structure of the bull's attention. The multi-head attention mechanism maps the dimensions of K, Q and V through a variety of different mappers, and then repeats the h attention operation, and obtains the output of the final multi-head attention mechanism after a linear change [9]. The specific calculation formula is shown in Eq. (2).

$$\begin{cases} MultiHead(Q, K, V) = Concat(head_1, \ldots, head_h)W^o \\ head_i = Attention(QW_i^Q, KW_i^k, VW_i^V) \end{cases} \tag{2}$$

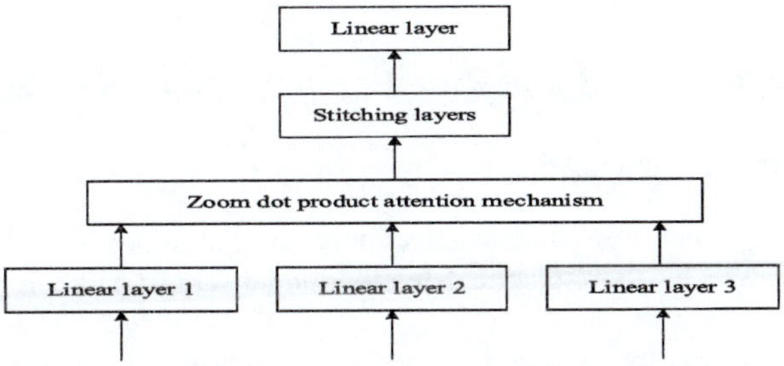

Fig. 2. Structure of the bullish attention mechanism

2.2 Waveform Mask Matching Principle

The underlying logic of the waveform mask matching technique is that the adjacent seismic waveforms recorded by the same station are similar. By superimposing the cross-correlation of template waveforms and continuous waveforms, missing events around the template can be identified [10]. The basic principle of the waveform mask matching method is as follows:

Suppose $\omega_{N,\Delta t}(t_0)$ is a continuous sampling of the N component of the nonzero time series $\omega(t)$:

$$\Omega_{N,\Delta t}(t_0) = [\omega(t_0), \omega(t_0 + \Delta t), \ldots, \omega(t_0 + (N - 1))\Delta t]^T \tag{3}$$

where t_0 represents the time of the first sampling point and Δt represents the sampling interval. Make the inner product of the two time series of the above equation $\omega_{N,\Delta t}(t_w)$ and $V_{N,\Delta t}(t_v)$:

$$[\vartheta(t_v), \omega(t_W)]_{N,\Delta t} = \left[\vartheta(t_v)_{N,\Delta t}, \omega(t_w)_{N,\Delta t}\right] = \sum\nolimits_{i=0}^{N-1} \vartheta(t_v + i\Delta t)\omega(t_W + i\Delta t) \tag{4}$$

The normalized correlation number can be expressed as:

$$C[\vartheta(t_v), \omega(t_W)]_{N,\Delta t} = \frac{[\vartheta(t_v), \omega(t_W)]_{N,\Delta t}}{\sqrt{[\vartheta(t_v), \vartheta(t_v)]_{N,\Delta t}[\omega(t_W), \omega(t_W)]_{N,\Delta t}}} \tag{5}$$

In Eq. (5), the value range of C is [−1,1]. The higher the value, the higher the degree of similarity between the two signals [11]. The number of correlations between the template and the continuous signal is expressed as $Cw(t)_{N,\Delta t}$, where $\omega(t)$ represents the seismic signal. Let the signal appear from the t_M moment:

$$C_\omega(t)_{N,\Delta t} = C[\omega(t), \omega(t_M)]_{N,\Delta t} \tag{6}$$

The basic workflow of the waveform template matching method includes: (1) preparing continuous waveforms, including preprocessing, shift start time, and bandpass filtering. (2) Use the seismic catalog to cut the template waveform and calculate the signal-to-noise ratio. (3) The continuous waveform and the catalog are cross-correlated to obtain the number of mutual relations. (4) Superimpose the number of correlations and calculate the average (CC) to measure the similarity of detected new events and templates.

3 Seismic Detection Model Based on Attention Mechanism

3.1 Model Network Structure

As shown in Fig. 3, the most advanced sequential model in the NLP domain is Transformer, based on which a new seismic detection model, TransQuake, is proposed in this paper.

Since seismic waves are continuous waveforms, in order to accommodate continuous waveform data, this paper proposes a new method to obtain the input of the encoder. The classification of seismic waves is based on whether the waveform contains seismic P-waves or not, and the results are obtained by softmax and linear layers based on the modified Transformer encoder. The shape of the reshaped waveform is shown in Fig. 3, after the STA/LTA process. The relationship between each position of the vectors is analysed using a fully connected layer before the PC is added to obtain the input to the encoder.

When reshaping each waveform, two factors need to be analysed. These two factors are: the size of the attentional model and the combination of the 3 components at each position. Therefore, when reshaping, each waveform is reshaped from 3 × 5000 to $k \times 3T$, where $k = (3 \times 5000)/3T$. Finally, they are fed in parallel to the PC layer and the FC layer. And in the FC layer, each neuronal cell is connected into a new new cell by Eq. (7).

$$y_m = \sum\nolimits_{n=1}^{3T} W_n x_n \tag{7}$$

In Eq. (7), y_m is value of the new cell, where the range of values of m is [0, d], d represents the dimension of the hidden cell; w_n is the weight to be trained; x_n is the value of the original cell.

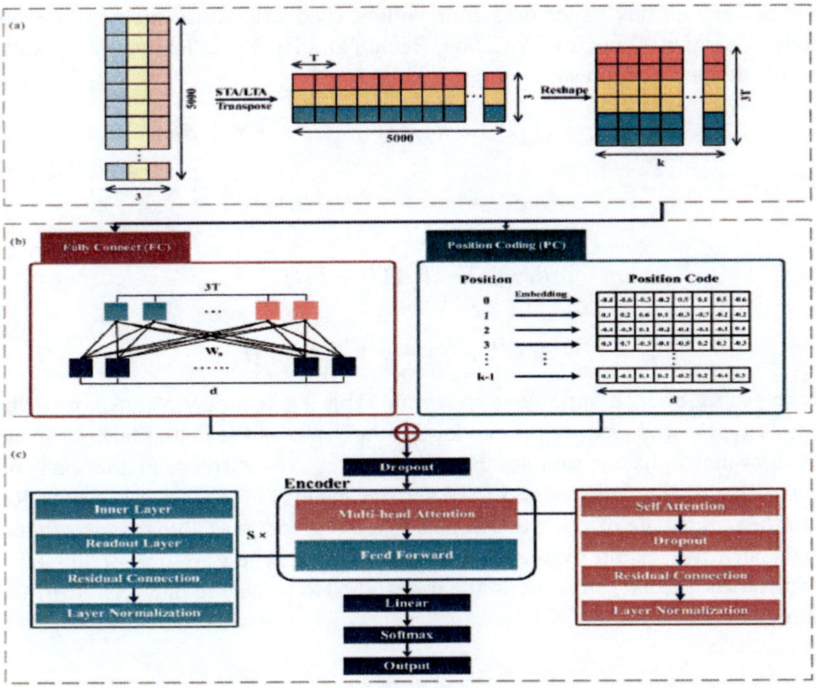

Fig. 3. TransQuake model architecture. (a) Data pre-processing; (b) Positional coding layers and fully connected; (c) Completed architecture of the model

In PC level, use the Xavier initializer to embed each position number of the reconstructed data into a vector whose length is also equal to the hidden unit. Since FC and PC are the same shape, we can add the elements corresponding to the two matrices together and take the new matrix as our input. Before entering the encoder, a dropout layer is used to reduce overfitting.

3.2 Introduce Attention Mechanisms

Attention-based transformers have lower total computational complexity when connecting more neurons, better model global features, and compute runs in parallel, with longer path lengths between long-range dependencies that can be learned from the network. The implementation principle of the attention mechanism is analyzed above.

4 Analysis of Experimental Results

The dataset used in this paper is from the rematch data of the "Aftershock Detection Competition" jointly organized by Alibaba Cloud and the Institute of Geophysics of China Earthquake Administration. This includes aftershock monitoring signals sampled at 100 Hz from July 1 to 31 recorded by 14 permanent seismic stations after the 2008 Wenchuan MW7.9 earthquake.

In experiments, this paper uses four widely used evaluation metrics to evaluate our method, namely Accuracy, Precision, Recall, and F1. Specifically, each evaluation indicator is defined as follows:

$$Accuracy = (TP + TN)/(TP + TN + FN + FP) \tag{8}$$

$$Precision = TP/(TP + FP) \tag{9}$$

$$Recall = TP/(TP + FN) \tag{10}$$

$$F1 = 2/(1/Precson + 1/Recall) \tag{11}$$

The meaning of each variable is shown in Table 1. Accuracy means how many of the sample types predicted by the model are the same as the actual labels. Accuracy refers to how many positive samples the model judges to be correct and how many of the waveforms that the model considers to be seismic events are true seismic events. Recall refers to the percentage of positive samples that the model was able to correctly detect and how many true seismic events it was able to retrieve. The F1-score is obtained from the Precision and Recall calculations and is designed to be able to find seismic instances accurately.

Table 1. The formula defined of TN, TP, FN, and FP

Output	Label	
	Positive	Negative
Positive	TP: True Positive	FP: False Positive
Negativ	FN: False Negative	TN: True Negative

In this paper, experiments are designed to verify the validity of the model in this paper, comparing it with traditional machine learning models, including logistic regression, support vector machines and random forests. These traditional machine learning models are widely used in earthquake detection.

The results of the experiment are shown in Fig. 4. From the experimental data, the following conclusions can be drawn in this paper. Comparing the experimental data, it can be concluded that TransQuake, the model in this paper, outperforms other models in terms of evaluation metrics. 10.4% higher than ConvNetquake and 5.2% higher than MSDNN.

Compared with other deep learning models in related studies, TransQuake can effectively model the dynamics and global information of seismic waves, and has a high interpretability for the results. In this paper, comparative experiments are conducted using aftershock data. Since machine learning is a data-driven approach, its performance depends heavily on its training dataset, so the better the model works when the training dataset is more relevant to the test dataset.

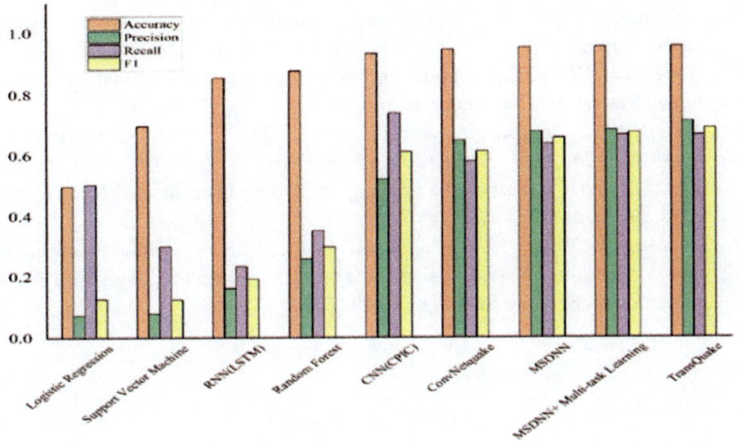

Fig. 4. Comparison of experimental results and effects of different methods

5 Conclusion

Earthquakes are natural disasters with destructive energy, usually the harm caused by strong ground motion is immeasurable, and seismic detection can provide emergency response time of seconds to tens of seconds before destructive energy reaches the warning area, greatly reducing casualties and economic losses. This paper first propose TransQuake. Based on the attention mechanism and combined with traditional STA/LTA algorithms, this network can effectively extract the dynamic and global features of seismic wave sequences. In order to verify the effect of the model, extensive experiments were carried out. Compared with a variety of cutting-edge seismic detection methods, the experimental results clearly verify the effectiveness of TransQuak, prove the modeling ability of Transformer structure on multi-dimensional seismic waveforms, and provide new method ideas for improving seismic detection performance.

References

1. Zhao, M., Chen, S.H.: (2021) Study on generalizability of deep learning-based seismic detection models in regional station networks. Earthquake **41**(01), 166–179 (2021)
2. Tang, J., Meng, T., Han, S.Y., et al.: A fault detection method for seismic data based on multi-resolution U-Net network. Petrol. Geophys. Explor. **56**(03), 436–445+409 (2021)
3. Yang, W.Y., Yang, J.R., Chen S.Q., et al.: Fault detection of seismic data based on U-Net deep learning network. Petrol. Geophys. Explor. **56**(04), 688–697+669 (2021)
4. Wu, L.H., Zhang, B.J., Lin, B.H., et al.: (2021) Application of BP neural network in seismic noise waveform detection. South China Earthq. **41**(03), 116–121 (2021)
5. Wang, Z.J., Wu, X.M., Du, Y.S., et al.: Deep learning-based seismic fault detection and fault combination. Oil Gas Geology Recov. **29**(01), 69–79 (2022)
6. Gong, J., Zhang, P., Zhang, Y., et al.: Azimuth detection of downhole seismometers in the seismic network of Jiangsu Province. Seismic Defense Technol. **17**(01), 181–187 (2022)

7. Qi, G.L., Yuan, C., Yu, M., et al.: Optimization of local seismic network layout based on seismic monitoring capability. J. Seismolog. **44**(03), 476–488 (2022)
8. Liu, L., Wang, H.F., Wang, Z.F.: Application of seismic detection methods based on long and short time window ratios. Electroacoustic Technol. **46**(09), 93–96 (2022)
9. Zhang, W.H., Yin, L.: Optimization algorithm for seismic detection of missed alarms based on CNN and BiLSTM. Intell. Comput. Appl. **13**(02), 169–173 (2023)
10. Xie, J., Cheng, X.Q.: A review on the application of deep learning in seismic localization. J. Earthq. Eng. **45**(01), 235–243 (2023)
11. Jiang, C., Zhang, P., White, M.C., Pickle, R., Miller, M.S.: A detailed earthquake catalog for Banda arc–Australian plate collision zone using machine-learning phase picker and an automated workflow. Seismic Rec. **2**(1), 1–10 (2022)

Research on Quantitative Trading Based on Deep Learning

Zhengyan Wang[1(✉)] and Yisong Zhao[2]

[1] Paris-Saclay University, 92330 Sceaux, France
zhengyan.wang@universite-paris-saclay.fr
[2] Sorbonne University, 75005 Paris, France

Abstract. Traditional quantitative trading strategies are widely used in stocks, futures and other financial markets, but the manual extraction method of features makes it lack the ability to effectively adjust strategies dynamically, and deep reinforcement learning can effectively simulate complex market environments and solve dynamic quantitative trading problems. Based on the development status of the financial industry, this paper introduces the deep reinforcement learning algorithm into the field of stock trading to build an intelligent trading model. Its goal is to discover the laws of the market in the learning of massive data, so as to carry out effective transactions, effectively avoid market risks and improve investors' returns. On the basis of the traditional DQN algorithm, corresponding to the actual requirements, we propose RB_DRL deep reinforcement learning algorithm model to improve the network structure. The experimental analysis results show that the improved model also shows good results in multi-group comparative experiments.

Keywords: Deep Learning · Neural Networks · Quantitative Trading · Data Mining

1 Introduction

The dramatic changes in financial markets in recent decades have been largely reflected in the gradual transformation of a market structure dominated by manual trading to one dominated by automated computerised trading. Developments in order generation, transmission and execution technology have facilitated this rapid transformation, greatly increasing the speed, capacity and complexity of the trading functions available to market participants. Trading markets facilitate capital formation and allocation by setting the price of securities and allowing investors to enter and exit positions in securities whenever and wherever they wish [1]. The great changes in the financial market in recent decades are mainly reflected in the gradual transformation of the market structure dominated by manual transactions into a market structure dominated by automated computer transactions [2]. The development of technology to generate, transmit, and execute commands has driven this rapid shift, greatly increasing speed, capacity, and complexity of trading features available to market participants. Traditional strategies typically use manual

J. C. Hung et al. (Eds.): FC 2023, LNEE 1132, pp. 23–30, 2024.
https://doi.org/10.1007/978-981-99-9538-7_4

extraction of financial features to remove data noise and uncertainty [3]. However, traditional machine learning mainly takes prediction as the direction of supervised learning, and cannot achieve direct transaction decisions or direct transaction costs. Therefore, this paper combines the current research status to enter the field of financial quantitative trading from the perspective of deep reinforcement learning [4]. While providing a new model for the field of financial quantitative analysis, it also expands another application scenario of deep reinforcement learning in the financial field, which has important academic research and practical application value.

2 Deep Reinforcement Learning

Deep reinforcement learning (DRL) is a product of the convergence of deep learning as well as reinforcement learning. Deep learning takes raw data such as images and feeds them directly into a multi-layer deep neural network, which gradually extracts higher-level features through non-linear transform learning and has a strong perceptual capability [5]. The latter has a bias towards learning problem-solving decision making capabilities, where the intelligence and the environment are constantly interacting and trial and error, using reward and punishment mechanisms to uncover optimal control strategies. The two complement each other's strengths, providing a general new way of thinking about solving complex problems, as shown in Fig. 1. In essence, deep reinforcement learning still follows the algorithmic thinking of traditional reinforcement learning [6]. The value-based approach approximates the state action value function through a neural network such as a DNN, which provides guidance for the selection of actions by the intelligence. Representative algorithms for such deep reinforcement learning methods include deep Q-networks [7]. Representative algorithms for this deep reinforcement learning approach include the Action Critic algorithm [8].

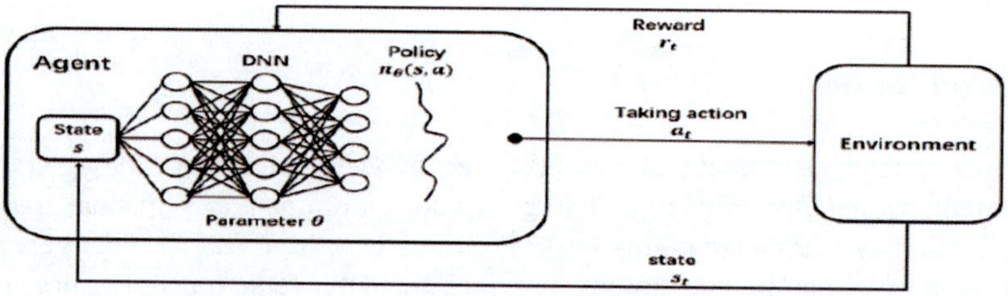

Fig. 1. Deep reinforcement learning process

Model-based deep reinforcement learning algorithms require that an intelligence be given a model of the environment or be asked to learn a model of the environment in order to perform a task in a particular environment. The most significant advantage of model-based deep reinforcement learning algorithms is that planning is considered in advance, where the intelligence tries out the available actions in advance of each step and then selects from the candidates, providing higher sampling efficiency compared to model-free deep reinforcement learning algorithms [9]. However, when the real environment is more complex, there is a gap between the model explored by the intelligences and the actual valid model, which can lead to poor performance of the intelligences in the real environment [10]. In contrast, although the model-free deep reinforcement learning algorithm gives up the potential gain in sample efficiency, it is easier to achieve the learning of the intelligent body to the optimal policy/optimal value function and to adjust the learning process of the intelligent body after a lot of interaction between the intelligent body and the real environment, with progressively better performance in a trial-and-error manner. The DQN algorithm's structure is illustrated in Fig. 2.

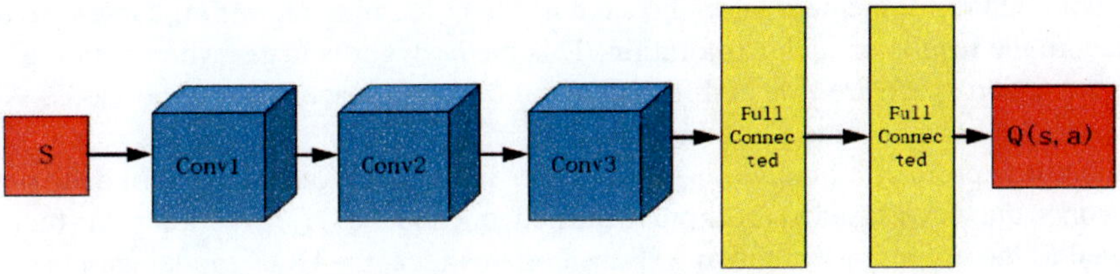

Fig. 2. Schematic diagram of the network structure of the DQN algorithm

Model-based DRL requires the construction of a model of the environment, relies on control theory and is often interpreted in terms of different disciplines. Model-free deep reinforcement learning learns policies based on information from direct interactions with an unknown environment, and so does not require knowledge of state transfer functions. Value function-based DRL algorithms learn an approximation of the optimal value function and indirectly learn deterministic policies [11]. The policy-based deep reinforcement learning algorithm learns an approximation of the optimal policy directly, outputting the probability of the next action and selecting the action based on the probability, but not the one with the highest probability, which is applicable to both continuous and discontinuous actions.

Model-based deep reinforcement learning algorithms require the construction of a model and the generation of a training set from the constructed model, so the complexity of the samples is not as high as model-free deep reinforcement learning algorithms, but the process of constructing the model is somewhat challenging. The classical algorithms for model-free deep reinforcement learning based on value functions are: the Deep Q-Network algorithm, and the Double Deep Q-Network algorithm. The Deep Q-Network algorithm is an offline algorithm based on value functions that combines the advantages of DNN and traditional Q-learning, differing from traditional Q-learning in that Deep Q-Networks use deep neural networks to fit Q-values and use an experience pool to store data obtained from interactions with the environment, whose empirical data can be reused.

3 Research on the Design of DRL Algorithm Based on Improvement

This study aims to improve the traditional deep reinforcement learning algorithm model and construct a new algorithm model. It learns the trading rules in the financial market through the model, so as to train an agent that can trade, and further build the corresponding quantitative trading system according to the trained algorithm model to carry out automatic quantitative trading.

3.1 Deep Reinforcement Learning Model Construction Based on Value Functions

In traditional reinforcement learning, such as the Q-learning algorithm, tables are used to record the results of each exploration. This method works better when both the state and the action space are low in dimension and the state space is not continuous. When states and actions become more dimensional, it is impossible to continue using tables, so the fitting problem of converting such tables into a function is raised. In deep neural networks, the state is used as the input to the network and the Q (s, a) action value function is used as the output of the network. Based on this idea, the DQN model adopts a two-network structure, that is, one network is used to calculate the current value function, and the other network is used to generate the target value. And use the current value, and the target value to build a loss function to optimize the deep neural network.

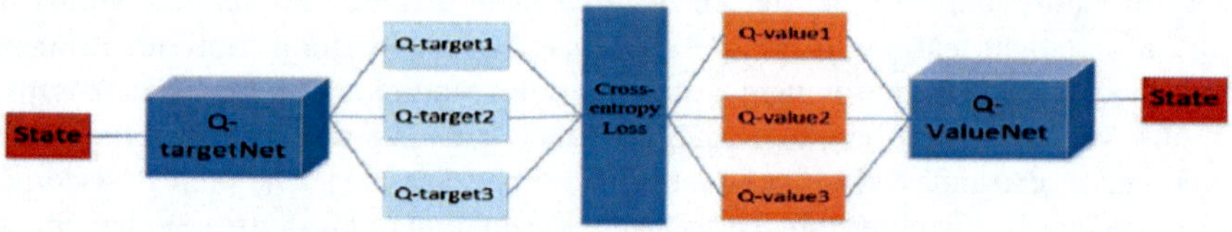

Fig. 3. RB-DRL model block diagram

Figure 3 is a structure diagram based on the improved DRL model, which consists of two parts in general. In the left half, a Q-targetNet is constructed to calculate the Q value of the target, and the input of the network uses the results of reinforcement learning exploration. The right half uses the same structure of Q-ValueNet to usher in the evaluation of the calculated action value function, and the last layer of the network uses the Softmax layer to input a 1 × 3-dimensional action value function vector, and build a cross-entropy loss function for network updates.

Figure 4 shows the overall flow chart of the RB-DRL algorithm model. In the model, on the one hand, the Q-ValueNet, is used to interact with the environment, carry out the exploration process of reinforcement learning, and store the results of each experiment in the memory playback unit. The target value network, that is, Q-TargetNet, generates the target Q value, which is used as the target of learning to evaluate the current value function. After the introduction of the target value network, the model can ensure that the target Q value is constant for a period of time during training. On the other hand, due to the existence of the memory playback unit, the exploration results of the experiment are persistently stored, and the data in the unit can be taken out for further training and learning every once in a while, which is convenient for further updating the network.

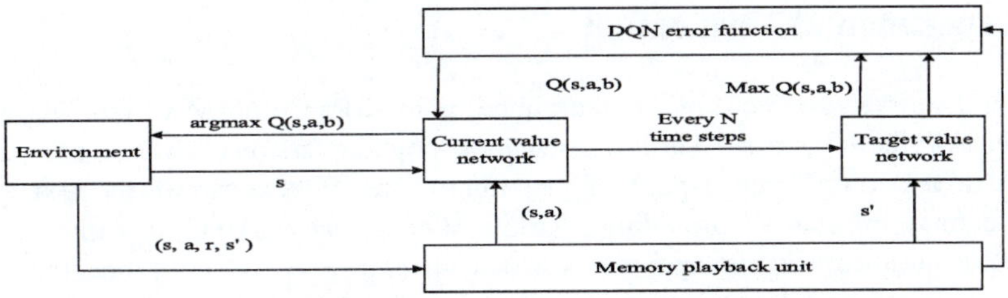

Fig. 4. RB-DRL algorithm flowchart

3.2 Design of Convolutional Neural Network with Residual Blocks

An improved convolutional neural network is introduced in the experimental study of this paper, and the structure diagram is shown in Fig. 5. The entire network is divided into 10 layers, including one input and output layer, three convolutional layers, two pooling layers, and three fully connected layers. The input layer inputs 20*20-dimensional sample data, and after a series of convolution calculations, outputs 3-dimensional result data in Softmax, which can also be considered to divide the samples into 3 categories. In the convolutional layer, the first layer uses a 3*3*64 convolution kernel for the input data for convolution, and the "SAME" method is used to fill the data edges during the convolution process to ensure that the overall dimension size remains unchanged during the convolution process. The sampling window of the pooling layer is 2*2, which adopts the maximum pooling method and fills the data like the convolutional layer. The fully connected layer accepts the output of the final pooling layer and calculates, uses the dropout method to randomly discard some parameters on the calculation results, and introduces the idea of residual in the fully connected layer, and the results of the first layer of the fully connected layer are transmitted to the third layer in addition to the second layer, so as to amplify the signal and enhance the fitting effect of the network model.

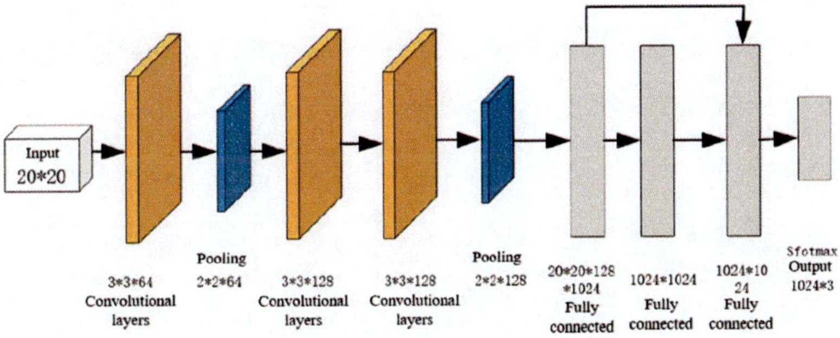

Fig. 5. RB-DRL network block diagram

4 Experimental Verification

The hardware environment of the experiment is Intel(R) Core(TM)i7-10210U, implemented in Python. Python 3.7 is the current more popular version of Python, compared to the previous Python 2 in all aspects of great improvement, Tensorflow is the next popular deep learning framework, providing a variety of neural network building interfaces.

The experiments in this paper were divided into two groups to verify the effectiveness of the improved model in this paper. Mainly select two stocks of Ping An and Shandong Gold for in-depth research and analysis. In terms of data, we select the data of the decade from 2010–01-01 to 2020–12-31, and divide it into training set and test set, and the training set selects the data of the eight years from 2010–01-01 to 2018–12-31, and the remaining two years are used as test set data to verify the learning results of the model.

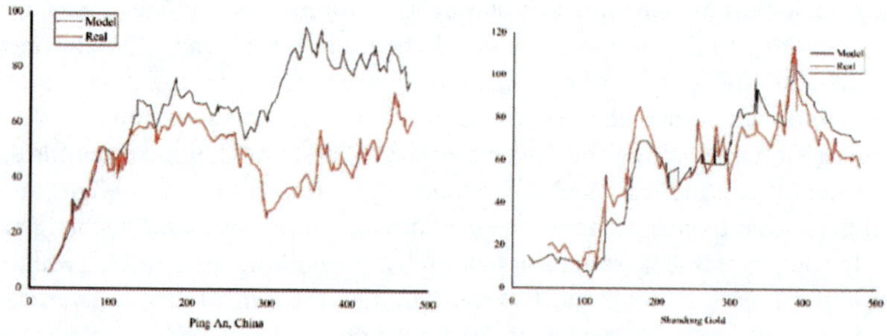

(a) (b)

Fig. 6. Model yield curve based on fully connected neural network

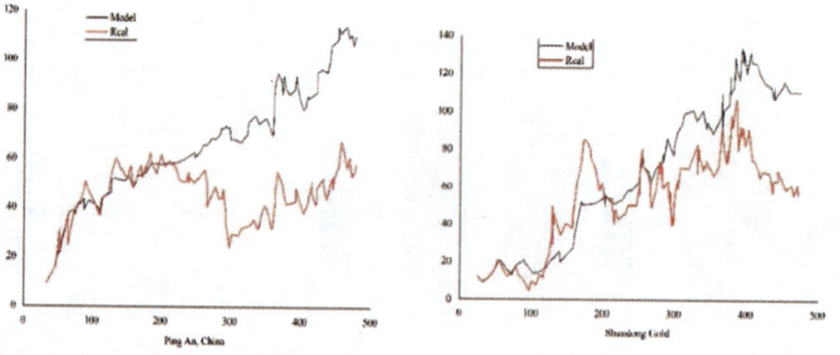

(a) (b)

Fig. 7. RB_DRL Model yield curve

Figure 6 is the model yield curve based on the fully connected neural network, and Fig. 7 is the experimental results of the improved RB_DRL deep reinforcement learning algorithm. Looking at the overall yield curve, the improved model performed significantly on both stocks compared to the no improvement experiment. The final returns were able to reach the level of the leading benchmark, the drawdown of individual stocks was reduced, and the risk level of the strategy model was further controlled. Ping An of China earned 126.26%, an increase of 47.76%, and Shandong Gold earned 116.70%, an increase of 48.73%. From the yield curve, we can also see that the fitting effect of the improved model has been improved, which can further seize the market opportunities. Buy in time to get a higher yield. And can sell stop loss in time during the decline process to reduce the risk of the model.

Experimental results show that the return after using the deep reinforcement learning algorithm can exceed the return of the benchmark buy and hold. Therefore, the experimental results show that on the one hand, the introduction of deep reinforcement learning in the field of stock trading is effective and has high research value. On the other hand, in the use of the model, it is also necessary to improve the model according to the actual results. Experimental results show that the improved deep reinforcement learning algorithm model (RB_DRL) in this paper is effective.

5 Conclusion

With the development of various algorithm technologies and the advent of the Internet 2.0 era, computers are gradually integrating into traditional industries with their powerful functions. In the financial field, the related algorithm technology has shined and achieved unprecedented achievements. On the basis of the traditional dDRL algorithm, RB_DRL deep reinforcement learning algorithm model is proposed, and the effectiveness of the model is verified through a series of comparative experiments. This paper proposes RB_DRL based on DRL, which solves the problems of insufficient fitting degree of financial data by traditional DQN algorithms, as well as the correlation between features and the importance of factors to a certain extent. However, in the field of combining artificial intelligence and finance, the current research is still in its infancy, and the follow-up research in this area will continue.

References

1. Ahmadi, E., Jasemi, M., Monplaisir, L., Nabavi, M.A., Mahmoodi, A., Jam, P.A.: New efficient hybrid candlestick technical analysis model for stock market timing on the basis of the support vector machine and heuristic algorithms of imperialist competition and genetic. Expert Syst. Appl. **94**, 21–31 (2018)
2. Cagliero, L., Garza, P., Attanasio, G., Baralis, E.: Training ensembles of faceted classification models for quantitative stock trading. Computing **102**, 1213–1225 (2020)
3. Huang, B., Huan, Y., Xu, L., Zou, Z.: Automated trading systems statistical and machine learning methods and hardware implementation: a survey. Enterprise Information Systems **13**, 132–144 (2019)
4. Abel, D., Dabney, W., Harutyunyan, A., Ho, M.K., Littman, M.L., et al.: On the expressivity of markov reward. Adv. Neural. Inf. Process. Syst. **34**, 7799–7812 (2021)
5. An, B., Sun, S., Wang, R.: Deep reinforcement learning for quantitative trading: challenges and opportunities. IEEE Intell. Syst. **37**(2), 23–26 (2022)

6. Houssein, E.H., Dirar, M., Hussain, K., Mohamed, W.M.: Assess deep learning models for Egyptian exchange prediction using nonlinear artificial neural networks. Neural Comput. **33**(11), 5965–5987 (2020)
7. A. Ntakaris, J. Kanniainen, M. Gabbouj, A. Iosifidis, (2018) "Mid-price prediction based on machine learning methods with technical and quantitative indicators", *Proc. SSRN*, pp. 1–40
8. Wang, Q., Xu, W., Huang, X., Yang, K.: Enhancing intraday stock price manipulation detection by leveraging recurrent neural networks with ensemble learning. Neurocomputing **347**, 46–58 (2019)
9. Kumar, G., Jain, S., Singh, U.P.: Stock market forecasting using computational intelligence: a survey. Arch. Comput. Methods Eng. **28**, 1069–1101 (2021)
10. Spooner, T., Savani, R.: Robust market making via adversarial reinforcement learning. In Proceedings of the Twenty-Ninth International Joint Conference on Artificial Intelligence, pp. 4590–4596 (2021)
11. Saini, A., Sharma, A.: Predicting the unpredictable: an application of machine learning algorithms in Indian stock market. Ann. Data Sci. **9**, 791–799 (2022)

A Realistic Training System for Maternal and Infant Health Care Based on MR Virtual Technology

Feng Sun(✉)

School of Public Administration, Guangdong Mechanical & Electrical Polytechnic, Guangdong 510440, China
2021011071@gdmec.edu.cn

Abstract. Maternal and infant nursing students must master sufficient knowledge and skills and have good professional competencies, especially before graduation, the assessment should be based on the requirements of maternal and infant health nursing professional competencies to assess the competencies that nursing students should have. The aim of this paper is to investigate the design and implementation of a realistic training system for maternal and child health care based on MR virtual technology. A complete imitation training system is constructed under the context-aware mechanism of maternal and infant health care, the workflow of the system is illustrated, and the design strategy of the maternal and infant health care imitation training system is proposed from the environmental context and task context. In the design strategy, a more detailed design guideline is proposed for the mother-infant health care imitation training system based on the multi-camera collaborative tracking technology of MR glasses. The experimental results show that the design solution is feasible.

Keywords: MR Virtual Technology · Maternal and Child Health Care · Realistic Training · Training System

1 Introduction

The Mixed Reality (MR) learning approach of wearing transparent smart glasses creates a new vision of wearable experimental education technology. This innovative approach provides active access to a variety of additional information and is more conducive to physical interaction with reality as well as conceptual learning and mastery [1, 2]. The integration of mother and child nursing teaching with XR (VR/AR/MR) technology is an important trend in modern education. VR has the distinguishing features of pure virtualisation and immersion, while AR emphasises the overlaying of virtual data based on real-world information. MR, on the other hand, combines the advantages of both very well. Mobile MR, supplemented by flexibility, can bring a natural perspective but with a novel experience to the teaching of maternal and child care [3, 4].

Clinical simulation is rapidly becoming an important tool in training students who are pursuing careers in nursing as a viable complement or alternative to practicing with

J. C. Hung et al. (Eds.): FC 2023, LNEE 1132, pp. 31–37, 2024.
https://doi.org/10.1007/978-981-99-9538-7_5

living patients. Laura K conducted an exploratory and descriptive study through a survey of nursing students in the Magdalena University practice laboratory using humanoid robots for nursing instruction. The results of the study showed that students in the project were positively motivated by the use of technology as a teaching strategy (specifically humanoid robots), suggesting a greater exploration of its use in educational nursing [7].Ijaz Ahmad presented a multimodal MRI interaction between real and virtual worlds. Previous related work has only focused on multimodal interactions between users on the real end and virtual beings, or visual and auditory interactions between users of VR experiences and users on the real end. As a result, they developed an MR game content that enables interaction between the real and virtual worlds by tilting the sitting position of the VR experience user according to the real side waiting and/or accompanying user's hand position [5]. Jason J. Saleem conducted an in-depth investigation into the relationship between maternal education and infant health. This study focused on the differential effects of maternal education on infant health outcomes across locations based on income status by using birth weight and low birth weight as measures of infant health. A substantial non-linear effect of maternal high school completion on infant health outcomes was found, which was masked when maternal education was specified as a linear variable. More importantly, the relationship between mother's education and infant health was concentrated in relatively deprived geographical areas. This is partly due to the fact that educated mothers living in poorer areas have substantially higher utilisation of health services compared to less educated mothers [6]. Therefore, it is relevant to study a realistic training system for maternal and child health care simulation based on MR virtual technology [7].

While most current research on maternal and infant health care training has been conducted from a single perspective of a professor, this paper combines contextual perception with the design of maternal and infant health care simulation training, exploring the interaction between the contextual needs of maternal and infant health care and the environment from a contextual perspective, providing a more detailed design direction for the design of maternal and infant health care simulation training systems that can enable users to autonomously perceive the situation of mothers and infants and provide them with This will enable the user to perceive the situation of the mother and baby autonomously and provide them with personalised health care guidance.

2 A Study of a Realistic Training System for Maternal and Child Health Care Based on MR Virtual Technology

2.1 Maternal and Infant Health Care Training Scenarios

The system's realistic training context contains the conditions, life patterns and behavioural preferences of mothers and infants. In the design of the system, the various types of information in the context are fully considered, and the situation of the mother and infant object from which the user conducts the realistic training is taken into account to consider how to improve maternal and infant health care and precise service capabilities [8, 9].

The environmental context contains the physical and social environment, and involves many environmental factors in the process of maternal and child health care.

The environmental context is considered in the maternal and infant health care simulation training system to provide design directions and strategies for the simulation training system based on the specific environmental elements mentioned above, and to enhance the maternal and infant health care capabilities of maternal and infant nursing students through the optimal integration of the environmental elements [10]. For example, the environmental scenario involves the residential and medical environment for mothers and infants [11].

Task contexts contain many health care behaviours for maternal and infant health care and the specific goals of these behaviours. Users should fully explore the opportunity points for experience optimization in each task context, improve the system task operation experience, and provide more complete and convenient health management services for the mother and infant subjects of the realistic training [12, 13].

2.2 Multi-Camera Collaborative Tracking Technology Based on MR Glasses

When users are wearing MR head-mounted displays for virtual experiments, the object with the Mark attached is facing the camera of the MR head-mounted display most of the time [14, 15]. However, the handheld Mark sometimes does not maintain a perfect orientation; on the other hand, some Marked objects placed on the table are themselves positioned with the Mark facing the other way, thus reducing the visual clutter caused by too many Marks appearing in the frontal view [16, 17]. Therefore, a multi-camera tracking feature would be helpful, and placing two user-facing auxiliary cameras on the desktop to track the back Mark would greatly improve the robustness and aesthetics of the maternal and child health care simulation training platform. In Eq. (1), f2 represents the satisfaction of the multi-camera collaboration feature and s2 indicates the degree of attribute adequacy. The feature satisfies the charm attribute.

$$f_2 = \beta e^{\lambda z s z} \quad \beta, \lambda_2 \in R \tag{1}$$

For the desktop MR maternal and child health care simulation training platform, the user needs to access the phenomenon of real-time MR experiments through the monitor. Three additional auxiliary cameras (ordinary RGB cameras) are set up to help track and identify the poses of the Mark, which are then transposed according to the positional relationship with the main camera. Ultimately, registration is performed according to the maximum confidence level obtained by calculating Eq. (2), with p denoting the camera and b denoting the Mark.

$$Con = \max(Con(p_1, b), Con(p_2, b), Con(p_3, b)) \tag{2}$$

For the mobile MR multi-camera practical training platform, the MR glasses come with one camera as the primary camera and only two additional secondary cameras are required. The difficulty in implementing the multi-camera function actually lies in the fact that the acquired poses need to be transposed according to accurate pre-calibrated multi-camera data and the virtual object is finally displayed on the screen of the head-mounted display in the correct pose [18].

3 Design of a Realistic Training System for Maternal and Child Health Care Based on MR Virtual Technology

3.1 WiFi Module

MR virtual technology-based maternal and infant health care simulation training system, need to be in the computer virtual simulation computing end, and a number of MR glasses operators to communicate with each other, different from the desktop virtual experiment platform of the single host mode, this platform to join the MR glasses constitute MR glasses - PC host of the client --server dual device model. Therefore, the dual devices are used to transmit the positional information of the virtual objects using WiFi to communicate under the local area network, and the direction is that the PC host processes the data and sends it to the MR glasses end.

3.2 MR Scene Data Storage and Loading Process Design

The information stored in the MR scene includes the position pose of the roving detector itself, augmented reality display, terrain information and manual label information. The specific data storage and loading process is shown in Fig. 1.

(1) When the server obtains the scene information from the roving probe, it first needs to form a terrain height map based on the DEM terrain and add an augmented scenario design.

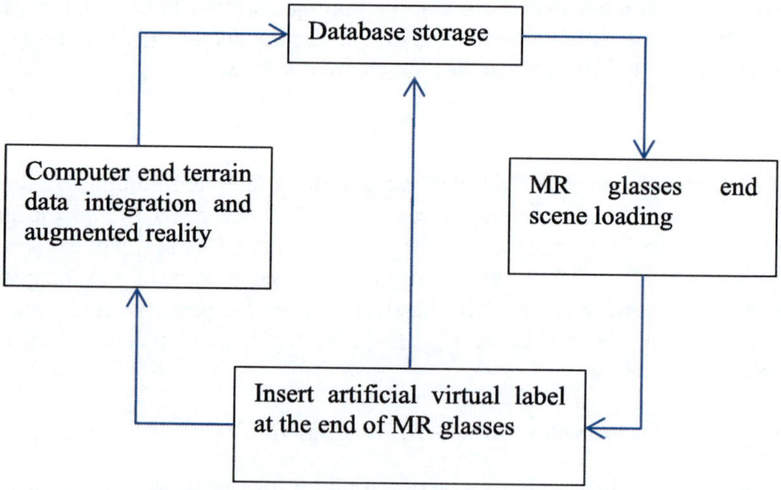

Fig. 1. Scenario storage and loading process

(2) Once the Unity scene is built, it is divided by square areas and the terrain height map, non-terrain obstacles and augmented reality location coordinates, category and size for each block of the database are uploaded.

(3) When using MR glasses to load a particular area map, it can then be loaded according to its coordinates. In the MR glasses side you can add labeled objects, move object positions and upload all scene information according to the final scene data at the end of the operation.

4 Implementation of a Realistic Training System for Maternal and Child Health Care Based on MR Virtual Technology

4.1 Implementing Human-Computer Interaction

The system uses MRTK as a mixed reality development component, using it to enrich the controls while using basic operations and simplifying the development effort. WindowsMixedReality runs as an extensible framework on Unity, providing developers with the ability to exchange core components. The basic MRTK file is first configured to use its corresponding component functionality at runtime. Its basic services include input methods, spatial awareness, camera, remote, 3D boundaries, etc. Through gaze and Tap gestures, spatial objects containing interaction methods with framework components can be manipulated for interaction methods, manipulation processing and object boundaries.

The interaction operations of mixed reality teleoperation in this paper are built on the basis of MRTK, adding suitable functional components to specific functions.

For example, the virtual tagging function is implemented. The tagging function is a 3D tagging of objects of interest to mother and child care students in the scene. The first step is to retrieve the prefabricated 3D virtual tag by clicking on the UI, add the interaction component ManipulationHandler for the spatial dragging operation, the tag size and direction adjustment component BoundingBox and the spatial position movement synchronisation component UpdatePosition. The functionality of the tabs is achieved by adding basic components.

4.2 User Satisfaction Masurement

After the completion of the system hardware and software design solution, a certain number of users were required to evaluate the design solution in order to check whether it was feasible. Fifteen users were selected to participate in this user satisfaction evaluation, and the users involved in the evaluation were given a detailed explanation of the functions and usage of the system, and a Likert scale was created from the perspective of several evaluation indicators, each item being divided into 5 levels (-2, -1, 0, 1, 2), in the order of extremely dissatisfied, dissatisfied, satisfied, more satisfied and very satisfied, and the users were asked to score through the satisfaction evaluation form. The scoring results are shown in Table 1 respectively. The average value of user satisfaction scores for each indicator is above 1, which shows that users are more satisfied with the system design solution, as shown in Fig. 2.

Table 1. System Scoring Results

Evaluation angle	evaluating indicator	mean value	Maximum	minimum value
Visual Design	color	1.72	2	1.66
	Icon	1.68	1.84	1.55
Function content	Comprehensive content	1.56	1.74	1.26
Operation	Convenient operation	1.61	1.78	1.38

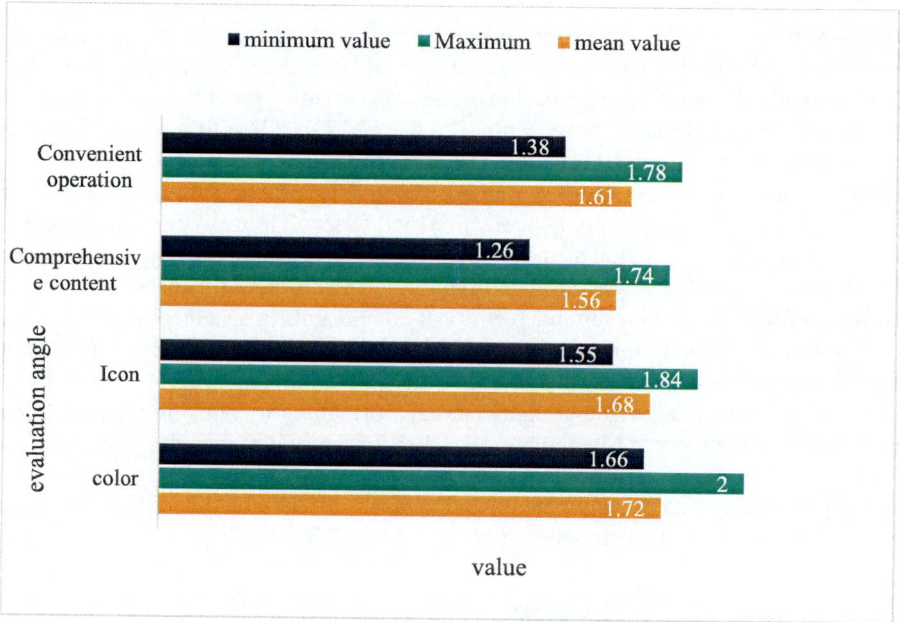

Fig. 2. Customer Satisfaction

5 Conclusions

Competence, knowledge and skills are both different and related to each other. Knowledge is the result, object and content of an individual's mental activity, while competence is the mental characteristic that an individual displays in order to complete an activity; skill is an automatic way of activity that has been consolidated through practice. The realistic training system can develop students' abilities, knowledge and skills. This paper proposes a design for a realistic training system for maternal and child health care based on MR virtual technology. In the process of system design, the situational tasks of maternal and infant health care are realistically reproduced, and the functions of system simulation are achieved, which proves the feasibility of simulation teaching as well as the broad application prospects. This study provides a theoretical basis for understanding students' competence and exploring scientific and effective laboratory management, which has certain social and practical significance. However, due to the limited capacity of the researcher, this study inevitably has many shortcomings and the system needs to be optimised based on the results of user satisfaction evaluation.

References

1. Hammady, R., Ma, M., Al-Kalha, Z., Strathearn, C.: A framework for constructing and evaluating the role of MR as a holographic virtual guide in museums. Virtual Real. **25**(4), 895–918 (2021)
2. Havenith, H.-B., Cerfontaine, P., Mreyen, A.-S.: How virtual reality can help visualise and assess geohazards. Int. J. Digit. Earth **12**(2), 173–189 (2019)
3. Dieker, L.A., et al.: Using virtual rehearsal in a simulator to impact the performance of science teachers. Int. J. Gaming Comput. Mediat. Simul. **11**(4), 1–20 (2019)
4. Alea, N., Bluck, S., Mroz, E.L., Edwards, Z.: The social function of autobiographical stories in the personal and virtual world: an initial investigation. Top. Cogn. Sci. **11**(4), 794–810 (2019)
5. Cirelli, L.K., Jurewicz, Z.B., Trehub, S.E.: Effects of maternal singing style on mother-infant arousal and behavior. J. Cogn. Neurosci. **32**(7), 1213–1220 (2020)
6. Ahmad, I., Asghar, Z., Kumar, T., Li, G., Manzoor, A., Mikhaylov, K.: Syed attique shah, marko höyhtyä, jarmo reponen, jyrki huusko, erkki harjula: emerging technologies for next generation remote health care and assisted living. IEEE Access **10**, 56094–56132 (2022)
7. Saleem, J.J., Wilck, N.R., Murphy, J.J., Herout, J.: Veteran and staff experience from a pilot program of health care system-distributed wearable devices and data sharing. Appl. Clin. Inform. **13**(3), 532–540 (2022)
8. Gillies, A., Smith, P.: Can AI systems meet the ethical requirements of professional decision-making in health care? AI Ethics **2**(1), 41–47 (2022)
9. Maibaum, A., Bischof, A., Hergesell, J., Lipp, B.: A critique of robotics in health care. AI Soc. **37**(2), 467–477 (2022)
10. Almagrabi, A.O., Ali, R., Alghazzawi, D., AlBarakati, A., Khurshaid, T.: A reinforcement learning-based framework for crowdsourcing in massive health care internet of things. Big Data **10**(2), 161–170 (2022). https://doi.org/10.1089/big.2021.0058
11. Kurazume, R., et al.: Development of AR training systems for humanitude dementia care. Adv. Robot. **36**(7), 344–358 (2022)
12. Fuchs, R., et al.: A system for real-time multivariate feature combination of endoscopic mitral valve simulator training data. Int. J. Comput. Assist. Radiol. Surg. **17**(9), 1619–1631 (2022)
13. Tuena, C., Riva, G.: Active Navigation training: an Innovative embodied-Based training system for spatial navigation in aging. Cyberpsychology Behav. Soc. Netw. **25**(1), 77–78 (2022)
14. Lattuada, M., Gianniti, E., Ardagna, D., Zhang, L.: Performance prediction of deep learning applications training in GPU as a service systems. Clust. Comput. **25**(2), 1279–1302 (2022)
15. Lavric, T., Bricard, E., Preda, M., Zaharia, T.B.: A low-cost AR training system for manual assembly operations. Comput. Sci. Inf. Syst. **19**(2), 1047–1073 (2022)
16. Fujs, D., Vrhovec, S., Vavpotic, D.: Towards personalized user training for secure use of information systems. Int. Arab J. Inf. Technol. **19**(3), 307–313 (2022)
17. Manabe, T., Omura, K.: Performance Evaluation of Bluetooth Low Energy Positioning Systems When Using Sparse Training Data. IEICE Trans. Fundam. Electron. Commun. Comput. Sci. **E105.A**(5), 778–786 (2022). https://doi.org/10.1587/transfun.2021WBP0007
18. Suda, H., Kotani, G., Saito, D.: INmfCA algorithm for training of nonparallel voice conversion systems based on non-negative matrix factorization. IEICE Trans. Inf. Syst. **E105.D**(6), 1196–1210 (2022). https://doi.org/10.1587/transinf.2021EDP7234

Cross-Border e-Commerce Payment Risk Based on Blockchain Underlying Technology

Zhijin Ding(✉)

Beijing Institute of Technology, Zhuhai, Zhuhai 519088, Guangdong, China
lkfwonderland@163.com

Abstract. In the financial field with the background of modern Internet, the core part of CBEC (Cross border e-commerc) system has become electronic payment. China's financial industry has serious problems in third-party platform payment, such as fund circulation and monitoring, lack of credit reporting system, privacy disclosure, etc., and should use various science and technology to fill the gap in Internet financial supervision and management. To ensure the safety of consumers and the companies to be consumed, eliminate the security risks in the gray zone, the standardization platform will firmly control the scale of personal information disclosure, create a healthy consumption and investment environment for consumers and investors, reduce and solve many risks and problems such as cross-border payment credit risk, technical risk, liquidity risk, privacy protection and supervision and management difficulties. This paper mainly studies how to use blockchain technology to discuss the risks of CBEC payment. At the same time, this paper first introduces the risks of the underlying blockchain technology and CBEC payment, and then applies the underlying blockchain technology to CBEC payment risk resolution, which reduces the risk of payment. Finally, it analyzes the actual application performance of the underlying blockchain technology. The experimental results of the article show that the questionnaire of CBEC companies and staff can show that there are great risks in e-commerce payment. Therefore, using the underlying technology of blockchain to reduce the risk of CBEC online payment is a topic worthy of study.

Keywords: Blockchain Underlying Technology · CBEC · Online Payment Risk · Blockchain Architecture

1 Introduction

In recent years, with the rapid improvement of science and technology and the rapid development of human society today, China's e-commerce has also brought some huge challenges and new beginnings with it. Especially in the current context of online finance, CBEC is booming because of demand, which has virtually changed our purchase model and also produced new trading methods. It is more convenient for us to purchase, but while the CBEC platform provides convenience, various problems occur quietly. How to control transactions and implement strict supervision on consumer transactions has become a problem. At present, there are few studies on solving the risk of CBEC. In this

J. C. Hung et al. (Eds.): FC 2023, LNEE 1132, pp. 38–45, 2024.
https://doi.org/10.1007/978-981-99-9538-7_6

case, Liu C has developed three types of supply chain game models: the basic model, the single trade credit model, and the trade credit and income sharing cooperation model. He used these models to analyze the optimal decision-making under centralized and decentralized conditions, and verified that a single trade credit cannot adjust the supply chain. However, partner agreements can coordinate the supply chain [1]. Akhmadeev R assessed the actual requirements for the implementation of VAT under the relevant conditions of participating in e-commerce payment, in order to develop and implement the target determination model. In the process of scientific and practical research on the specific conditions of the implementation and payment of indirect tax, it is determined that there are differences due to the applicable indirect tax [2]. In the current CBEC environment, Song B put forward the view that because the multi-variety and small-batch transaction mode of CBEC prohibits the customs from comprehensively inspecting the quality of goods, the laws and regulations of different countries may conflict, which may cause significant risks to CBEC [3]. However, the above research has not combined with reality to solve the risk of CBEC payment, and the evaluation and comparison are few, which need to be improved.

This paper first describes the risk and risk management of CBEC payment, discusses the prevention of payment risk of CBEC companies, and analyzes the risk of payment methods, payment channels and electricity recovery, thus laying a theoretical foundation for CBEC to avoid online payment risk.

2 Blockchain Underlying Technology

Information layer: distributed ledgers, the core technology of blockchain, need to use cryptography technology to ensure the integrity and security of data [4]. Network layer: also known as point-to-point technology, in the blockchain network, nodes are equal, and each node determines the transaction information and exchange rights through consensus algorithm;

Protocol layer: also known as distributed protocol, this layer mainly provides a set of standard distributed protocols for the blockchain system to achieve the scalability, security and consistency of the data after being linked. Incentive layer: incentive mechanism refers to the code of conduct that ensures the continuous and healthy operation of the blockchain system through the design of incentive mechanism in the system [5]. Contract layer: The contract layer is the core function of the blockchain system, that is, the smart contract is used to execute the consensus agreement, ensure transaction security and control node permissions, and ensure data verification and tamper-proof.

Its structure is shown in Fig. 1:

The big data layer is composed of blockchain big data blocks, timestamp, hash function, blockchain big data organization mechanism and asymmetric encryption algorithm [6]. The network layer includes P2P network, transmission mechanism and data verification mechanism. The protocol layer includes the existing blockchain of the Protocol. The public identification layer includes the existing blockchain of the public identification mechanism. The typical common identification mechanism mainly includes workload evidence, proof of interest, authorization certificate and interest. The main contract layer of the encouragement layer mainly includes script language, algorithm

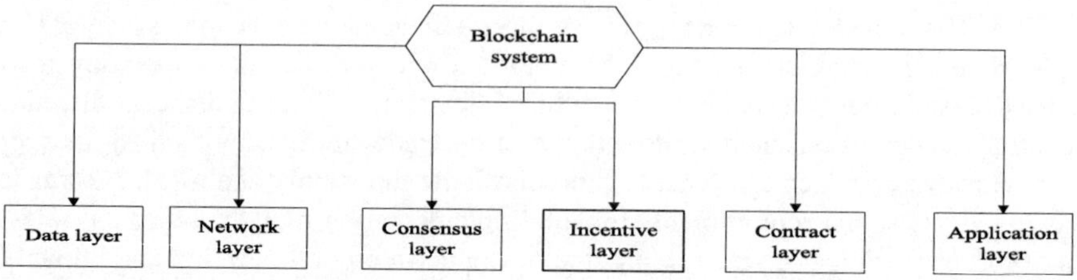

Fig. 1. Blockchain system architecture

mechanism, and jointly recognized contracts, of which the contract is the core, The application layer includes application site blockchain, programmable currency, programmable finance, programmable currency, programmable currency, programmable currency, programmable finance, programmable society, and the specific structure is shown in Fig. 2:

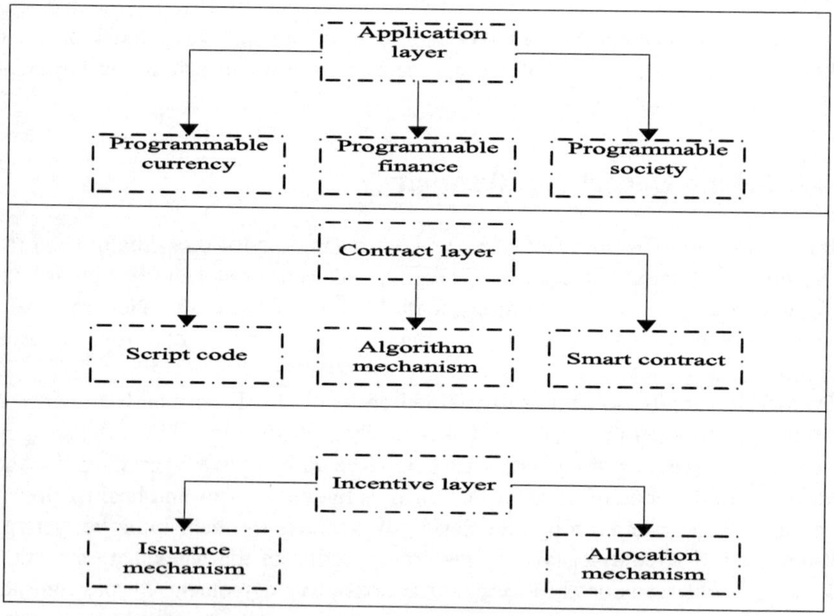

Fig. 2. Specific architecture classification of data layer

2.1 Blockchain Underlying Technology Algorithm

Block interval method: The block interval needs to be dynamically adjusted according to the network load at any time according to the real-time situation [7]. When the network idle time interval becomes longer and the network busy time interval becomes shorter, the dynamic contraction can be realized. The interval calculation method of Block is to make statistical calculation based on the number of recent transactions. The more transactions, the shorter the interval, and the less transactions, the longer the time.

The block interval calculation formula is as follows:

$$Block\_interval = min\_interval + \frac{max\_interval - min\_interval}{1 + transaction\_num} \quad (1)$$

where, min_ Interval is the minimum block interval, max_ Interval is the maximum time interval. Transaction_ Num is the average transaction in the last m blocks, where m can be an integer greater than 1, m, min_ Interval and max_ The interval is preset by consensus calculation or set by smart contract.

Workload proof method: By calculating the value, the hash value of the content after transaction data collation reaches the specified limit [8]. When a node successfully finds a target that meets the standard, it will immediately send the broadcast packet to all networks, and the nodes in the network will immediately verify after receiving it. The number of required preambles 0 is called the difficulty value. The Nonce value is iterated until the target value is valid. The process of finding the nonce value of the target value that meets the appropriate conditions is called drilling. The algorithm formula is as follows:

$$target = Hash(index, previousblock\ hash, timestamp, block\ data, nonce) \quad (2)$$

$$t\arg et < difficulty//target\ new\_difficulty = prev\_difficulty* \quad (3)$$

The first 2015 block generation time/1209600 (generally the number of seconds required to generate a block every 10 min and 2016 blocks)// The CPU required to adjust the difficulty expression is the workload proof. To repair a block, and the proof of the block and all subsequent block workloads must be recreated.

2.2 Advantages and Disadvantages of Blockchain Underlying Algorithm

The blockchain underlying algorithm technology has become more and more popular with the continuous expansion of development and application, and this powerful development impetus from the needs of all walks of life gives the blockchain underlying algorithm an opportunity to develop and change. Therefore, although blockchain has made more and more achievements in each industry, each algorithm has its advantages and disadvantages, and the advantage of the lower algorithm is decentralization [9]. Because blockchain uses peer-to-peer network technology to store data and uses distributed computing and storage, it does not store centralized hardware or management architecture, and the rights and obligations of each node are equal.

At the same time, because the system is open, the customer's personal information is encrypted, and the system is maintained synchronously by the node and all the maintenance function nodes, so that everyone can query the blockchain data through the public interface and develop relevant applications [10]. The whole system is highly transparent and creates a trusted database. However, at the same time, the lack of privacy of blockchain algorithm is one of the disadvantages. The shared algorithm chain allows each user to obtain a complete data backup. The transaction data is public. For the accounts and transaction information of some commercial institutions, the algorithm has no privacy.

3 E-Commerce Payment Risk Experiment

This paper takes the staff and sales customers of CBEC companies in Zone C as the survey objects, and uses the network survey method to investigate the potential payment risks of e-commerce in Zone C. In the questionnaire survey, there was no communication between participants to ensure the objectivity and fairness of the questionnaire.

3.1 Questionnaire Design

A network survey was conducted on the current payment mode and existing problems of e-commerce companies in Zone C. Problems: online risk, credit risk, digital signature risk, accounting risk, gateway risk, etc. On this basis, an anonymous questionnaire was used to investigate all the questions to be investigated.

3.2 Questionnaire Distribution and Recovery

First of all, the designed anonymous questionnaire was produced and distributed to the staff of CBEC companies and sales customers in the form of online 60 copies each. The group distributed this time has relatively high cultural quality. A total of 120 questionnaires were distributed this time. A total of 102 questionnaires were collected this time, including 64 questionnaires for staff of the Power Supply Department and 38 questionnaires for customers. Summarize the collected questionnaires and eliminate the ones with unclear or unclear answers. At last, 92 valid questionnaires were collected, including 48 for internal staff and 44 for power customers.

3.3 Statistical Data Results

The collected internal staff questionnaires are counted to determine the impact of the following risks on online payment of e-commerce, where A represents a great impact, B represents an impact, C represents a little impact, and D represents no impact. The data obtained during statistics are shown in Table 1 and Table 2:

Table 1. Risk impact feedback from staff of CBEC companies

risk	A	B	C	D
Risk of online	55%	33%	7%	5%
Credit risks	36%	24%	29%	11%
Digital signature risk	47%	36%	16%	1%
Accounting risk	49%	35%	15%	1%
Gateway risk	54%	22%	15%	9%

From the above data, it can be seen that the five risks have a great impact on the transaction rate of CBEC payment. Therefore, this paper needs the existence of an early warning and prevention mechanism, and the lack of risk of CBEC payment management mechanism.

Table 2. Sales customer feedback risk impact

risk	A	B	C	D
Risk of online	55%	34%	2%	9%
Credit risks	54%	30%	6%	10
Digital signature risk	45%	38%	4%	13%
Accounting risk	54%	30%	5%	11%
Gateway risk	64%	12%	5%	19%

4 Results and Discussion of CBEC Payment

4.1 Risk Status of CBEC Payment

First of all, the order turnover of companies that use blockchain technology and do not use blockchain technology in Zone C is studied. From Fig. 3, it can be seen that the order turnover of companies that use blockchain technology is generally higher than that of companies that do not use blockchain technology, which can indicate that blockchain technology helps companies achieve the goal of economic growth while improving payment security.

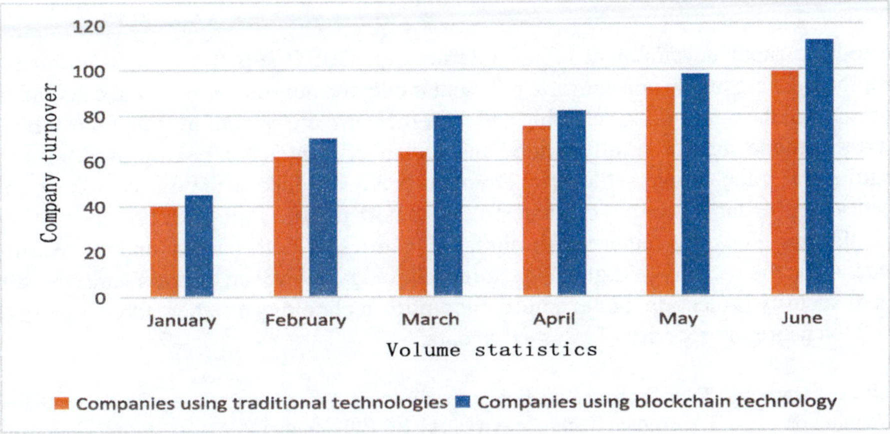

Fig. 3. Data Sheet of Video Software Usage Algorithms in Zone A in the First Half of 2021

4.2 Risk Analysis of CBEC Payment

(1) Importer credit risk

In the B2B export transactions of CBEC, some importers with bad credit use the convenience and timeliness of the network to cheat, or deliberately default when the trading market has adverse changes. For example, on the Made in China website, there are some buyers who take advantage of the loopholes in the online enterprise information review system, use false enterprise information to buy and sell, cheat the export enterprises' goods, or threaten the export enterprises with bad comments, force the export enterprises to reduce prices or make other concessions, and other acts that endanger the economic interests of foreign trade export enterprises.

Information risk of CBEC platform.

Foreign trade enterprises need to publish their basic information and supply information on the cross-border platform in order to negotiate and sign contracts with importers. For this reason, the data update of the algorithm will become slow.

Cross-border exchange collection of risks.

The payment for CBEC B2B export by foreign trade enterprises is collected through the network platform. Restricted by the cross-border payment environment and the third-party payment system, there are huge risks in the cross-border payment process of foreign trade enterprises.

Cross-border customs clearance risk.

With the change of China's CBEC B2B export customs clearance mode, the information transfer between the relevant departments such as customs, tax, banking, commodity inspection and CBEC logistics enterprises has become one-way, making it impossible to effectively share information between various departments.

5 Conclusion

In today's rapid development of the Internet era, CBEC still faces many problems, mainly focusing on the non-uniform payment mode, the non-uniform capital calculation system, and the non-uniform clearing and settlement system. Therefore, the future CBEC payment mode should be fully unified and socialized, which can be improved through multilateral trade capital settlement. Payment is the key link of the transaction, so the security and reliability of payment are crucial to both parties. From the current situation, the platform does not require high digital signatures of both parties, and its security depends on the security of digital signature technology. However, data transmission still needs security protection. For example, encryption technology or other technologies are used to ensure the security of transaction data.

Acknowledgment. Fund Project: Research on the Financial System of Cross-border E-commerce in Zhuhai based on Blockchain Technology (2021YBC120) Zhuhai 2021–2022 Philosophy and Social Science Planning Project.

References

1. Su, W., Wang, Y., Qian, L., et al.: Creating a sustainable policy framework for cross-border E-commerce in China. Sustainability **11**(4), 943 (2019)
2. Lipych, L., Mokhniuk, A.: E-business in ukraine: peculiarities, tendencies, prospects. Economic J. Lesia Ukrainka Eastern Euro. National Univ. **1**(21), 74–79 (2020)

3. Liu, C., Chen, K., Li, M., et al.: Trade credit and revenue sharing of supply chain with a risk-averse retailer. Math. Probl. Eng.Probl. Eng. **2021**(6), 1–15 (2021)
4. Akhmadeev, R., Morozova, T., Voronkova, O., et al.: Targets determination model for VAT risks mitigation at B2B marketplaces. J. Entrepreneurship Sustain. Issues **7**(2), 1197–1216 (2019)
5. Song, B., Yan, W., Zhang T.: Cross-border e-commerce commodity risk assessment using text mining and fuzzy rule-based reasoning. Adv. Eng. Inform. **40**, 69–80 (2019)
6. Bhagi, A., Sarkar, A., Sethuraman, V., et al.: Blockchain technology for immunisation data storage in India: opportunities for population health innovation. BMJ Innovations **8**(1), 1–3 (2022)
7. Alharbi, T.: Deployment of blockchain technology in software defined networks: a survey. IEEE Access (99), 1–1 (2020)
8. Du, W.D., Pan, S.L., Leidner, D.E, et al.: Affordances, experimentation and actualization of FinTech: A blockchain implementation study. J. Strategic Inform. Syst. **28**(1), 50–65.3 (2019)
9. Younus, A.M., Abumandil, M.: Factors affecting the adoption of blockchain technology for the development of australian logistics. Design Eng. (Toronto), **2021**(7), 9133–9141.5 (2021)
10. Weiss, M.B., Werbach, K., Sicker, D.C., et al.: On the application of blockchains to spectrum management. IEEE Trans. Cognitive Commun. Netw. PP(99), 1–1 (2019)

Brand Visual Identification System for Urban Brand Image Design

Ying Li(✉)

Department of Design, Xianyang Normal University, Xianyang 712000, Shaanxi, China
ly1375@sina.com

Abstract. Brand visual identity system is not an isolated system, it is a very systematic whole. The block vision system is closely connected with the city and belongs to a part of urban planning. The overall and unified VRS reflects the good operation of all departments in a city, and also shows the neat style of the city. In this paper, the design of brand visual identity system based on urban brand image design, taking the construction of urban commercial block visual identity system as an example, discusses the process of urban brand design and 2 matters needing attention in the design of brand visual identity system. By means of questionnaire survey, this paper analyzes the satisfaction of temporary residents and long-term residents of city a to the city's brand image and their awareness of the city's regional culture. The test results show that the most satisfactory of the temporary residents is that the "cultural atmosphere" accounts for 47.2%, the "urban environment" is nearly 28% lower than the "cultural atmosphere", and the "cultural heritage" accounts for more than 80% of the temporary residents and long-term residents, of which the "local folk customs" account for 11.4% and 9.3% respectively, and the "industrial culture" accounts for less than 6%. From this, we can see that the most attractive place of a city is the urban culture, that is, the design of urban brand image is very important; A the cultural heritage, local folk customs and industrial culture in urban regional culture have distinct characteristics. When shaping the visual image of urban brand, we should organically combine the three. Only in this way can we truly reflect the spiritual outlook, value orientation and deep cultural connotation of a city. It is very necessary to design a brand visual identity system based on urban brand image design.

Keywords: Urban Brand · Brand Image Design · Visual Identification System · Identification System Design

1 Introduction

Taking the construction of urban commercial street as an example, this paper designs the brand visual identification system based on the design of urban brand image. Excavate and highlight the personality of commercial blocks, combined with the city brand image. Commercial block visual recognition should consider the integration with the whole city, not only with the city image and urban material form, but also integrate the design of commercial block visual recognition system(VRS) into the overall link of urban

J. C. Hung et al. (Eds.): FC 2023, LNEE 1132, pp. 46–52, 2024.
https://doi.org/10.1007/978-981-99-9538-7_7

planning, so as to achieve systematic and integrated design. The history and culture of the city make the city present a unique face. The VRS of commercial blocks can also reflect the character of the city from the elements of block image, scale, materials, color and so on. This not only ensures the overall consistency, but also imperceptibly promotes the city brand image.

Many scholars at home and abroad have studied the design of brand visual identification system based on urban brand image design. Tamari t takes the main stadium of the 2020 Tokyo Olympic Games as an example to analyze the controversial design of Tokyo stadium related to Japan's national brand initiative. Through the investigation of Zaha Hadid, a world-famous architect, this paper considers why Hadid's aesthetics and narration of Olympic venues can not fully shape the new image of Japan, which can be regarded as a national cultural heritage [1]. Ponzini D tested some of these hypotheses by comparing two case studies: the Agbar tower in pobrennu, Barcelona and the Doha tower in the West Bay area of Doha, Qatar. The planning and development of these two cities are different from each other. Based on this evidence, it criticizes the simple assumptions circulated internationally and calls for a deeper understanding of urban design, planning and policy factors [2].

The connotation of regional culture is the internal characteristic of most domestic commercial blocks. The innovation of this paper in designing the brand visual identification system of urban brand image lies in taking the local cultural context as the design starting point. Vivid graphic symbols are used to spread impressive effects, so as to have a certain identity and appeal. This can be easily recognized and accepted by the public, so as to arouse their resonance and cultural identity. Their own characteristics are often unique and innovative, with characteristics and uniqueness; The unique visual design has set up a new image for the commercial district and played a positive role in publicizing the characteristics of the city [3, 4].

2 Design of Brand Visual Identification System for Urban Brand Image Design

2.1 Urban Brand Design Principles

People oriented principle: people oriented principle is a design idea that focuses on people's spiritual and emotional needs from the perspective of people and on the basis of meeting people's basic needs. Commercial block VRS is a visual and spatial multi-dimensional three-dimensional design, especially needs to reflect the care for people, which is also a concentrated embodiment of humanistic spirit. When designing the VRS of commercial blocks, we should combine the principles of ergonomics and make appropriate design according to people's perception system, people's behavior and activity law, people's static and dynamic scales and people's safety needs. Therefore, when implementing the specific design, designers should pay particular attention to the grasp of color, graphic symbols, materials and other elements. From a visual point of view, the clean and beautiful neighborhood environment with soft color matching can make people pleasing to the eye and produce a comfortable feeling [5].

Normative principle: the VRS should be implemented according to certain design criteria and adhere to the principle of unity and standard to standardize the operation in order to play its function. Of course, on this basis, we also need to carry out expansionary innovative design to show the charm of design on the basis of ensuring strong visual recognition. Logo is the first image element in visual recognition and the first impression of the image of commercial block in the hearts of the audience. The sign of the commercial block represents the visual symbol of the block. It not only reflects the concept and spiritual connotation of the block, but also the symbolic pattern of the whole block, which has the function of positioning and explanation. Therefore, the logo design of commercial blocks needs to use concise graphic symbols to convey specific information and meaning, so that the public can perceive and obtain information at the first time. Therefore, the shape and color of the logo should be easy to identify and remember [6, 7]. The signs should be used strictly and reasonably, which will lead to confusion in the process of commercial use. Otherwise, it is easy to cause confusion in the process of commercial use.

On the basis of meeting the use function, we should also meet people's spiritual needs and meet people's basic aesthetic needs. In terms of basic elements, vivid and interesting modeling images are adopted, which can be remembered by those who have seen it, and has a good communication effect. The design of basic elements should be careful in ingenious conception and refinement. The composition can be seen as perfect as possible, and then constantly revised and refined to extract the essence of its dross, enrich and develop the scheme. Secondly, in terms of shopping environment, a good shopping environment in commercial blocks can not only attract consumers to stop, stimulate shopping desire, but also shape a good image of commercial blocks. The shopping environment includes the external environment of the block, the internal environment of the store, the facade of the window, etc.; It also includes the natural environment and artificial environment of the block. Reasonably arrange the elements of the VRS of the commercial block into the block environment to form a harmonious and beautiful whole and create a strong shopping atmosphere [8, 9].

2.2 Brand Visual Identity System Design

(1) Tap their own cultural characteristics

In the process of positioning the visual core of the commercial block, we should analyze according to the local traditional cultural history, economic level and the characteristics of the target population of the block, excavate the representative graphic symbols and design them. Accurate and meaningful positioning of commercial blocks can make people refreshing and easy to remember.

(2) Establish a perfect guidance facility system

A reasonable commercial block guidance facility system should be an integral part of the urban guidance facility system, which echoes with the urban guidance facility system. Similar to the urban guidance facilities, the design of the guidance facilities of the commercial block should consider the characteristics of the block itself, as well as people's perception of the block, and tap the cultural characteristics of the block itself as the innovation of the design. Starting from the cultural characteristics of the block, its characteristics are refined and simplified into a representative graphic symbol, so that users can understand the meaning and characteristics of the graphic at a glance. In addition to graphic symbols, text, color, material, modeling and technology are all factors involved in the design of guiding facilities. People perceive the city through various hints such as color, current situation, words, symbols, hearing and even smell. Maps, house numbers, road signs, landmark buildings and other guiding facilities make it easy for people to find the way [10].

The design of commercial block guidance system should pay attention to systematicness, which is not only the inheritance and development between block guidance and city guidance, but also the guidance facilities of commercial block itself are a system. According to the function, the guiding facilities of commercial blocks can be divided into block traffic guidance, commercial area guidance, cultural tourism guidance, etc.; According to different traffic types, it can also be divided into bus guidance, subway guidance and pedestrian guidance. In the design process, we should pay attention to the relationship between these classifications to achieve unity and change. The guiding information to be provided can also be divided in order through different colors to determine what information pedestrians see first and then [11].

(3) Provide humanized public facilities

Public facilities are essential elements in the space environment of commercial blocks. In the design process, we should not only reflect the aesthetic principles, but also meet the functional requirements. In particular, the scale of public facilities should comply with the principle of ergonomics, so that the size is reasonable and comfortable. As a part of the city, the commercial block should skillfully apply some technical elements in the urban VRS to the design of public facilities of the commercial block.

Generally speaking, the public facilities of commercial blocks should be in the same line with their city brand image. In a modern city, especially in the design of large-scale facilities, it is necessary to consider reasonable, uniform, material saving, reasonable and simple production and processing functions, and the minimization of cost control. However, we should also avoid too patterned modeling styles, add some unique and interesting designs, and achieve the integration of modern science and technology culture and historical culture. This shows that the design of public facilities should be combined with the cultural characteristics of the block and the surrounding environment, integrated with the environment, and have strong visual appeal [12].

3 Investigation and Research on Brand Visual Identification System of Urban Brand Image Design

Using the online questionnaire survey method, this paper investigates and analyzes the satisfaction of temporary residents and long-term residents of city a on the brand image of the city, as well as the regional culture (awareness) of long-term residents and short-term residents of city A.

In this paper, SPSS 22.0 software is used to count and analyze the questionnaire results, and t-test is carried out. The t-test formula used in this paper is as follows:

$$t = \frac{\overline{X} - \mu}{\frac{\sigma X}{\sqrt{n}}} \tag{1}$$

$$t = \frac{\overline{X_1} - \overline{X_2}}{\sqrt{\frac{(n_1-1)S_1^2+(n_2-1)S_2^2}{n_1+n_2-2}(\frac{1}{n_1} + \frac{1}{n_2})}} \tag{2}$$

where formula (1) is a single population test, AA is the average number of samples, s is the standard deviation of samples, and N is the number of samples. Formula (2) is a double population test, s_1^2 and s_2^2 are the two sample variance, n_1 and n_2 are the sample size

4 Investigation and Analysis of Urban Brand Image

By means of questionnaire survey, this paper investigates and analyzes the satisfaction of temporary residents and long-term residents of city a on the brand image of the city (including urban environment, citizen quality, urban construction, job opportunities, public security management, cultural atmosphere, convenience of life and others). The test results are shown in Table 1 and Fig. 1

Table 1. Satisfaction of residents in different aspects of city a

	urban environment	Citizen quality	urban construction	Job opportunities	Public security management	cultural atmosphere	Convenient life	other
Temporary residents	19.8%	3.7%	13.5%	1.6%	2.8%	47.2%	11.3%	7.4%
Long term resident population	15.4%	2.4%	17.6%	5.7%	4.9%	37.8%	12.3%	6.2%

From the above test results, it can be seen that the most satisfied temporary residents are the "cultural atmosphere" accounting for 47.2%, the "urban environment" is nearly 28 percentage points lower than the "cultural atmosphere", the "urban construction" 13.5% and "convenient life" (11.3%) are about 10 percentage points, the selection rate of the remaining options is less than 10.0%, and the selection rate of "job opportunity" is the lowest. It can be seen that the most attractive place of a city is the urban culture, that is, the design of urban brand image is very important.

Next, test the awareness of long-term residents and short-term residents of city a to the regional culture (cultural heritage, local folk customs and industrial culture) of the city, and conduct a questionnaire survey. The test results are shown in Fig. 2.

In various surveys of regional culture, the data on the regional cultural awareness of city a shows that the proportion of "cultural heritage" of temporary residents and long-term residents has reached more than 80%, of which "local folk customs" account for 11.4% and 9.3% respectively, and "industrial culture" accounts for less than 6%. This data shows that the citizens of the city are very attractive to cultural heritage. In addition, local folk customs such as folk culture and urban scenery are loved by the majority of

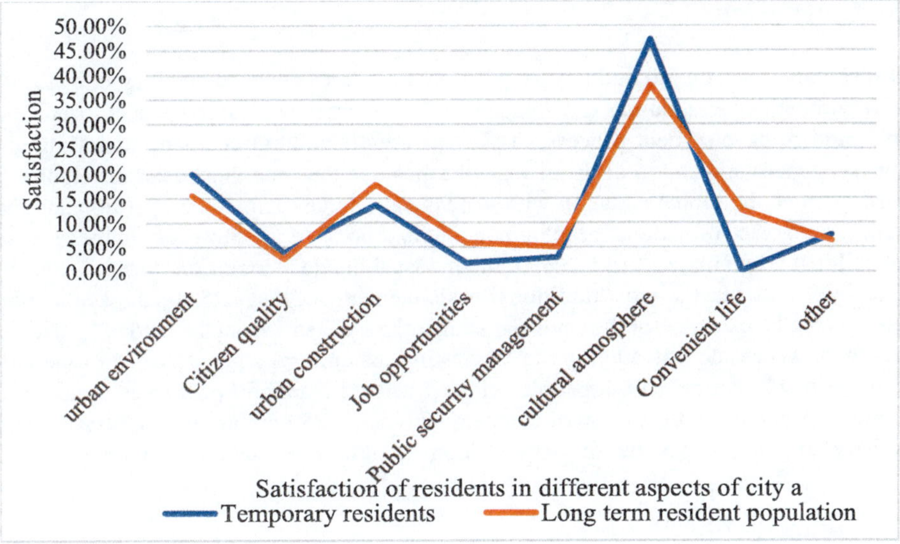

Fig. 1. Satisfaction of residents in different aspects of city a

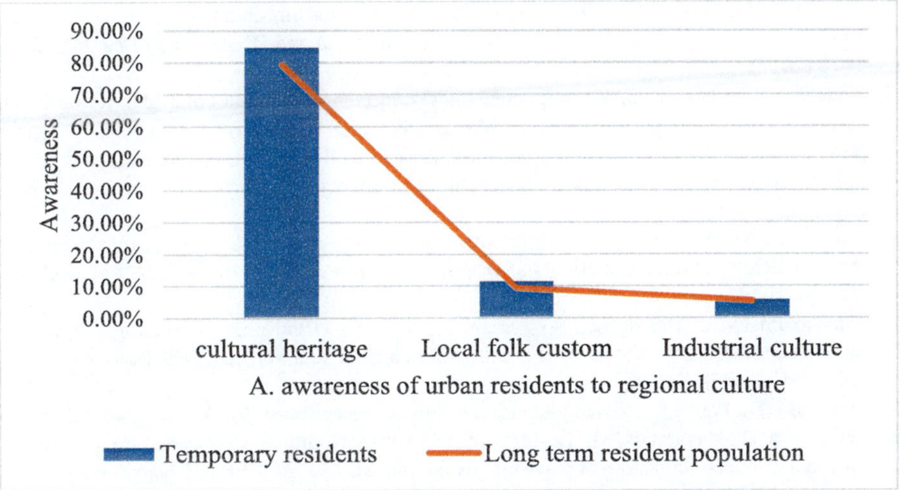

Fig. 2. A awareness of urban residents to regional culture

young people. Due to the vague understanding of the connotation of industrial culture and the lack of a consistent sense of industrial culture cognition, it is difficult to organically connect industrial culture with urban brand building. It can be seen from the above data that people have a low understanding of industrial culture and local folk customs, and there is still a certain gap in building urban brands with industrial culture and local folk customs.

5 Conclusions

It is essential for a city to find its own position, fully tap its own advantages, enhance its core competitiveness, build a city image that can reflect the characteristics of its ancient and modern development process, display its unique cultural character and highlight the advantages and charm of the city in all aspects in the changing social development environment, the global situation with increasingly fierce regional competition and the accelerating pace of national development in the process of urbanization. In the process of building urban image, the design of urban visual image recognition system is the top priority. No matter the most traditional basic elements in the VRS, or the application elements of block space, or the environmental interface elements in the field of landscape research, we should pay attention to excavating the cultural characteristics of the city and the block, adhere to the people-oriented and standardized principles around the same core, and make the design of commercial block VRS combining functionality and artistry, so as to promote the further development of commercial block and city.

References

1. Tamari, T.: Star architects, urban spectacles, and global brands: exploring the case of the Tokyo Olympics 2020. Int. J. Jpn. Sociol. **28**(1), 45–63 (2019)
2. Ponzini, D., Arosio, P.M.: Urban effects of the transnational circulation of branded buildings: comparing two skyscrapers and their context in Barcelona and Doha. Urban Des. Int. **22**(1), 1–19 (2017)
3. Changhyun, J., Moonsun, Y., Jungyong, L.: The influence of brand color identity on brand association and loyalty. J. Prod. Brand Manag. **28**(1), 50–62 (2019)
4. Kim, K.I., Kim, J.W.: A study on urban tourism factors affecting urban brand image & revisit intention of chinese tourists: focusing on Chinese tourists visiting busan. J. Tour. Leisure Res. **31**(8), 39–58 (2019)
5. Cecilia, Cassinger, Jorgen, et al. The narrative rhythm of terror: a study of the Stockholm terrorist attack and the "Last Night in Sweden" event. International Journal of Tourism Cities, 2018, 4(4):484–494
6. Vinayakumar, R., Alazab, M., Srinivasan, S., et al.: A visualized botnet detection system based deep learning for the internet of things networks of smart cities. IEEE Trans. Ind. Appl. **56**(4), 4436–4456 (2020)
7. Dzyabura, D., Peres, R.: Visual elicitation of brand perception. J. Mark. **85**(4), 44–66 (2021)
8. Pedeliento, G., Kavaratzis, M.: Bridging the gap between culture, identity and image: a structurationist conceptualization of place brands and place branding. J. Prod. Brand Manag. **28**(3), 348–363 (2019)
9. Budiman, A., Wijaya, P.W.: Perancangan Media Komunikasi Visual Pada Museum Uang Sumatera. AKSA Jurnal Desain Komunikasi Visual **3**(1), 356–369 (2020)
10. Bonakdar, A., Audirac, I.: City branding and the link to urban planning: theories, practices, and challenges. J. Plan. Lit. **35**(2), 147–160 (2020)
11. Bae, S., Jung, T.H., Moorhouse, N., et al.: The influence of mixed reality on satisfaction and brand loyalty in cultural heritage attractions: a brand equity perspective. Sustainability **12**(7), 2956 (2020)
12. Tien, N.H., Anh, D.B.H., Ngoc, P.B., et al.: Brand building and development for the group of Asian international education in Vietnam. Psychol. Educ. **58**(5), 3297–3307 (2021)

Data Quality Evaluation Method Based on Density Clustering Algorithm and Its Application

Limin Zhao, Guangcai Liu, Peng Wei(✉), Wenbin Zhang, Li Sun, and Peihao Qiao

North Automatic Control Technology Institute, Taiyuan, Shanxi, China
wp7786@126.com

Abstract. Data quality assessment is mainly to analyze, evaluate and calculate the collected original information, and then to obtain the decision results. At present, there are many research methods for data quality. Through density clustering of data, we can get some parameters. This paper proposes a new data quality evaluation model to improve the performance of products and services. This paper mainly analyzes the data quality evaluation model by using experimental comparison and density clustering algorithm analysis. The experimental data shows that because the weight of the precision rule set in the evaluation is 0.18, the precision evaluation value in the final evaluation value is 14.99.

Keywords: Density Clustering Algorithm · Data Quality · Evaluation Method · Application Research

1 Introduction

There are two main methods of data quality evaluation, namely, classification based on density clustering and comprehensive scoring. Data quality is an important part of information security. The results obtained by density clustering algorithm are more accurate and reliable than traditional methods. The quality of data directly affects the user experience. Data quality assessment is a complex process, which requires statistical analysis of problems in the system, and then takes effective measures to improve the system performance. Density clustering algorithm is a data quality evaluation method. This method can further process the original sample information.

Many scholars have studied data quality evaluation methods and density clustering algorithms. For example, some experts choose the model combination method to simulate the change characteristics of time series data [1, 2]. Some experts also claimed that the key to solving this problem is to reduce the dimension, group the existing high-dimensional data, and eliminate the noise in high-dimensional data to a certain extent [3, 4]. In addition, some experts claim that on most data sets, dense clustering can obtain better clustering results than traditional clustering algorithms, and there is almost no need to manually set parameters in the clustering process [5, 6]. The research of data quality evaluation mainly focuses on the classification of data of different types and

J. C. Hung et al. (Eds.): FC 2023, LNEE 1132, pp. 53–60, 2024.
https://doi.org/10.1007/978-981-99-9538-7_8

grades, and summarizes the corresponding models, including cluster analysis, density function and statistical description.

This paper first studies the density-based clustering algorithm and improved methods, and analyzes its basic classification and characteristics. Secondly, the necessity of data quality detection and evaluation methods is discussed, and the factors affecting data quality evaluation are proposed. Then the comprehensive evaluation of data quality is simulated. Finally, relevant data and conclusions are drawn through experimental comparison.

2 Data Quality Evaluation of Density Clustering Algorithm

2.1 Density-Based Clustering

Clustering is a method of data processing, which classifies the samples and then merges the samples with similar degree into a group. According to different classification objects, a new class is formed by combining similar properties in the data. A sample space can contain multiple categories at the same time. In a prototype-based cluster, a cluster is a collection of objects, where each object is closer to the prototype of the cluster than the prototype of other clusters. This technology is simple and can be used for different types of data. Although it usually needs to be executed several times, it is also relatively effective. It is relatively quiet and can handle clusters of any shape and size. It can find many clusters. Grid-based clusters can determine the grid size of each object and the number of grids for each object in a single data scan. It obtains the final clustering result based on the statistical analysis of the original set and its characteristic information. The density clustering algorithm is an unsupervised data set [7]. It consists of all meaningful points in a sample space set. Density clustering algorithm is a data quality evaluation method. It can divide a sample into multiple clusters according to certain rules, and then classify these monomers. Density clustering algorithm is a multi-attribute data analysis method. The density clustering algorithm has good discrimination in data classification.

Density clustering algorithm is a data mining method. It mainly analyzes the differences between the distance and dimensions between sample points, and then draws conclusions. In this case, a density measurement function is needed to calculate it. After the sample set is obtained by density clustering algorithm, a tree model is built according to its attributes. The density clustering algorithm is used to estimate a data from the data of the sample set. First, the sample set is clustered, and then all objects in each group are divided into the same cluster according to the classification map obtained. Then use different methods to calculate the distance between each cluster. Data compression technology mainly includes two processes. The first is to delete some invalid information from the memory. The second is to estimate the effective signal.

The clustering algorithm based on object density and direction determines the central object of the density attraction point chain according to the definition of density accessibility [8]. In the algorithm, we use Gaussian function to formally simulate the influence of each data point.

$$g_{guass}(a,b) = f^{\frac{c(a,b)^2}{2\lambda^2}} \tag{1}$$

represents the distance between object a and object a_i.

$$\text{density}(a_i) = \sum_{k=1}^{m} f^{\frac{c(a_i,b_k)^2}{2\lambda^2}} \tag{2}$$

The maximum local data point of the global density function in the data space. The neighborhood radius is as follows:

$$R = \frac{\mathrm{j}(C)}{\mathrm{m}^{coefR}} \tag{3}$$

where j (c) is the average distance of all objects. According to the cluster definition, the characteristics of the object can be obtained at the class boundary. If the density value of most data points is close to 1, increase the value of the density parameter. If most values of the density parameter are close to m, the density parameter value must be reduced [9, 10].

2.2 Data Quality Assessment

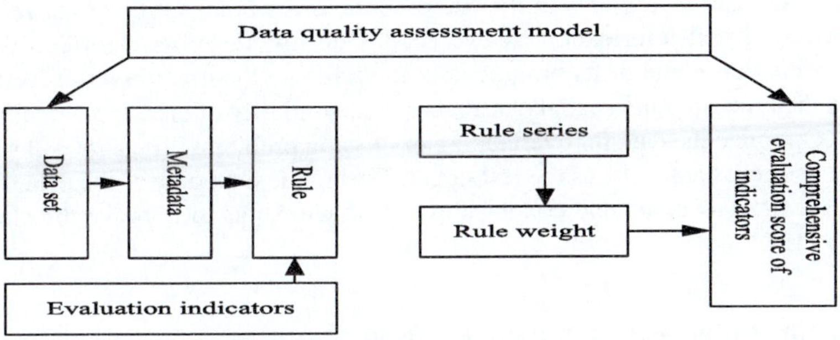

Fig. 1. Data Quality Assessment Model

The evaluation of data quality is the process of analyzing, processing and evaluating the data used by users. In real life, because it is impossible for the system to customize personalized needs for everyone, there are differences between different groups. Therefore, in order to better serve people, it is necessary to sort out and utilize a large number of individual characteristics information, and then establish a corresponding database to collect relevant useful information resources. The quality of data is determined by many factors, including acquisition methods, equipment and algorithms. This paper mainly introduces the data quality evaluation model based on density clustering. The first module is to classify randomly selected samples from different perspectives. The second module is to determine the importance of the selected data type, and grade these variables according to their importance to show their advantages and disadvantages. Data quality assessment is a dynamic process. It is to obtain data through analysis, collation and statistics within a specific time interval. Data quality assessment refers to the estimation, analysis and inspection of relevant information obtained by users using computer systems under specific application environments through certain methods. Data quality evaluation is to evaluate and analyze the results of a test, and give reasonable suggestions according to its advantages and disadvantages [11]. In order to ensure the reliability of data, the original information obtained is analyzed to make correct decisions. Data quality evaluation analyzes, calculates and gives corresponding results for each indicator. The main influencing factors of data quality assessment are as follows. Different users have different understanding of the same content. The impact of different types of algorithms on the quality evaluation results is also very different. The data quality assessment model is shown in Fig. 1:

Data quality assessment is a complex system with uncertainty and multi-objective optimization problems. In the process of practical application, we need to evaluate these indicators. The density clustering algorithm, as one of the unsupervised learning methods, can solve this kind of problem well. Density clustering can divide features into different types when classifying massive data. This division method has high efficiency. The application of density clustering algorithm in data quality evaluation is mainly to calculate by establishing functions.

The traditional data quality assessment method is to quantify user behavior and perceived information, and then transform it into a decision tree. However, in real life, people often have large differences among different groups. The evaluation results can not reflect the user's problems in the use process as a whole, so it can not reflect the different use of text information by everyone. Density clustering algorithm is a data quality evaluation method. Its basic idea is to divide a global graph into several clusters, and then measure the spatial distribution and similarity of each individual in these clusters. Optimize the data for overlapping or missing problems. Then we will get more accurate and reliable results as the test target. During the experiment, the data obtained are statistically processed and conclusions are drawn, so as to improve the clustering accuracy [12].

3 Comprehensive Evaluation of Data Quality

3.1 Test Process

This test record selects the fields of some tables in the database to be tested. The examination rules are determined by the interview and examination. Collect and synthesize the information obtained from these two methods to obtain different series of rules required by the system. The implementation of tests in this system depends on the rules and test data of each rule set.

3.2 Single Factor Set Test Evaluation Set

Based on the precise evaluation model of data quality of single factor set, select the precise evaluation management module of single factor set from the evaluation tools, and evaluate the above eight item tables from the perspective of factor set 2 (context data quality). The test data comes from the data of the company management system in the database. This system is an enterprise management system based on B/S architecture of

NET program language based on Windows Server and SQL Server database. The main service of the system is to provide services for laboratory staff and manage the internal workflow of the laboratory. Key functions include business process management, human resource management, authority control and process control.

3.3 System Function Realization

The back-end code of the system is implemented in Java language. The front-end is implemented in JavaScript language, and the front-end and back-end interact with RESTful APIs. The system structure is shown in Fig. 2:

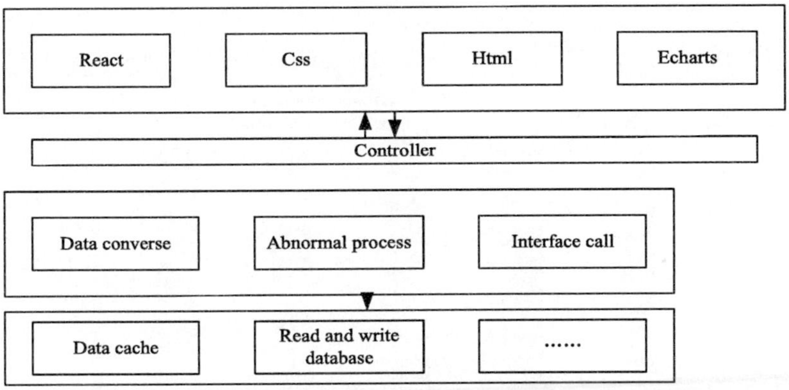

Fig. 2. Structure of the System

Regarding the system architecture, the system uses MVC architecture. The "data source" section considers the relational and non-relational data stored in the system. Data layer (DAO) The framework used by the system is MyBatis. Add a cache layer to the system according to the running speed of the system. Since the evaluation rules and other information rarely change, the data from the evaluation rules will be cached when the system starts to improve the system running speed.

4 Experimental Results

4.1 Evaluation Score Display

Go to the main module of data quality evaluation, select the quality evaluation module, click the evaluation button, run the test, get the score of each evaluation series, get the evaluation value according to the weight of the evaluation series, and save each evaluation data in the database for future reference. The evaluation scores are shown in Table 1:

As shown in Fig. 3, in the accuracy evaluation, there are 40540 verified data sets, of which 40520 are qualified and 20 are inconsistent. After reviewing and verifying the enterprise management system database and data files, there are records indicating the accuracy, integrity, consistency and credibility of data quality indicators. Does not match. As shown in the sample results, the system allows full automatic testing in terms of accuracy, integrity, consistency and credibility, with high accuracy and time saving.

Table 1. Number of Data Evaluation

	Total quantity	Qualified quantity	Unqualified quantity
1	40540	40520	20
2	36270	36250	20
3	30120	30108	12
4	10570	10560	10
5	45220	45205	15

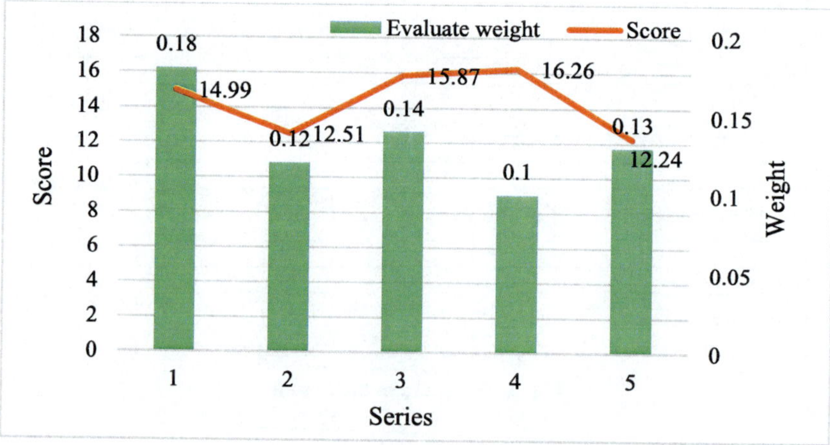

Fig. 3. Data Evaluation Score and Weight

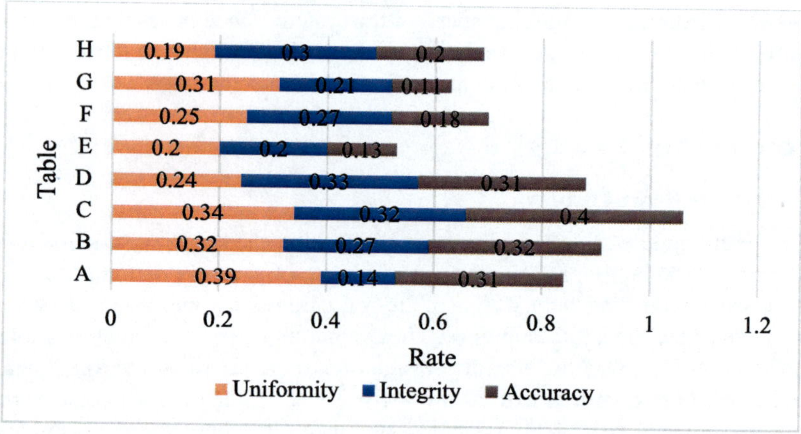

Fig. 4. Visual Analysis of Test Results

4.2 Visual Analysis of Test Results

When applying the accurate single-factor evaluation model, the 8 evaluation tables are automatically evaluated. Click here to open the analysis result visualization module, which will automatically fill in data. As shown in Fig. 4, we can see that the system module can not only clearly see the results of each test item table and dynamically display different indicators.

5 Conclusion

In this paper, we mainly use the density clustering algorithm to evaluate the data quality. The data quality is studied based on the density clustering algorithm, which is mainly aimed at the problems that may occur in different applications. Experiments have verified that the method has high accuracy and performance advantages. This method can greatly reduce the time and cost required in the process of data quality assessment. The density of the data is divided by clustering algorithm and divided into three categories, namely, mean, standard deviation and extreme value. The density function fitting parameter estimation method is used to improve the clustering effect, and some statistics are introduced in the data quality clustering process to reduce its computational complexity. Do density clustering analysis on the original data, and compare the clustering results with the original classification to obtain the optimal solution.

References

1. Purushothaman, R., Selvakumar, S., Rajagopalan, S.P.: Hybrid grasshopper and chameleon swarm optimization algorithm for text feature selection with density peaks clustering. Int. J. Comput. Intell. Appl. **21**(3), 2250018:1–2250018:18 (2022)
2. Phamtoan, D., Vovan, T.: Automatic fuzzy clustering for probability density functions using the genetic algorithm. Neural Comput. Appl. **34**(17), 14609–14625 (2022)
3. Khader, M., Al-Naymat, G.: Density-based algorithms for Big Data clustering using MapReduce framework: a comprehensive study. ACM Comput. Surv. **53**(5), 93:1–93:38 (2021)
4. Jaya Lakshmi, B., Madhuri, K.B., Shashi, M.: Efficient hybrid algorithms for density based subspace clustering to deal with density divergence for improved quality and conciseness. Evol. Syst. **12**(2), 489–501 (2021)
5. Bali, B., Singh, B.M.: Unification of firefly algorithm with density-based spatial clustering for segmentation of medical images. Int. J. Comput. Appl. Technol. **65**(4), 316–324 (2021)
6. Wiharto, Wicaksana, A.K., Cahyani, D.E.: Modification of a density-based spatial clustering algorithm for applications with noise for data reduction in intrusion detection systems. Int. J. Fuzzy Log. Intell. Syst. **21**(2), 189–203 (2021)
7. Devarakonda, N., Kavitha, D., Kamarajugadda, R.: Escape the traffic congestion using brainstorming optimization algorithm and density peak clustering. Ingénierie des Systèmes d Inf. **26**(3), 285–293 (2021)
8. Karkosková, S.: Data governance model to enhance data quality in financial institutions. Inf. Syst. Manag. **40**(1), 90–110 (2023)
9. Dastoorian, R., Wells, L.J.: A Hybrid Off-Line/On-Line Quality Control Approach for Real-Time Monitoring of High-Density Datasets. J. Intell. Manuf. **34**(2), 669–682 (2023)

10. Romero, G.C., Esteban, M.P.E., Sáez, M.D., Such, M.M.: A shape expression approach for assessing the quality of linked open data in libraries. Semant. Web **14**(2), 159–179 (2023)
11. Leithner, M., Fikar, C.: A simulation model to investigate impacts of facilitating quality data within organic fresh food supply chains. Ann. Oper. Res. **314**(2), 529–550 (2022)
12. Naake, T., Huber, W.: MatrixQCvis: shiny-based interactive data quality exploration for omics data. Bioinform. **38**(4), 1181–1182 (2022)

Optimization of Computer Network Connection Enhancement Relying on Nonlinear Active Queue Management Algorithm

Bixia Wu(✉)

Minbei Vocational and Technical College, Nanping 353000, Fujian, China
yx900312@126.com

Abstract. The relationship between the packet loss rate and queue length of the main queue management in the computer network connection process, through the analysis of the main queue management algorithm in the computer network packet loss process unreasonable problems, in order to effectively solve this problem, this paper through the construction of data information processing model, the use of the new model of computer network transmission process packet loss rate analysis, according to the quadratic function model. The new packet loss rate calculation formula is composed, and this formula is used to upgrade and optimize the network blocking process queueing management process packet loss algorithm, which can ensure that the main queueing management algorithm is more reasonable for packet loss management, which can make the fairness and stability of the system in the application process improved and the stable operation of the system is stabilized.

Keywords: Nonlinear Active Queue Management Algorithm · Packet Loss Rate · Queue Length · Quadratic Function

1 Introduction

With the continuous progress of science and technology, the development of new industries such as computer Internet has gradually been widely used in people's lives. People enjoy the benefits brought by the development of computer networks at the same time, but also is the many research scholars gradually focus on the problem of network routing optimization, because the use of modern computer networks continue to expand the field [1], in the process of network data transmission will appear delay, data packet loss and bandwidth limitations and many other problems, serious impact on the normal operation of computer networks [2]. Therefore, an optimization method needs to be proposed to achieve network management and resource optimization. The optimization process of computer network routing problem is actually a special kind of NP-hard combination optimization [3]. This type of problem can be categorized as a hard-to-solve combinatorial optimization problem, which has gained the attention of many research scholars because of the relatively large range of problem usage involved [4, 5]. The traditional

J. C. Hung et al. (Eds.): FC 2023, LNEE 1132, pp. 61–66, 2024.
https://doi.org/10.1007/978-981-99-9538-7_9

approach taken is based on statistical-mathematical principles, based on obtaining accurate data or constructing a model before the optimization problem can be solved more quickly, but in the actual field of application, the use of complex mathematical models constructed in different fields [6], but it is generally difficult to obtain accurate data information, especially in the more restrictive environment can effectively solve the computer network connection enhancement optimization problems [7].

Based on the above analysis, this paper proposes to apply the nonlinear active queue management algorithm to the computer network routing connection process, conducts a detailed analysis of the data problem of routing optimization, and also proposes to use the nonlinear active queue management algorithm to take enhancement measures in the state transfer rules and data information model, which can well solve the slow convergence speed of the computer network queue optimization process and easily fall into the local The problem of minimal values. Finally, the experimental research results show that the algorithm proposed in this paper has good practicality. It can effectively solve the computer network optimization problem.

2 Nonlinear Active Queue Management Algorithm

The use of nonlinear active queue management algorithm in the computer network connection enhancement optimization process can effectively analyze the packet loss rate and queue length in detail, and can effectively solve the phenomenon of anomalies at the maximum and minimum thresholds in the network packet loss process [8].

The computer network structure used in this paper, can be more than N connected users in the computer network, can appear in the shared bandwidth using C as the bottleneck link, then in the computer network connection enhancement optimization will have the same characteristics of the user information loop delay using d to represent. Then the computer network traffic model can be expressed as follows.

Translated with www.DeepL.com/Translator (free version)

$$\dot{x}_i(t) = \frac{1}{d^2} - \beta x_i(t)x_i(t-d)p\big(\lambda(t-d),\ \widetilde{C}(t-d)\big) \quad (1)$$

$x_i(t)$ in the expression can be in the t moment appear in the ith sender in the network stable operation corresponding to the transmission rate; $\lambda(t) = \sum_{j=1}^{N} x_j(t)$ said in the computer network routing optimization of the bottleneck queue appeared in the input rate; $\widetilde{C}(t)$ said the virtual link bandwidth of the network; then using the function $p(\cdot)$ can be in the computer network connection process, in t-d moment can calculate the network packet marking probability; if in the $\beta < 1$ network environment, you can use $(1/d)\sqrt{3/2p^*}$ to represent the throughput according to the network in stable operation.

Translated with www.DeepL.com/Translator (free version)

3 Computer Network Connection Enhancement Optimization

The basic operating process of the nonlinear active queue management algorithm is mainly based on the probabilistic lost parts, while marking them to reduce the time of network queuing in order to effectively avoid the problem of computer network blockage. The non-linear active queue management algorithm used can be divided into two processes, the first step is to calculate the mean value of the queue length, which needs to be based on the packet loss rate of the computer network, the second step is to set the packet loss rate, which will be based on the partially discarded data, which can effectively reduce the length of the queue, in the optimization process of the computer network connection enhancement, which can be based on the mean value of the queue fast packet loss rate size value [9]. The weighted sliding mean square can be used to calculate the expression as:

Translated with www.DeepL.com/Translator (free version)

$$avg_L(t_{n+1}) = \left(1 - w_q\right) \times avg_L(t_n) + w_q \times q(t_n) \tag{2}$$

In the expression, $avg_L(t_n)$ represents the average queue length size value, w_q represents the weighting factor, and $q(t_n)$ represents the actual length value of the real-time queue in the network operation.

In the process of computer network operation, if the phenomenon of network congestion occurs, for the occurrence of continuous packet loss or global synchronization problem in the process of computer network operation, this paper uses marker probability expression as follows:

$$P = \frac{P_a}{1 - count \times P_a} \tag{3}$$

In order to can effectively improve the nonlinear active queue management algorithm in the computer network operation process convergence problem, this paper uses the data information update method, the method not only needs to consider the local impact of data information factors, but also needs to update the global information problem. In the process of computer network k need to pass through the path (i, j), but also need to combine the expression (2) for data timely update; if in the computer network connection enhancement optimization k to achieve from the starting point to the end of the path selection, according to the expression (4) all the path data information element for global timely update, the expression is as follows.

$$\tau_{ij} = (1 - \rho_1) \times \tau_{ij} + \rho_1 \times \frac{1}{Q} \tag{4}$$

According to the data pheromone that does not appear on the path is updated in time according to expression (5) expressed as:

$$\tau_{ij} = (1 - \rho_1) \times \tau_{ij} \tag{5}$$

The expression ρ_1 denotes the retention factor corresponding to the data pheromone, $Q = AQ_1 + BQ_2$, Q_1 denotes the path (i, j) of the computer network connection, the sum of the delay and delay jitter corresponding to node i, Q_2 denotes the energy consumed corresponding to path i, and B denotes the total bandwidth of the network.

The computer network connection enhancement optimization process that can be implemented according to the above analysis of the offender method is shown in Fig. 1.

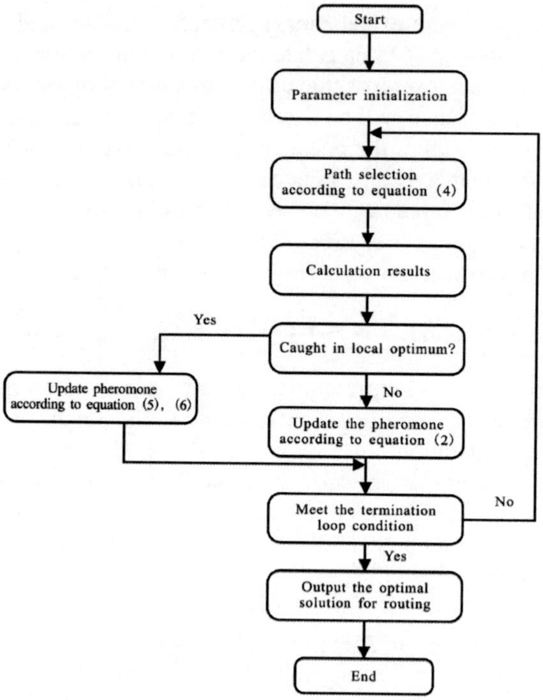

Fig. 1. Computer network connection enhancement optimization process

4 Simulation Test and Analysis

In this paper, we use simulator ns-2 to verify the performance of the proposed nonlinear active queue management algorithm, the computer network topology is shown in Fig. 2, where $si(i = 1, 2, \cdots, N)$ indicates that the application transmitter D can be used as a UDP application transmitter port, for the bottleneck link between nodes n1 and n2, the propagation delay size between nodes n2 and n3 is d = 180 ms, and the propagation propagation delay between other nodes The packet length of the computer network is calculated as 500 Byte, and the capacity of any queue is 500 packet; not counting the bottleneck link, then the other links queue management process uses the nonlinear active queue management algorithm.

Setting the buffer space within the router to between 10–100 packets, then the message size can use 1000 B. In such a computer network environment, a comparison of the results of the three processing algorithms used is shown in Fig. 3.

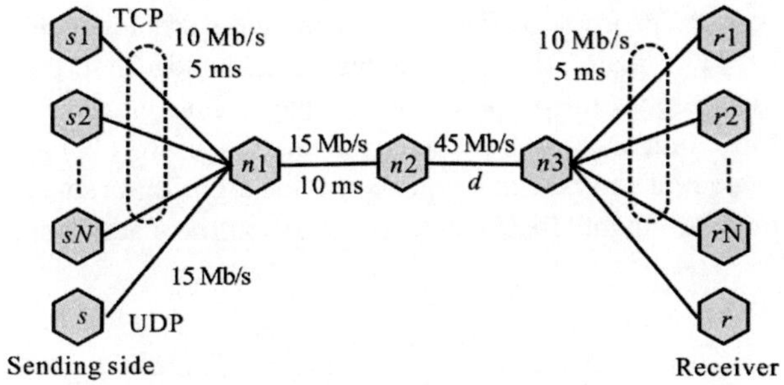

Fig. 2. Topology of a computer network

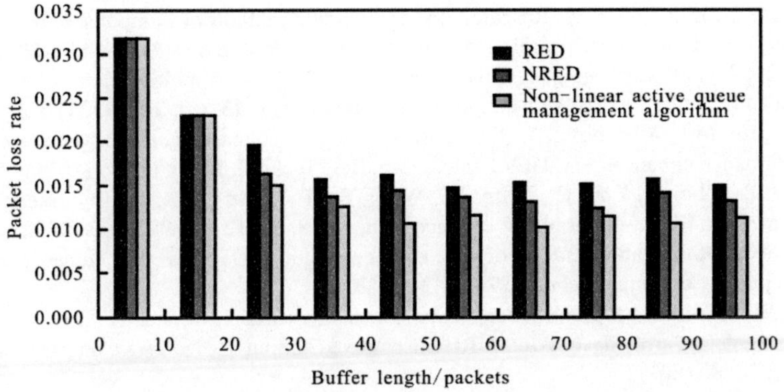

Fig. 3. Relationship between buffer space and packet loss rate

The test results in Fig. 3 show that the nonlinear active queue management algorithm outperforms the RED and NRED algorithms if the buffered queue is larger than 20 packets, and the performance gap between the advantages is maintained between 20% and 28%. Whether analyzed from the principle or from the experimental test results, the nonlinear active queue management algorithm outperforms the traditional algorithm in terms of packet loss rate on the computer network, and the algorithm has better stability in the process of computer network connection.

5 Conclusions

With the continuous progress of science and technology, the scale of computer network operation has been expanded by fast reading, which also leads to the problems of delay and data packet loss that constantly appear during the transmission of network data information. In order to effectively improve the stability of the computer network data transmission process and reduce the energy consumed by network operation, the computer network routing problem can also be solved. In this paper, we use the nonlinear active queue management algorithm to analyze the problem of distributed parallel characteristics, and improve the rules of node transfer and data information update in the queue management process to effectively improve the efficiency of traditional queue management [10, 11]. Finally, the experimental results show that: the nonlinear active queue management algorithm proposed in this paper can optimize the most suitable routing data information according to the characteristics of the changing network environment, the problem of too fast convergence of this algorithm reduces the problem of computer networks falling into local extremes in the nonlinear active queue management process.

References

1. Theodoropoulos, P., Spandonidis, C.C., Fassois, S.: Use of convolutional neural networks for vessel performance optimization and safety enhancement. Ocean Eng. **6**(3), 177–181 (2022)
2. Lu, H., Li, M., Zhang, Y.: Research on optimization method of computer network service quality based on feature matching algorithm. J. Phys. Conf. Ser. **3**(5), 16–23 (2021)
3. Lei, D.: A pansharpening generative adversarial network with multilevel structure enhancement and a multistream fusion architecture. Remote Sens. **131**(3), 727–732 (2021)
4. Aririguzo, M.I., Mbachu, C.B.: Transport control protocol based computer wireless network performance enhancement. IJERT-Int. J. Eng. Res. Technol. **11**(6), 69–75 (2021)
5. Chen, B., Zhu, L., Zhu, H., Yang, W., Wang, S.: The loop game: quality assessment and optimization for low-light image enhancement, vol. 8, no. 12, pp. 901–912 (2022)
6. Liu, X.C.: Optimization strategy of volte voice connection delay and enhancement of customer perception. Commun. Technol. **27**(1), 43–62 (2019)
7. Wei, W.Y., Shi, Y., Yang, J.X., Meng, H.X.: Artificial neural network and convex optimization enable antenna array design. Int. J. RF Microwave Comput. Aided Eng.Comput. Aided Eng. **11**(2), 72–80 (2021)
8. Wang, X., Yang, D., Jian, X., Chen, M., Guna, J.: Performance analysis and optimization for coverage enhancement strategy of narrow-band internet of things. Futur. Gener. Comput. Syst.. Gener. Comput. Syst. **101**(11), 434–443 (2019)
9. Cui, P., Wang, H., Wang, Y.: Optimization path analysis of computer network database security management technology. China Comput. Commun. **10**(1), 203–218 (2019)
10. Wang, W.: Deployment and optimization of wireless network node deployment and optimization in smart cities. Comput. Commun. Commun. **155**(7), 117–124 (2020)
11. Ly, A., Jl, B., Sz, B.: Network optimization of hospital computer processing system and clinical effect of alendronate in the treatment of premature ovarian failure. Microprocess. Microsyst. Microsyst. **7**(2), 103–114 (2020)

Establishment of Ambient Media Advertising Order Snatch System Considering Price Stepping (PPA) Algorithm

Yueyang Zhao(✉)

Design Department, Dankook University, Yongin-si 16890, Gyeonggi-do, South Korea
uiw821@126.com

Abstract. With the rapid development of modern information society, all types of advertisements emerge in an endless flow, where ambient media that emphasizes the multi-sensory experience of users rise as the times require. In this paper, based on the price stepping algorithm, a new ambient media advertising model based on O2O that integrates offline large-screen resources is proposed to further promote the development of ambient media in modern society. The practice demonstrates that the penetration of ambient media advertising into the living environment allows consumers to remember them more quickly.

Keywords: Price Stepping · Ambient Media Advertising · Order Snatch System

1 Introduction

With the rapid development of modern information society, various advertisings are emerging. Major brands use various methods to promote their products and compete for eye-catching. In this environment, the influence of traditional outdoor advertising is gradually declining, and ambient media that emphasizes the multi-sensory experience of users emerges. In daily life, buses are frequently-used means of transportation. The bus station advertising is not restricted by time and region. It faces a broad audience, and its information spreads widely [1, 2]. Once the bus stop advertising is established, it will be effective for a long time. Over time, people's impressions will deepen, the effect of communication will be better, and it will play a direct communication role with consumers. Studies have shown that 74% of consumers go out to travel by buses, the average frequency is 20 times a month, at least 16 h: waiting room advertisings are divided into ordinary consumers with a 50% share of the market except for TV The most exposed advertising media outside [3, 4]. At present, many brands are not satisfactory in the promotion of advertisings. Many advertisings only focus on the visual and sound and ignore the established fact that people have five senses [5]. However, there are still many excellent ambient media advertising cases that not only stop at visual effects. They combine multiple senses to offer users a refreshing feeling [6, 7]. The concept of "ambient media" originated in Europe. Mark Austin and Jim Aitchison summarized the definition of "ambient media" in the book "Is anyone out there", i.e., something suitable for spreading advertising information, which can be used for writing, coloring, and hanging, and anything that can be used to convey brand connections [8, 9].

J. C. Hung et al. (Eds.): FC 2023, LNEE 1132, pp. 67–74, 2024.
https://doi.org/10.1007/978-981-99-9538-7_10

To exert the role of ambient media in life more effectively, based on the price stepping algorithm, a new ambient media advertising model based on O2O that integrates offline large-screen resources is proposed in this paper to further promote the development of ambient media in modern society. The penetration of ambient media advertising into the living environment allows consumers to remember them more quickly.

2 Price Stepping Algorithm

The ideal advertising budget of advertisers is directly proportional to the advertising transaction volume, and the advertising transaction volume is closely related to the advertising period and the advertising area. To predict the advertising transaction rate in a certain period or a certain region, we established a linear regression model based on historical data to fit the advertising transaction probability p(t) in each period and the advertising transaction probability p(d) in each region.

$$h(x) = \sum_{i=0}^{n} p(t)_i x_i = p(t)x \tag{1}$$

where $x_1, x_2, \cdots, x_n$ is the total daily transaction volume, and h(x) represents the transaction volume during the t period of the day.

$$g(x) = \sum_{i=0}^{n} p(d)_i x_i = p(d)x \tag{2}$$

where $x_1, x_2, \cdots, x_n$ is the daily transaction volume, and g(x) represents the transaction volume in the d area of the day.

The fitted p(t) and p(d) are only used to predict the next day's advertising order transaction. When a new advertising transaction occurs, the advertising transaction rate needs to be continuously adjusted in accordance with the new data to calculate the continuous Ad transaction rate.

The value of advertising orders in different delivery periods is different, and advertisers' demand for delivery during prime time is higher than that of unpopular periods. The system adopts a relatively high pricing strategy during prime time when competition is high, which can better match the value of this time period and bring higher returns to equipment owners. This paper defines that the competition degree measurement based on the delivery period can reflect the gap between the delivery amount in this period and the average delivery amount per unit period. The difference is shown in Eq. (3):

$$\partial(t) = \frac{tv(t) - \overline{tv}}{\overline{tv}} \tag{3}$$

where tv(t) is the delivery volume of the period, and represents the average delivery volume of the period. As the value of tv(t) is not available when the order is generated, the transaction prediction probability p(t) at this time period needs to be used, as shown in Eq. (4):

$$\partial(t) = \frac{p(t) * \overline{v} - \overline{tv}}{\overline{tv}} \tag{4}$$

where $\overline{v}$ represents the average daily delivery volume. At this point, $\partial(t)$ represents the change trend of the degree of competition with time. When the delivery volume of this period is close to the average delivery volume of the period, the value is 0, the high competition period value is greater than 0, and the low competition period value is less than 0.

In the competitive relationship of the same delivery period, various delivery areas also have different delivery volumes. Taking a relatively high pricing strategy in hot areas with high competition can balance the market.

3 Overall Design of the Client

Through the demand analysis of ambient media advertiser clients, the functions of advertiser clients are divided into the following five modules: advertising module, order management module, user management module, statistics module and setting module. To realize the functions required by the user, the client also needs the support of the network module and the storage module. The overall structure is shown in Fig. 1.

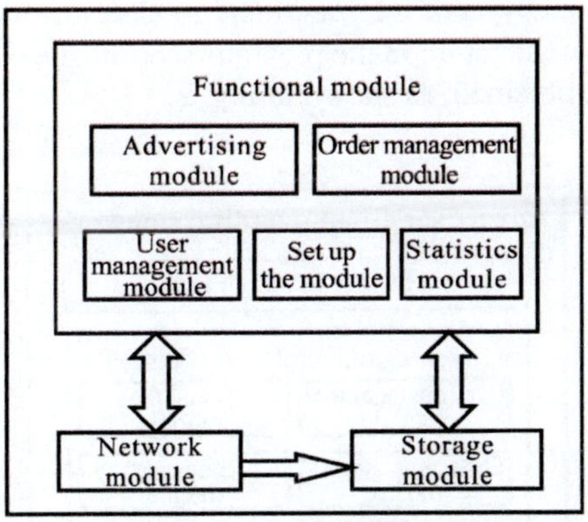

Fig. 1. Overall architecture of the ambient media advertiser client

The user management module provides account management functions such as registration, account password login, and third-party login. After the advertiser logs in, the user management module can also be used to add or modify personal information and modify the password.

The ad serving module provides users with an entrance to publish advertising orders, including uploading advertising resources and publishing orders. Advertisers can choose to upload image ads or upload video ads. Uploading image ads supports album selection and instant shooting [10]. Multiple pictures can be uploaded simultaneously, and the upload progress can be viewed. The upload of image ads also requires the settings of the image rotation time, uploaded video ads support album selection, and recording, and it can only upload a video during the upload process. Advertisers also need to fill in the title, introduction, delivery time, delivery area, and other information of the advertising.

The order management module displays the details of all orders, including order information and advertising information. The order information includes order status, order release time, etc. The advertising information includes the display of thumbnails of the advertising resources, the advertising area, and the advertising period. The order management module can also manage the delivery status of orders and the local storage status of orders.

The statistics module is used to display the information collected by the set-top box, such as the flow of people. The advertiser client uses a list to display the flow of people for each successful order. It is convenient for advertisers to quantify the advertising revenue.

The setting module can clear the cache, view the client version number, developer information, and update the client function.

The storage module is used for the analysis and local cache of user information, advertising information, order information, etc. It provides data interface support for the functional modules.

The network module is responsible for communicating with the server, providing functions such as uploading advertising resources, uploading orders, and obtaining statistical data.

Through the above analysis of the functional modules of the advertiser client, and the analysis of the iOS architecture in the previous section, the overall framework of the advertiser client can be obtained, as shown in Fig. 2.

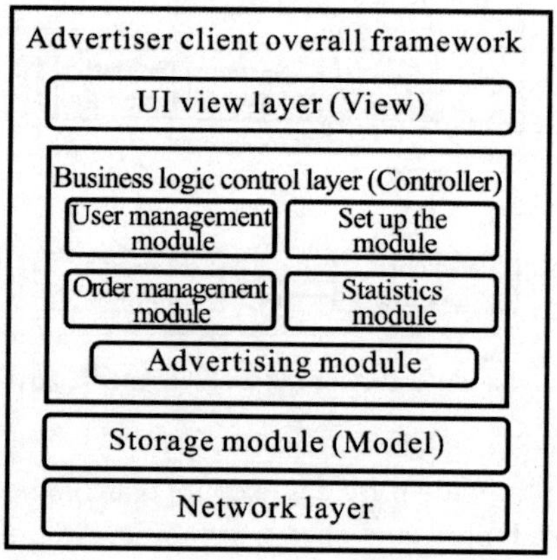

Fig. 2. Four-layer framework diagram of ambient media advertiser client

As shown in Fig. 2, the advertiser client's framework is divided into four layers, the view layer implements the client's UI presentation, and user interaction response, the business logic control layer implements the logical processing of five business function modules. The storage layer mainly performs data processing. Parse and encapsulate different persistence methods for different data. The network layer is mainly responsible for communication with the server and implements network status monitoring and data encryption.

4 Implementation of the Ambient Media Advertising Order Snatch System

Advertisers place advertisements on the advertiser's client. Advertisers can upload images to the server by selecting or recording pictures or video resources, and enter information such as the name of the advertising, the profile of the advertising, the area where the advertising is served, and the time of delivery.

Applications can use cameras and photo libraries through image pickers. The image picker interface is implemented by the UI Image Picker Controller controller class, which inherits from the UI Navigation Controller class and can be used to select photos and record videos.

The XS Upload Controller class is a control class for selecting or recording all advertising resources. It must implement the UI Image Picker Controller Delegate protocol and the UI Navigation Controller Delegate protocol. It connects the image picker, the XS Upload View, XS Request Model, and XS Request Data Manager, the interactive model of the ad delivery network, to control the entire process. This controller class creates an instance of 5 songs, 8 £ 6, 1 ^ 11 € 0, and 0111", and specifies the commission. Based on the user operation, the image source is specified, and at the same time, the user wants to select whether the image or video is selected, and controls the image. The picker selects or records, and saves to the album after recording.

The XS Upload View class is a view class for advertising. This class draws the UI layout presented to the user, such as the button btn Choose Pic for uploading advertising resources, the prompt information label, the text summary for the advertising introduction, the date picker, etc. Some methods of operating the control are designed, such as the click event of the select time button design click trigger event show Date Picker (), this method implements the click time button, the time selector pops up. When the time is selected on the time picker, the date Picker Value Changed () method is triggered. XS Upload Controller refers to the XS Upload View class to implement the methods provided.

The XS Request Model class is the data model for the entire advertising. Its attributes include advertiser ID, ad Name, ad Abstract, delivery start time begin time, delivery end time end time, delivery address Name, etc. The information entered by the advertiser when the advertising is placed can be matched with the model object. The XS Upload Controller class uses this data model as an attribute, processes the user input information obtained from XS Upload View, and stores it in this attribute.

The XS Request Data Manager class is a network class that inherits AF Netwoking and is responsible for the implementation of the server interface. This class designs a method for uploading advertising orders to the server, which uses POST to communicate with the server. The XS Upload Controller class implements the singleton of the network class as an attribute, calls the method of the network class, and uses the value in the attribute request based on the XS Request Model class as the parameter of the interface method to perform network communication. If the network condition is poor, an exception is thrown, and a pop-up window prompts the user.

When the advertiser needs to upload a large number of advertising resources, the network interruption during the upload process needs to be considered. To improve user experience and avoid unnecessary waste of network traffic, the client implements a breakpoint resume function. When using the HTTP protocol for network requests, the protocol will automatically upload large files in pieces, but if the network is interrupted during the upload process, we cannot get a breakpoint. After the network connection, you need to upload it again. Hence, when uploading locally, the large files are divided into pieces, such as 1M/piece, and the pieces are divided and uploaded to the server in sequence. After the server receives them, these small pieces are merged into the original file. At this time, the upload status of each piece is recorded locally, and when the network is interrupted and connected again, the local upload starts from the breakpoint to realize the continued transmission of the breakpoint.

Use the UI Image Picker View to obtain files and save them in the XS Ad File class. The attributes of this class include file type, file path, file name, file size, the total number of slices, and thumbnails. Use the read Data With Chunk method to fragment the file, and each fragment records the status of the upload, waiting means uploading, loading means uploading, and end means uploading has been successful.

Table 1. Ambient media advertising on bus platforms

	Vision		Hearing	Tactile	Sense of smell	Taste	Five senses
	Morphology	Color	Voice	Material	Smells	Taste	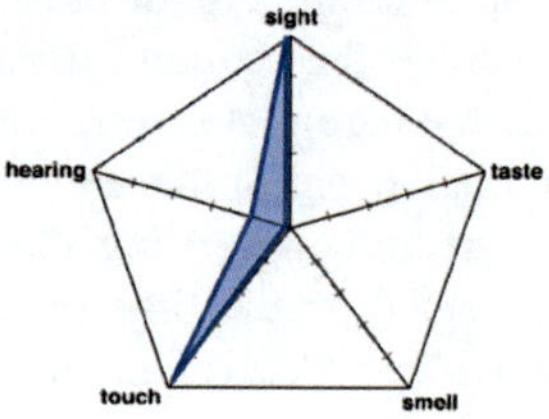
Approach	●	●	○	○	○	○	
Entrance	●	●	○	●	○	○	
Interactive	●	●	●	●	○	○	
Leave	●	●	○	●	○	○	

The process of uploading to the server is as follows:

1) The client sends an upload request to the server. The request parameters include the uploaded file name, file size, and other information.
2) After the server approves to upload, it generates an identifier to identify the upload session and returns it to the client.
3) After the information is received, the client starts sharding and records the upload status in the sharding. Information such as the identifier returned by the server, the starting position of the fragment, and the size of the fragment are appended to the upload information.
4) After the fragment is received, the server validates the fragment ID and size. If the verification is passed, the server stores the fragments and returns the correct response to the client. If the verification fails or the message is not received for a long time, the error message is returned to the client.

5) After the correct response is received, the client changes the upload status of the shard to finish and repeats steps (c) and (d) to upload the remaining shards in multiple threads. If you receive an error message from the server, an exception is thrown. Wait for the next upload.
6) When upload again, the client skips the shard whose upload status is finish, and only uploads the shards in the wait and load status to realize the resuming of the breakpoint.
7) After all the file fragments are received, the server merges the fragments in order, restores the source file, encrypts the file, and calculates the verification code. The server returns the message that the file upload was successful to the client. The client clears the shards and completes the upload of large files. Table 1 shows the ambient media advertisings are placed for bus platform order snatchs.

5 Conclusions

In this paper, based on the price stepping algorithm, we proposed a new ambient media advertising model based on O2O that integrates the functional modules of advertising master client and device master client. The frameworks of the two master clients are designed respectively. At the same time, the ambient media advertising order snatch system is constructed. The offline large-screen resources are integrated to further promote the development of ambient media in modern society [11]. The penetration of ambient media advertising into the living environment allows consumers to remember them more quickly.

References

1. Wei, L.: Establishment of business English corpus based on foreign economic demand dependent on automatic generation algorithm of conceptual semantic network. In: Macintyre, J., Zhao, J., Ma, X. (eds.) SPIoT. LNDECT, vol. 97, pp. 581–589. Springer, Cham (2021). https://doi.org/10.1007/978-3-030-89508-2_75
2. Golubkova, E.: The urban environment transformation through ambient media in the estimates of the Yekaterinburg advertising community. KnE Soc. Sci. **5**(2), 364–373 (2021)
3. Zhang, G.J., Xiao, C.Y., Ojo, O.O.: Dissimilar friction stir spot welding of AA2024-T3 /AA7075-T6 aluminum alloys under different welding parameters and media. Defence Technol. **17**(2), 14 (2021)
4. Yang, J.: Implantable network advertising video marketing based on FPGA and Sobel Algorithm. Microprocess. Microsyst. **80**, 103347 (2020)
5. Casado-Aranda, L.A., Martinez-Fiestas, M., Sanchez-Fernandez, J.: Neural effects of ambient advertising: an fMRI analysis of voice age and temporal framing. J. Ambient Manage. **206**(15), 664–675 (2018)
6. Park, D.J.: United States news media and climate change in the era of US President Trump. Integr. Ambient Assess. Manage. **14**(2), 202–204 (2018)
7. Brulle, R.J., Aronczyk, M., Carmichael, J.: Corporate promotion and climate change: an analysis of key variables affecting advertising spending by major oil corporations, 1986–2015. Clim. Change. Change **159**(1), 87–101 (2020)
8. Erlygina, E., Vasilyeva, A.: Ambient media in a market system. Bull. Sci. Pract. **5**(11), 280–283 (2019)

9. Po-Chin, Y., Wui-Chiang, L., Hao-Yen, L., et al.: Use of Facebook by hospitals in Taiwan: a nationwide survey. Int. J. Ambient Res. Public Health **15**(6), 1188–1194 (2018)
10. Angela, C., Peter, S., Tony, S., et al.: Implicit messages regarding unhealthy foodstuffs in Chinese television advertisements: increasing the risk of obesity. Int. J. Ambient Res. Public Health **15**(1), 70–78 (2018)
11. Sheehan, K., Morrison, D.: Advertising leadership and climate change: the efficacy of industry professionals to address climate issues. Advertising Soc. Q. **19**(1), 1–8 (2018)

Face Recognition System Based on Data Box Security Mechanism

Linru Yue[1], Danhong Chen[1](✉), Jun Zhao[2], Kehui Li[1], and Meilin Li[1]

[1] School of Economics and Management, Shenyang Aerospace University, Shenyang, Liaoning, China
icatci22139@163.com

[2] School of Marxism, Shenyang Aerospace University, Shenyang, Liaoning, China

Abstract. Face recognition began in the 1960s, by virtue of its rapidity, it has been widely used in various fields. However, face recognition technology still has a large room for progress, at the same time, the exposed data security problem is also one of the key defects concerned by the public. With the development and application of face recognition technology, the problem of secure storage of face data has been paid more and more attention. The leakage, tampering and damage of facial information caused by system vulnerabilities will cause huge losses to the personal and property security of users. Therefore, this paper proposes a face recognition system which is based on data box security mechanism to realize the secure storage of face information. At the same time, this paper also puts forward a variety of face recognition algorithm fusion, can further reduce the error of face recognition, improve the accuracy of face recognition.

Keywords: Face recognition system · Data security · Security mechanism

1 Introduction

Computer network and communication technology advances by leaps and bounds,as a result of which, authentication is becoming increasingly significant. With its rapid and accurate characteristics, face recognition quickly enters the service market and becomes a common way of identity verification in hotels, residential areas, dining places and other places where a variety of personnel are concentrated [1]. At the same time, the derived face payment, face check in and other functions have greatly improved the convenience and sense of science and technology of life. But as a result of face recognition has identification process, autonomy, personal relationship and risk prevention and control complexity, imprecise information protection system is easy to crack, resulting in facial information disclosure, tampering and destruction, box of security mechanisms in this paper, based on the data of face recognition system, which can realize the secure storage of the face information, help the healthy development of face recognition [2].

J. C. Hung et al. (Eds.): FC 2023, LNEE 1132, pp. 75–82, 2024.
https://doi.org/10.1007/978-981-99-9538-7_11

2 Related Works

In the face recognition system development status, and the past face recognition technology is different, now face recognition technology has been improved, but the comprehensive application has not reached the best state [3]. One deficiency of face recognition is that the stability and dependability of this information is not high enough. Compared with biometric features such as fingerprints and irises, faces contain less information and lack of complexity. In the process of obtaining, the stability of face information is affected by the light in the environment, the pose of the person, the expression of the person, and whether there is occlusion [4]. The recognition errors and image quality problems caused by the lack of samples are also inevitable [5]. How to better recognize faces in different lights and angles? How to determine the identity clearly and accurately is still a technical pain point that needs to be solved [6]. In addition, with the further expansion of the application range of face recognition, the storage security of facial information has become a key problem that needs to be solved urgently [7].The commonly used data security protection methods in the early years can be divided into the following categories [8]. Although the following methods can protect the security of data in a large scope, they generally have the drawbacks of long backup cycle and high construction cost [9]. As show in Table 1.

Table 1. Traditional data protection methods

type	role
RAID protection	Soft RAID is to use software to achieve RAID function, without any hardware to achieve, this kind of program can run in the bottom of the operating system, from the host SCSI or IDE controller submitted to the physical hard disk, virtual into a variety of modes of virtual disk, and then submitted to the upper layer of the program interface
Conventional backup	Through this method, data is retained in specific way in order to be rebuilt after system breach or other extreme cases
Copy	Instantaneously transfer data from one place to another over a network (LAN/WAN) to form one or more copies
Snapshot	A rapid data protection technique in which a snapshot is a mirror image of a data set at a specific time, also known as instant copy, and is a complete available copy of the data set
CDP	Continuous Data Protection (CDP) technology is a revolutionary breakthrough to the traditional data backup technology. The system continuously monitors and records the changes of source data, thus automatically protecting data. When a disaster occurs, the system simply selects the point in time to recover the data. In principle, data can be recovered to any point in time

3 Methods

3.1 Face Recognition System Based on the Data Box Security Mechanism

Resolve to solve these problems of safe storage of face information. In this paper, a data box security mechanism is proposed, which uses the data box model to encrypt the face information, adopts the traditional key cryptography system in the copyright and privacy protection of data in the data box, and incorporates the opt-in and opt-out technologies. In this way, data can be stored securely.

The key is divided into two kinds: symmetric key and asymmetric secret. Symmetric key encryption whice be called as private key encryption at the same time. In this way, data only be encrypted and decrypted when the sender and the receiver of the information use this key [10]. The significant preponderance of symmetric key encryption is its efficient encryption and decryption, which can successfully encrypt a lot of data. Asymmetric key encryption which be called as public key encryption at the same time. It needs to use a pair of keys to perform data decryption and encryption. One is publicly released, also be named as public key, is controlled by the sender to encrypt the data, the other is private key which is controlled by the receiver to decrypt the date.

Opt-in refers to obtaining the consent of the data subject before sharing and exploiting the face information [11]. Opt-out refers to when face data becomes outdated, irrelevant, and out of scope, and can be forgotten. So as to ensure the security of face information.

3.2 Realization of Data Security Box Mechanism

3.2.1 Pipeline is Introduced

To improve the security of face recognition, a new method has been proposed, P2SGrad. This method calculates the gradient length just utilize \theta_{i, j} in the training face recognition model. $\theta_{i,j}\theta_{i,j}$ Formally, the gradient length work out by P2SGrad is per-parametric, independent of the number of C classes, and also independent of the specific definition of logit F_i, y_i f_{i, y_i}F_i, y_i. P2SGrad don't use a stipulated formula for the loss deviation, which is not only concise but also rigorous. When j =y_i, , the length of the proposed gradient is positively related to} \theta_{i, j}}, and when j!$\theta_{i,j}\theta_{i,j}$=y_i, , they are inverse relationship. More ingenious, the gradient length in P2SGrad depends only on} \theta_{i, j}} so it is conforms to the test metrics of face recognition arithmetic. $\theta_{i,j}\theta_{i,j}$ It can effectively improve the security of face recognition system.

3.2.2 SoftMax Passenger

Softmax Passenger method is introduced to improve the accuracy of face recognition. In the operation of face recognition,the cosine Softmax cross-entropy loss has two part formula, Cross-entropy loss and Softmax function. Apply the following formula.

Assume the facial features are denoted by vector_xi and then let's substitute logit Fi,j f_{i,j}$f_{i,j}$ into the softmax function:

$$f_{i,j} = s \cdot \frac{\langle \vec{x}_i, \vec{w}_j \rangle}{\|\vec{x}_i\|_2 \left\| \overrightarrow{W}_j \right\|_2} = s \cdot \left\langle \widehat{x}_i, \widehat{W}_j \right\rangle = s \cdot \cos\theta_{i,j} \quad (1)$$

In the formula above, s is the hyperparameter and is the classification score (logit) that class j is assigned by x_i which is the weight vector of class j. $f_{i,j}\overrightarrow{W}_j\overrightarrow{x}_i$ and; $\overrightarrow{W}_j$ Normalized vectors of the respective. $\overrightarrow{x}_i\overrightarrow{W}_j\theta_{i,j}$ Is the Angle between the feature and the class weights. $\overrightarrow{x}_i\overrightarrow{W}_j$. The probabilities $f_{i,j}$ is given by entering logits into the following formula:

$$P_{ij} = \text{Softmax}() = f_{i,j}\frac{e^{f_{i,j}}}{\sum_{k=1}^{C} e^{f_{i,j}}} \tag{2}$$

In the formula above, C is used to represent the number of this classes and the result $P_{i,j}$ is the probability of $\overrightarrow{x}_i$ which is substituted into class j.If $\mathrm{j} = \mathrm{y_i}$, P_{ij} is the probability that $\overrightarrow{x}_i$ is assigned to the corresponding y_i.

3.2.3 Introducing Face Recognition Metric Graph

In the face recognition algorithm, the most widely applied metric learning method is triplet loss function, which was first applied into face recognition. Triplet loss can separate the distance between positive and negative example pairs with a certain margin. Mathematically speaking, each triplet i should meet the following formula:

$$\|f(x_a) - f(x_P)\|_2^2 + \alpha < \|f(x_a) - f(x_n)\|_2^2 \tag{3}$$

The case of binary classification using SoftMax loss

$$||x||(||W_1||cos\theta_1 - ||W_2||cos\theta_2) = 0 \tag{4}$$

In the above formula, x represents the feature vector. W_1 and W_2 is the weight to each class. θ_1 and θ_2 represent the angles x forms with respect to W_1 and W_2, respectively. As show in Fig. 1 and Fig. 2, these two decision bounds can be made more stringent by adding a multiplicative margin into the above formula:

$$||x||(||W_1||cosm\theta_1 - ||W_2||cos\theta_2) = 0 for class 1 \tag{5}$$

$$||x||(||W_1||cos\theta_1 - ||W_2||cosm\theta_2) = 0 for class 2 \tag{6}$$

This way can greatly improve the degree of differentiation between categories and the compactness within each category.

4 Experimental Process

4.1 Step 1

Get the set S which containing M face images. During the presentation, we have 25 face images, as shown in the following figure. As show in Fig. 3, every image can be transform into an n-dimensional vector, next, we put the M vectors into a set S, as follows:

$$S = \{\Gamma_1, \Gamma_2, \Gamma_3, \ldots\ldots, \Gamma_M\} \tag{7}$$

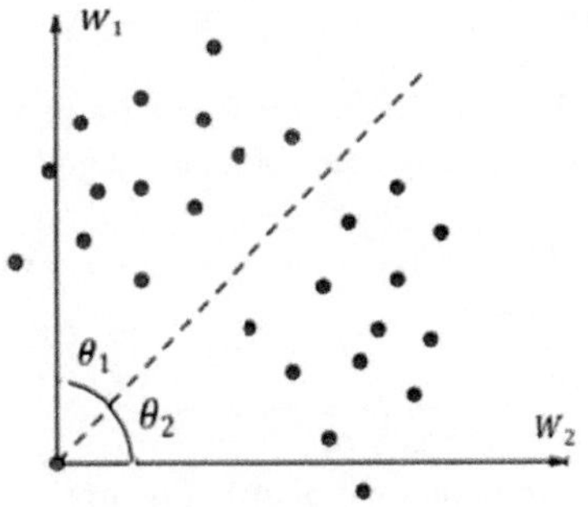

Fig. 1. Algorithm diagram of face recognition process (1)

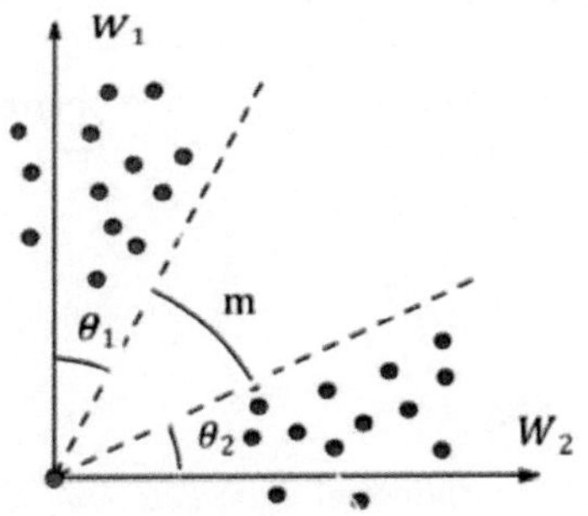

Fig. 2. Algorithm diagram of face recognition process (2)

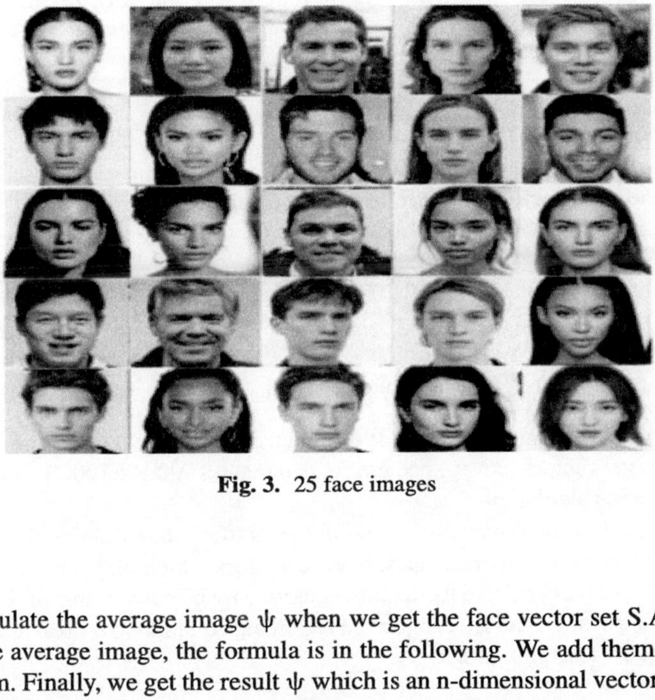

Fig. 3. 25 face images

4.2 Step 2

We can calculate the average image ψ when we get the face vector set S.As to how to calculate the average image, the formula is in the following. We add them up and then average them. Finally, we get the result ψ which is an n-dimensional vector. If we bring it into the image form, you get the following result which is called "average face".

$$\psi = \frac{1}{M}\sum_{n=1}^{M}\Gamma_n \tag{8}$$

4.3 Step 3

Subtracting the average value obtained in Step 2 from each factor in set S,and then we can get the difference value between the normal image and the average image.

$$\Phi\text{i} = \Gamma\text{i} - \psi \tag{9}$$

4.4 Step 4

Find M orthogonal unit vectors u_n which is used to depict the distribution of Φ.

k (k = 1,2,3...) in u_n are computed by the following equation:

$$\lambda_k = \frac{1}{M}\sum\nolimits_{n=1}^{M}\left(u_k^T\Phi_n\right)^2 \tag{10}$$

u_k can only be determined when λ_k is minimum. By the way, u_k also satisfies the following equation:

$$u_l^T u_k = \delta_{lk} = \begin{cases} 1 \; if \; l = k \\ 0 \; otherwise \end{cases} \tag{11}$$

u_k is a unit orthogonal vector in the formula above. Calculating u_k above is essentially calculating the eigenvectors of the following covariance matrix:

$$\mathrm{C} = \frac{1}{\mathrm{M}}\sum\nolimits_{\mathrm{n}=1}^{\mathrm{M}}\Phi_{\mathrm{n}}\Phi_{\mathrm{n}}^{\mathrm{T}} = \mathrm{AA}^{\mathrm{T}} \tag{12}$$

Among them

$$\mathrm{A} = \left\{\Phi_1, \Phi_2, \Phi_{3,\ldots,}\Phi_{\mathrm{n}}\right\} \tag{13}$$

In an N × N (for example, 100 × 100) dimensional image, it is too large to directly calculate its eigenvectors (covariance matrix can reach 10000 × 10000), so we have the following simple calculation.

If the amount of the dimension of the image is more than training images, such as (N > M^2). Under this circumstances, feature vectors which play a role, are only M-1 instead of N^2.So we can solve the feature vectors only by calculating an N × N matrix. Setting this matrix as L, and we can use the following equation to represent the elements in the m row and n column of L:

$$L_{mn} = \Phi_m^T\Phi_n \tag{14}$$

The eigenvectors u_l of the covariance matrix can be described as follows when we have found the M eigenvectors v_l of the L matrix.

$$u_l = \sum\nolimits_{k=1}^{M} v_{lk}\Phi_k \; l = 1, \ldots, M \tag{15}$$

These eigenvectors, if reduced to a pixel arrangement, are called eigenfaces (see Fig. 4). There are 25 feature faces in the picture.

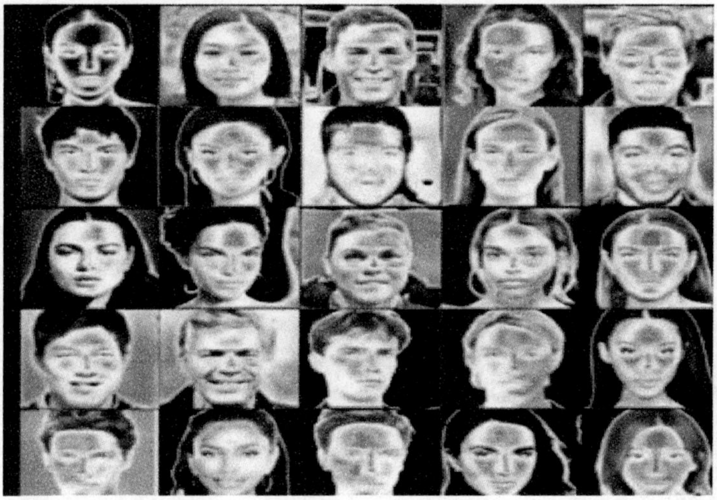

Fig. 4. 25 features

5 Results and Discussion

After further modification and improvement of the data security box and face recognition, we carried out experiments on the system. Our experiment found that the face recognition system based on data box security mechanism can improve the detection speed by about 30% compared with other systems. At the same time, the security of the data box storage mechanism is checked. It is found that the data security box can resist the external attacks to the maximum extent to ensure the safe storage of information. The combination of scientific algorithms and tight protection mechanisms magnifies the advantages of face recognition technology and makes it easier to gain public trust. In the future, first of all, we plan to further innovate the algorithm, improve the accuracy of face recognition, and ensure that the face data is not infringed in the process of storing the data security box. Secondly, we will expand the capacity of the face database to ensure that the system can meet the large-scale application conditions.

6 Conclusion

On the basis of analyzing face recognition algorithm and exploring data security protection method, this paper proposes a face recognition algorithm based on data box security mechanism. The fusion of multiple face recognition algorithms can not only ensure the accuracy of recognition, but also improve the speed of recognition. At the same time, the data box security mechanism can maximize the security of face information. In general, the face recognition system based on the data box security mechanism from the point of view of security, for the user to consider all aspects. In the most efficient way, it meets the requirements of users in the aspect of personal information protection, and has a good development prospect.

Acknowledgments. This research was supported by the college students' innovative entrepreneurial training plan of Shenyang Aerospace University under grant X202210143046.

References

1. Pillai, S.S., et al.: Image assisted data security using key encrypted file. Int. J. Eng. Res. **V4**(05), 1–4 (2015). https://doi.org/10.17577/IJERTV4IS050194
2. Thapa, C., Camtepe, S.: Precision health data: requirements, challenges and existing techniques for data security and privacy. Comput. Biol. Med. **129**, 104130 (2021). https://doi.org/10.1016/j.compbiomed.2020.104130
3. Hkirdov, D.A., Sagatov, E.S., Dmitrenko, P.S.: Trap method in ensuring data security. Inf. Technol. Nanotechnol. 189–198 (2019)
4. Sugumar, R., Iman, S.: Data security in public cloud storage environment. Int. J. Eng. Res. Technol. **V4**(06), 1–6 (2015). https://doi.org/10.17577/IJERTV4IS060127
5. Khosla, P.K., Amandeep, K.: Big data security solutions in cloud. Adv. Parallel Comput. **29**, 80–106 (2018)
6. Saoji, S.U., Kumar, N., Jayakumar, Gupta, V.: Attendance management system using deep learning and facial recognition technique. Int. J. Innov. Technol. Explor. Eng. **8**(10), 4097–4101 (2019). https://doi.org/10.35940/ijitee.J9423.0881019
7. Abdulhussain, S.H., Mahmmod, B.M., AlGhadhban, A., Flusser, J.: Face recognition algorithm based on fast computation of orthogonal moments. Mathematics **10**(15), 2721 (2022). https://doi.org/10.3390/math10152721
8. Singh, S., Singh, M., Tarar, S.: Improved facial recognition based authentication approach to secure big data. Int. J. Innov. Technol. Explor. Eng. **8**(6c), 14–20 (2019)
9. Bagrov, N., Konushin, A.S., Konushin, V.S.: A semi-automatic method of collecting samples for learning a face identification algorithm. Program. Comput. Softw. **45**(3), 133–139 (2019)
10. Hub, M., Prihodova, A.K.: Impact of globalisation on data security – authentication issues. SHS Web Conf. **92**, 05009 (2021). https://doi.org/10.1051/shsconf/20219205009
11. McCaldin, P.E.: Towards building a facial identification system using quantum machine learning techniques. J. Adv. Inf. Technol. **13**(2), 198–202 (2022)

Development Strategy of Smart Cities Based on Artificial Intelligence

Chonghui Li(✉)

School of Information and Business Management, Chengdu Nfeusoft University, Chengdu, Sichuan, China
lichonghui@nsu.edu.cn

Abstract. The rapid development of technologies such as big data, cloud computing and AI has laid the foundation for solving contemporary urban governance problems, and at the same time, has enabled the development of "smart cities" in a deeper way. The "wisdom" of a smart city is mainly expressed in the interconnection of various factors within the city, making the city an organic whole and enabling more accurate and intelligent decision-making. In today's information age, smart cities are seen as a new way of thinking about urban governance and the future direction of urban construction. The aim of this paper is to investigate smart city development strategies based on artificial intelligence. From a "people-centred" perspective, the development strategy of smart cities and artificial intelligence technologies are analysed. The project's practical results show the effectiveness of the platform design and implementation.

Keywords: Artificial Intelligence · Smart City · Development Strategy · Video Surveillance

1 Introduction

Smart city will bring the evolution and update of management mode to the development of modern industrial city, and also bring unexpected changes to the living and production mode of urban residents. Smart city is a new form of urban development that has continued since the industrial revolution. It is established by utilizing the achievements of human civilization. Relying on technological media such as big data and artificial intelligence, smart city can effectively integrate the information of urban life and urban development, and promote the informatization and wisdom of human society through the city. It is also an important option to solve problems in urban development [1, 2]. At the same time, systematically applying smart city to specific cases in theory, defining its concept and grasping the specific forms of smart city application are conducive to strengthening the understanding of smart city development mechanism, enriching urban construction theories and smart city development models, and promoting relevant theoretical research of smart city [3, 4].

Given the increasing global population, the construction of smart cities has become critical, aiming to improve urban flow management with efficient information and communication technologies (ICTs) [5]. Vehicle-mounted sensor networks (VSNS) play a

J. C. Hung et al. (Eds.): FC 2023, LNEE 1132, pp. 83–90, 2024.
https://doi.org/10.1007/978-981-99-9538-7_12

crucial role in keeping smart cities running efficiently. Of course, there are a number of challenges that need to be addressed before virtual storage networks can be introduced on a large scale, including the concept of accurate topological analysis methods and beneficial collaborative mechanisms in the city-wide information sharing process. Hedi Haddad constructed a VSN-assisted smart city model and evaluated a range of smart applications in public services and urban flow management. Then, the information source selection algorithm of complex network and the city information sharing mechanism based on reinforcement learning are considered, supplemented by a series of open challenges [6]. With the rise and development of the Internet of Things as a new form of sustainable development, the concept of a "smart city" has gained considerable development. Smart cities are based on autonomous and distributed infrastructure, including heterogeneous network infrastructure for intelligent information processing and control systems, as well as ubiquitous sensing involving millions of information sources. However, due to the continuous growth in the amount of data and the number of Internet-of-Things devices, high latency, bandwidth bottlenecks, security and privacy, and scalability appear in the current smart city network architecture [7]. To address the limitations of today's smart city networks, an efficient, secure and scalable distributed architecture needs to be designed by bringing computing and storage resources close to endpoints. Siddhant Jain has proposed a new hybrid network architecture for smart cities, taking advantage of emerging software-defined networking and blockchain technologies. To achieve efficiency and address current limitations, their architecture is divided into two parts: the core network and the edge network. By designing a hybrid architecture, they propose an architecture that inherits the advantages of both centralized and distributed network architectures. They also proposed a proof of work scheme in the model to ensure security and privacy. In order to evaluate the feasibility and performance of the proposed model, they evaluated it based on various performance indicators. The evaluation results show the validity of their proposed model [8].

At present, scholars at home and abroad have conducted a large number of researches on smart cities, but most of the achievements focus on the "wisdom" level and pay insufficient attention to the human factor. In contrast, this paper explores and analyzes the development principles and strategies of smart city construction from a new perspective of "people-oriented", and proposes a video and image application platform for the construction of a new smart city in M County.

2 Artificial Intelligence Technology and Smart City Development Strategy

2.1 People-Oriented Development Principle

The understanding of the connotation of "people-oriented" should start from "people" and "this". Only by fully understanding the meaning of "people" and "this" can we better understand the meaning of "people-oriented" [9].

The "people-oriented" mentioned in this paper refers to the "people" living in the urban ecosystem. Smart city provides all-round services for "people", so that urban residents can enjoy efficient, convenient, safe and green urban life. This requires that the construction of new smart cities must take the actual needs of "people" as the main body to increase their happiness and satisfaction. Through the innovative application of modern information technology, urban information exchange and business cooperation can be realized, so that citizens can better understand the city and improve the comfort and efficiency of urban life. To be specific, urban development should be achieved through public participation, and attention should be paid to meeting the needs of citizens both materially and spiritually, paying attention to the personality and characteristics of cities.

2.2 Development Strategies of Smart Cities

(1) Strengthen data resource management and promote data sharing and openness

By establishing a data resource management platform, we can promote the storage and sharing of massive data in cities, and realize the convergence, governance, analysis and opening of data. Relying on the underlying big data platform to provide hierarchical storage space for data, the data is unified and aggregated to guide people within the scope of system management [10, 11]. Analyze data quality indicators according to the transportation industry data standard system, and provide data cleaning tools, views and other data governance tools to continuously improve data quality. By comparing and analyzing the data stored, associated and collected, the system of data consanguinity and data relationship is established. Finally, data permission management is used to orderly open the data. Combined with intelligent transportation storage data resources, expand the range of data collection, exchange and sharing [12, 13].

(2) Develop innovative technologies such as digital retina to effectively extract key data

The intelligent level is improved through video parsing pre-processing and end-to-end cloud collaborative transformation. Different from the way that traditional cameras process video at the back end, digital retina identifies the feature images of "people, cars, non-images" obtained from video data stream based on multi-service requirements. Then, the encoded video stream is stored locally and uploaded to the cloud on demand, and all compact feature streams are synchronized to the cloud in real time [14]. In this way, not only can the retrieval and storage capabilities of the system be improved, manpower input in the later stage be saved, but also the delay caused by system response in emergency events can be greatly shortened [15].

2.3 Artificial Intelligence Technology

(1) Video management platform

In intelligent video surveillance, it is based on network and centered on image processing. The construction of video surveillance system greatly depends on the integration and upgrading of management platform, especially in the large system, the role of platform is increasingly prominent. And video surveillance system also gradually from the scope of security surveillance, extended to other fields.

(2) AR technology

AugmentedReality (AR) is an emerging technology that integrates image acquisition and processing with human-computer interaction. Its realization process is to enhance the real environment by effectively matching the virtual information generated by computers with the real environment.

One of the key processes of the augmented reality system to achieve information enhancement is the recognition and matching of the target image, one of the key processes of the augmented reality system to achieve information enhancement is the recognition and matching of the target image, SIFT algorithm for the rotation of the image has a good adaptability, for the outside light changes and radiation also has a high stability.

If I(x, y) is an initial given image and G(x, y, δ) is a Gaussian function that can change accordingly in the scale space, then the scale space of the initial given image can be expressed as:

$$L(x, y, \delta) = G(x, y, \delta) \times I(x, y) \tag{1}$$

In the above Eq. (1), δ represents the size of the scale space, and the value of δ is inversely proportional to the blur degree of the image. The larger the value of δ, the general appearance of the image; The smaller the value of delta, the clearer the detail of the image. The Gaussian function G(x, y, δ) is:

$$G(x, y, \delta) = \frac{1}{2\pi\delta^2} e^{-\frac{(x^2+y^2)}{2\delta^2}} \tag{2}$$

After a series of scale space transformation steps and a series of operations such as double downsampling, the final Gaussian pyramid can be obtained. Gaussian difference (DOG) can find a series of stable and invariant extremum point σ in the scale space. Gaussian difference (DOG) function function is generally defined as:

$$G(x, y, \delta) = L(x, y, k\delta) - L(x, y, \delta) \tag{3}$$

where, kδ and δ represent the smooth scale of two consecutive images. Feature points are generally composed of some local extreme points in DOG space.

3 The Realization of Smart City Development Strategy

3.1 Operating Environment Design

The running environment of the central system platform should first be built on a server with a well configured environment. If you choose IE8 or above, you can realize the operation of the system platform. In addition to Oracle, Tomcat and Java Jdk need to be installed and configured. These software are used to support the operation of the system platform.

3.2 Video Image Application Platform Design

Video and image application platform is a unified entrance, which can provide various government departments and units with video intelligent application services based on video, pictures, face, human body, vehicle structured data and model data, including basic video application, query and retrieval application, distribution and alarm application, etc. The functional architecture of the video image application platform is shown in Fig. 1.

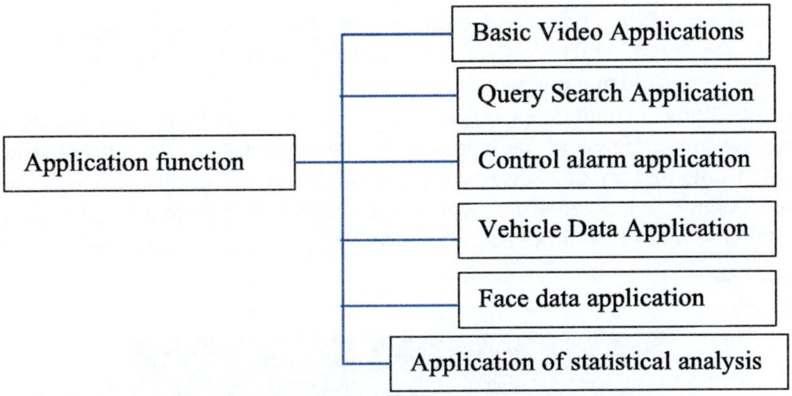

Fig. 1. Functional architecture diagram of video image application platform

3.3 Functional Architecture Design of Video Image Information Database

Data acquisition is mainly to provide a standard interface for the video surveillance system of each functional unit to synchronize the collection of image data. By using mature cloud computing technology, a large number of valuable data is stored and shared, and it is divided into subject libraries to provide support for related commercial applications. Data management is the classification, data processing, file sharing, data seriation processing of the data stored in the image database. In the aspect of system data query, rapid query of the data in the system using advanced full-text retrieval technology, in the aspect of system management, including user management, log management, dictionary management, and status monitoring of the system.

4 Practice Verification

4.1 Background

In 2022, M County was listed as the pilot county for the innovation and reform of grassroots social governance mechanism. Driven by AI deep learning, the project completed the integration of comprehensive governance information network, public safety video surveillance network, comprehensive governance visual network, people hotline integration and comprehensive emergency response in Tonglu smart governance Information Center in accordance with the construction and management norms of comprehensive governance center. To build a people-oriented smart city center system featuring "integration of civil air defense + technical defense, and online + offline linkage".

4.2 AR Live Command

The panoramic camera and AR/VR video map command combat platform are adopted. Based on the 4K video images of the real road, various traffic information such as video surveillance and bayonet are superimposed to realize the functions of AR panoramic real-time video, AR panoramic video linkage and joint control, and visual management of distribution control and alarm, so as to build a new experience of three-dimensional and precise command and combat system.

The functions of cloud image AR real scene command include multi-picture layered display, real-time preview of panoramic video, linkage and joint control of panoramic video, real-time display of alarm data, joint linkage of GIS map, automatic patrol of AR scene, high-low linkage of AR scene, high-altitude linkage of AR scene, automatic target tracking and other functions. The effect of panoramic video linkage control is shown in Fig. 2.

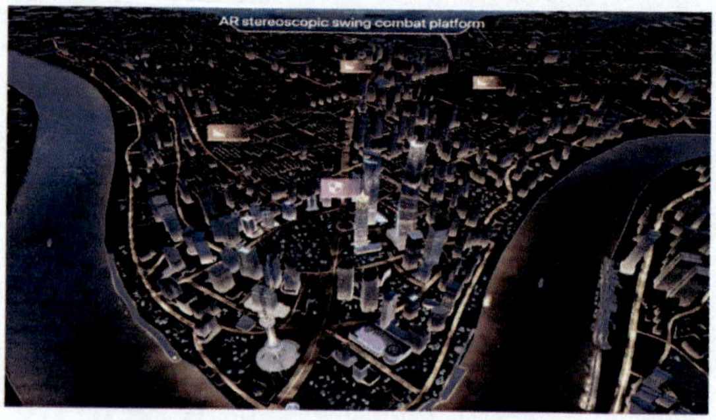

Fig. 2. Schematic diagram of panoramic video linkage and joint control effect

4.3 Urban Traffic Operation Monitoring

Through the study of this project, the intercity transport system such as passenger station, toll station, airport and so on can be comprehensively monitored. Urban traffic operation monitoring can be the urban traffic high-speed toll station, passenger station, train station, airport location, and it with passenger flow, traffic flow, video and other relevant information displayed on the map. By analyzing the data, the annual and monthly passenger data of urban traffic can be displayed, and the changes of monthly traffic trips can be studied. The monitoring of urban traffic operation is shown in Table 1.

Table 1. Monthly Passenger Traffic Volume (Unit: 10000 person times)

Time	Civil aviation	High speed	Highway	Railway
January	57	89	104	65
February	44	88	101	52
March	46	80	94	50
April	41	79	80	44
May	58	92	112	60

5 Conclusions

In the process of smart city construction, video surveillance system is an essential core part. This paper aims to analyze the existing problems of smart city construction in M County from a new perspective of "people-oriented", and put forward corresponding countermeasures and suggestions for implementing the "people-oriented" concept and improving the new smart city. The establishment of video surveillance system is inseparable from the development and accumulation of computer technology, network technology, storage technology and chip technology. IT is because of the development and accumulation of these basic IT technologies step by step, to ensure the upgrading of video surveillance system supporting equipment, so as to promote the upgrade of the whole video surveillance system. Similar to the application of traditional communication products, the wireless transmission of video surveillance system is conducive to the wider and more flexible application of the system. The wireless transmission technology based on ordinary SIM and UIM cards can further promote the wireless penetration of video surveillance system.

References

1. Fontaine, P., Minner, S., Schiffer, M.: Smart and sustainable city logistics: design, consolidation, and regulation. Eur. J. Oper. Res. **307**(3), 1071–1084 (2023)
2. Gordon, E., Harlow, J., Teng, M., Christoferetti, E.: Toward a collaborative smart city: a play-based urban living laboratory in Boston. Int. J. Human–Comput. Interact. **39**(2), 302–318 (2023). https://doi.org/10.1080/10447318.2021.2012384
3. Fadda, M., Anedda, M., Girau, R., Pau, G., Giusto, D.D.: A social internet of things smart city solution for traffic and pollution monitoring in Cagliari. IEEE Internet Things J. **10**(3), 2373–2390 (2023)
4. Rahman, M.A., Shamim Hossain, M., Showail, A.J., Alrajeh, N.A., Ghoneim, A.: AI-enabled IIoT for live smart city event monitoring. IEEE Internet Things J. **10**(4), 2872–2880 (2023). https://doi.org/10.1109/JIOT.2021.3109435
5. Sachan, A., Kumar, N.: SDN control-enabled and time-quantum-driven max-pressure approach for intersection management in smart city. IEEE Syst. J. **17**(1), 1694–1702 (2023)
6. Haddad, H., Bouyahia, Z., Horchani, L., Jabeur, N., Gharrad, H.: A connected mobility scheme for taxi supply-demand balancing in a smart city context. Comput. Inform. **41**(2), 550–570 (2022)
7. Abbas, S., Khan, M.A., Athar, A., Shan, S.A., Saeed, A., Alyas, T.: Enabling smart city with intelligent congestion control using hops with a hybrid computational approach. Comput. J. **65**(3), 484–494 (2022). https://doi.org/10.1093/comjnl/bxaa068
8. Jain, S., ShashankGupta, K.K.S., Rodrigues, J.J.P.C.: Fog computing in enabling 5G-driven emerging technologies for development of sustainable smart city infrastructures. Cluster Comput. **25**(2), 1111–1154 (2021). https://doi.org/10.1007/s10586-021-03496-w
9. Sharma, R., Arya, R.: A secure authentication technique for connecting different IoT devices in the smart city infrastructure. Clust. Comput. **25**(4), 2333–2349 (2022)
10. Conde, J., Munoz-Arcentales, A., Choque, J., Huecas, G., Alonso, Á.: Overcoming the barriers of using linked open data in smart city applications. Computer **55**(12), 109–118 (2022)
11. Kostic, Z., et al.: Smart city intersections: intelligence nodes for future metropolises. Computer **55**(12), 74–85 (2022)

12. DeHart, J.: Understanding equity and diversity in smart city infrastructures: Chicago, IL, USA. XRDS **28**(3), 60–61 (2022)
13. Kataria, A., Puri, V.: AI- and IoT-based hybrid model for air quality prediction in a smart city with network assistance. IET Netw. **11**(6), 221–233 (2022)
14. Jadric, M., Mijac, T., Cukusic, M.: End-user approach to evaluating costs and benefits of smart city applications. Int. J. E Serv. Mob. Appl. **14**(1), 1–15 (2022)
15. Mohanty, A., et al.: Identification and evaluation of the effective criteria for detection of congestion in a smart city. IET Commun. **16**(5), 560–570 (2022). https://doi.org/10.1049/cmu2.12344

Modeling of Financial Risk Control Imbalance Dataset Based on Benchmarking Management Optimization Algorithm

Yichen Liu[1(✉)] and Jun Yu[2]

[1] School of Economics, Henan University of Economics and Law, Zhengzhou 450000, China
15039578750@163.com

[2] Economic and Management College, Civil Aviation University of China, Tianjin, China

Abstract. Imbalance modeling of financial risk control refers to a situation where the sample in the data set is unbalanced due to various factors, such as the credit level and repayment ability of different financial customers. In order to evaluate and control risks more accurately, it is necessary to model unbalanced data. This article mainly designs models for unbalanced data sets of financial risk control, and uses different algorithms to compare and analyze the computational capabilities of the algorithms. It studies the financial risk control of benchmarking management optimization algorithms. Experimental data show that the maximum AUC value and accuracy value obtained by the benchmarking management optimization algorithm in the risk control of different financial enterprises exceed 0.85. When creating a model, it is necessary to consider the characteristics of unbalanced data and apply appropriate algorithms and techniques to ensure the reliability and stability of the model. At the same time, it is necessary to continuously optimize and improve the model to adapt to different risk scenarios and customer needs.

Keywords: Benchmarking Management · Optimization Algorithms · Financial Risk Control · Balanced Dataset

1 Introduction

The continuous development of data mining technology, as well as the rapid updating of data acquisition equipment and processing software, have promoted the research of risk control algorithms in the financial industry. Financial risk control technology is based on data sources by analyzing, processing, and controlling collected data. Data mining technology is an important branch of information processing, occupying an increasingly high position in the financial field. The imbalance in financial risk control is an important issue in the allocation of data resources. Based on benchmarking management optimization algorithms, this paper studies the modeling of benchmark databases, the construction of risk early warning models, and the design and implementation of decision support systems in the financial industry.

J. C. Hung et al. (Eds.): FC 2023, LNEE 1132, pp. 91–99, 2024.
https://doi.org/10.1007/978-981-99-9538-7_13

The application of data mining technology in the financial field has become increasingly common. Financial benchmarking can help companies identify risks. Many scholars have conducted relevant research on benchmarking management optimization and financial risk control. For example, some scholars introduced the modeling principles of financial risk control issues, and combined with data mining related theories, studied and analyzed a new technology for optimizing models based on multiple attribute decision-making methods [1, 2]. Some scholars have proposed an optimization model for solving the problem of large data volumes and severe redundancy [3, 4]. In addition, some scholars have introduced the concept of financial risk control and analyzed dataset modeling from both theoretical and practical perspectives [5, 6]. Financial risk control imbalance dataset identification refers to the collection, screening, and analysis of data with different attributes to detect anomalies or deviations from true values, and to classify them to achieve optimal control.

This paper first studies the benchmarking management optimization algorithm, expounds its basic theory, and analyzes its advantages and disadvantages. Secondly, it analyzes the concept and control elements of financial risk control. Then, the process of modeling unbalanced data sets is discussed. Finally, through empirical analysis of benchmarking management optimization algorithms, the AUC values and accuracy of different algorithms are obtained.

2 Financial Risk Control of Benchmarking Management Optimization Algorithm

2.1 Benchmarking Management Optimization Algorithm

Benchmarking Management Optimization (BMO) is an algorithm used to solve unbalanced dataset problems. Unbalanced datasets are classification issues where the number of samples varies by category. For example, in the field of financial risk control, the number of debt customers is often far lower than the number of good customers.

The BMO algorithm is based on a new idea of extracting reference samples from the original dataset, and then generating a new training set by resampling the reference samples [7]. The steps are as follows: Reference Sample Selection: Select some positive and negative samples from the original dataset as reference samples. Generally, the selected reference sample must be a sample that can represent both positive and negative examples of the original dataset. Resample: Resample the basic sample to generate a new training set. During the resampling process, certain strategies are used to balance the number of positive and negative samples, such as oversampling, undersampling, and SMOTE. Training Model: Use the new training set to train the classification model. Test Model: Use a set of tests to evaluate the performance of the model [8].

The advantage of the BMO algorithm is that it ensures that the newly sampled driver set better represents the original dataset, thereby improving the performance of the classifier. In addition, the BMO algorithm effectively avoids the problem of over fitting and under fitting because the newly sampled learning set is more balanced, allowing the classifier to learn functions that better represent the dataset [9]. However, BMO algorithms also have some drawbacks, such as the need to select appropriate reference samples, resampling strategies, and classifiers, which can lead to excessive or insufficient matching issues. At the same time, the BMO algorithm also has certain computational complexity, and it is necessary to consider the requirements of actual application scenarios [10].

2.2 Financial Risk Control

Financial risk control refers to the process of identifying, evaluating, managing, and monitoring financial sector risks through the use of various management techniques and methods. Its purpose is to ensure the normal operation of financial institutions and the legitimate interests of their customers.

The main elements of financial risk control are: risk assessment: determining the customer's risk level by assessing the customer's credit status, repayability, historical behavior, etc., and providing a basis for subsequent risk control [11]. Risk control: Take a series of measures, such as setting credit limits, adjusting interest rates, and increasing collateral, to control and manage risks and reduce the risk of loss for financial institutions. Risk monitoring: Monitor the activities and behaviors of financial institution customers in real time, quickly detect abnormal situations, and take appropriate measures to correct abnormal situations to ensure the safety of financial institution operations [12]. Anti fraud: Use various technical means, such as manual auditing, automatic auditing, and big data analysis, to verify and compare customer information to prevent fraud. Compliance management: Comply with relevant laws, regulations, and industry standards to ensure that the operations of financial institutions meet the legitimate interests of customers. Financial risk control typically uses data analysis and artificial intelligence to improve the accuracy and effectiveness of risk assessment and control [13].

Establish a sound risk management framework, clarify risk management responsibilities and processes, and clarify risk control measures for each link. Evaluate the customer's credit history, transaction history, and other relevant information, predict the likelihood of customer failure, and minimize losses and risks [14]. Monitor and analyze customer behavior, trading patterns, and capital flows, early identify potential fraud and risks, and take measures to avoid losses. Create and optimize risk models using data mining, machine learning, and other technologies to predict and identify potential risks. Verify customer identity and transaction information to prevent illegal transactions and money laundering [15]. Use encryption technology, network security technology, and other means to ensure the security and privacy of financial data. Develop a comprehensive emergency plan, respond to and manage emergencies in a timely manner, and reduce losses and risks. Financial risk control technology includes various methods and technologies that need to be widely applied to improve the efficiency and accuracy of risk control. At the same time, with the development of financial technology, financial risks will continue to evolve, requiring continuous updating and optimization of risk control technologies.

2.3 Modeling Unbalanced Datasets

Unbalanced data sets are common in financial risk control. The number of debt customers is often far lower than the number of good customers. In this case, traditional classification models may encounter errors or computational errors. Therefore, unbalanced data modeling is an important issue in financial risk control. The experimental process for modeling unbalanced data is shown in Fig. 1:

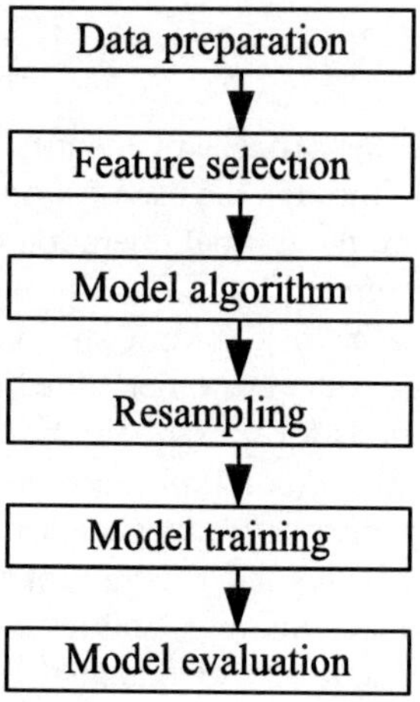

Fig. 1. Experimental Flow of Unbalanced Data Modeling

Data Preparation: Select an unbalanced record, such as a credit default record, from the financial risk control records. The data set is divided into training sets and test sets. Feature engineering: feature extraction and dataset preprocessing. Some classic feature selection methods, such as correlation analysis, information collection, and chi-square testing, can be used to select features that are highly relevant to the classification target to be modeled. Model selection: Select classification algorithms suitable for modeling unbalanced data, such as decision trees, logical regression, support vector machines, and random forests. Methods such as cross validation can be used to compare the performance of different models. Resampling: Due to data set imbalance, some resampling methods can be used to balance data sets, such as oversampling, undersampling, and remote. It should be noted that resampling may lead to model distortion or fitting problems, which need to be adapted to actual situations. Model training: Use training sets to train models and perform operations such as hyperparametric adjustment to improve model performance. Model evaluation: Use a set of tests to evaluate model performance, such as accuracy, recall, F1 value, and other indicators. You can draw ROC curves and PR curves to evaluate the classification effect of the model. In practical applications, unbalanced data modeling must also consider the specificity of business scenarios, such as identifying and controlling debtor customers when controlling financial risks. Therefore, it is necessary to adjust and optimize it according to the actual situation.

3 Empirical Analysis of Benchmarking Management Optimization Algorithms

3.1 Experimental Design

Determine comparative objects and research objectives, and determine evaluation indicators and data sources. Collect and organize data needed for the experiment, including sample data, reference data, and scoring index data. The quality and reliability of data must be ensured. Design an experimental plan, including selecting comparative algorithms or models, experimental procedures, and experimental parameters. The reproducibility of the experiment and the interpretability of the results must be considered. Conduct experiments according to the experimental design, and record the experimental process and results. The control of test conditions and the accuracy of data collection must be observed. Conduct statistical analysis and visual representation of experimental results, compare the performance of different algorithms or models, and draw conclusions and recommendations. Analyze the advantages and disadvantages of each algorithm or model, and propose suggestions for improvement and optimization. Based on experimental results, the best algorithm or model is selected and applied to actual scenarios to improve management efficiency and decision-making quality.

3.2 Identification of Financial Risk Control Imbalance Dataset

The unbalanced dataset in financial risk control refers to a situation where the number of negative samples in the dataset is significantly greater than the number of positive samples. This uneven data distribution may cause machine learning algorithms to focus too much on negative samples and ignore positive samples during training, thereby reducing the prediction accuracy and reliability of the model.

Some methods for identifying unbalanced financial risk control datasets:

Oversampling method: Compensates for the number of positive and negative samples by replicating positive samples or generating new positive samples. Under sampling method: Balance the number of positive and negative samples by randomly discarding some negative samples. Threshold adjustment method: By adjusting the classification threshold, the model pays more attention to positive samples. Integrated learning method: Improve the prediction accuracy and robustness of models by integrating the prediction results of multiple models. Sample weighting method: By setting different weights for different samples, the model pays more attention to positive samples.

Identifying unbalanced data sets used to control financial risk requires extensive use of various methods and techniques to improve the accuracy and reliability of model predictions. At the same time, it is necessary to select the most suitable method according to the actual situation, and optimize and adjust the model to adapt to different data distributions and different application scenarios.

3.3 Experimental Environment and Data

Experimental environment: Windows 11, 8GB RAM, 2.80GHz processor, Python 3.7.3 based algorithm. The experiment used a Python machine learning tool, sicikt learn, using default settings. The dataset selected for the experiment is from the credit default records of five different financial companies (F1, F2, F3, F4, F5). The sample ratio is the ratio between the majority of samples and the minority of samples, which can indicate the degree of imbalance in a dataset. In order to ensure the accuracy of the experimental results, the cross validation method was used in the experiment. The obtained dataset information is shown in Table 1:

Table 1. Data Set Information

	Sample number	Number of attributes	Sample ratio
F1	460	20	1.1
F2	390	14	1.3
F3	210	40	1.2
F4	760	10	1.9
F5	360	30	2.0

In order to verify the effectiveness of benchmarking optimization algorithm, decision tree algorithm structure and the compatibility between logistic regression algorithm and random forest algorithm, the benchmarking optimization algorithm is compared with decision tree algorithm, logistic regression algorithm and random forest algorithm. Decision trees are easy to implement and efficient, so they are selected empirically selected as classifiers. These three algorithms were tested on multiple datasets. Using AUC as an indicator to evaluate the performance of the classification algorithm, the following formula:

$$AUG = \frac{1 + S_{\text{rate}} - E_{\text{rate}}}{2} \tag{1}$$

E_{rate} is the rate misclassified in most classes, with the formula:

$$E_{\text{rate}} = \frac{E}{E + D} \tag{2}$$

D_{rate} is the correctly classified ratio in most classes, expressed as:

$$D_{\text{rate}} = \frac{D}{D + E} \tag{3}$$

4 Comparative Analysis of the Experimental Results

4.1 The AUC Results of the Algorithm

As shown in Fig. 2, we can find that the AUC value of BMO algorithm is the highest compared with other algorithms. The AUC value of BMO algorithm can reach 0.88, and the AUC value of decision tree algorithm is the lowest, both below 0.7. However, the AUC value of the random forest algorithm is above 0.7 in the classification of the unbalanced data set.

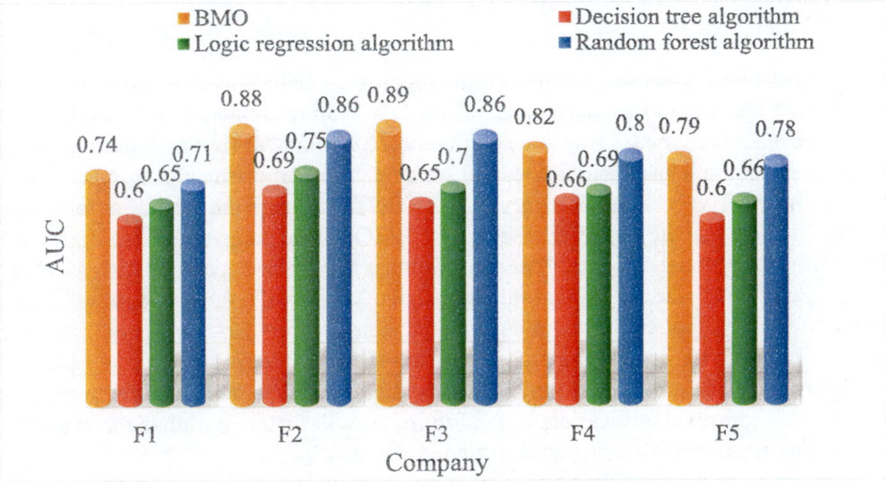

Fig. 2. The AUC Results of the Algorithm

4.2 Accuracy Results of the Algorithm

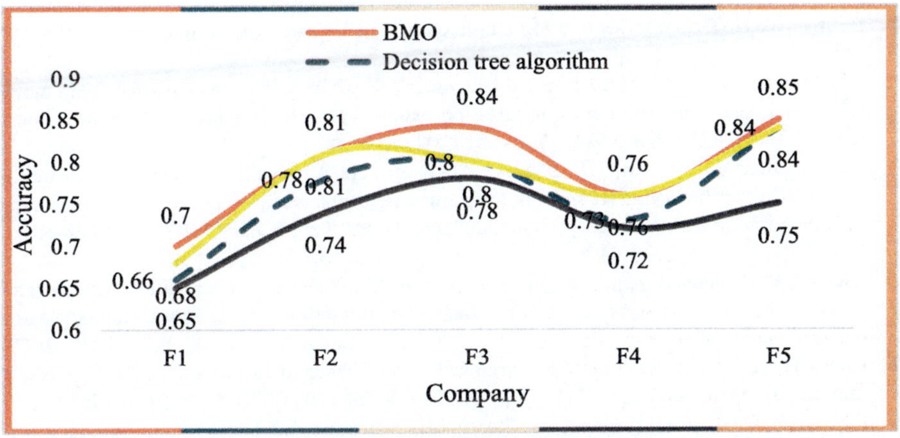

Fig. 3. Accuracy Results of the Algorithm

As shown in Fig. 3, we can find from the figure that the accuracy of the BMO algorithm is above 0.7 in the unbalanced data sets of 5 different financial companies, among which the highest accuracy can reach 0.85. The logistic regression algorithm had low accuracy in the cases and the lowest accuracy among the four algorithms.

5 Conclusion

In the experiment, we used credit default records as unbalanced records. First, we designed the features of the dataset and selected the features related to the classification target to be modeled. Then, we used a classifier based on the BMO optimization algorithm to train and test the dataset and compare it with other classification algorithms. Finally, we used the AUC values and accuracy of the ROC to evaluate the model performance. The results show that the classifier based on the BMO optimization algorithm works well in modeling the unbalanced data. This algorithm has the highest accuracy with AUC values compared with other classification algorithms. Meanwhile, the proposed algorithm can be effectively applied in the field of financial risk control. In conclusion, the modeling of financial risk control imbalance data based on BMO optimization algorithm is an effective classification algorithm that can improve the accuracy and effectiveness of financial risk control classification. In the future, we will further optimize the algorithm to cope with more complex financial risk control scenarios.

References

1. Lubbock, A.L.R., Lopez, C.F.: Microbench: automated metadata management for systems biology benchmarking and reproducibility in Python. Bioinformatics **38**(20), 4823–4825 (2022)
2. Di Martino, S., Peron, A., Riccabone, A., Vitale, V.N.: Benchmarking management techniques for massive IIoT time series in a fog architecture. Int. J. Grid Util. Comput. **12**(2), 113–125 (2021)
3. Albahri, O.S., et al.: Multidimensional benchmarking of the active queue management methods of network congestion control based on extension of fuzzy decision by opinion score method. Int. J. Intell. Syst. **36**(2), 796–831 (2021)
4. RajeevKumar, M., Ansari, T.J., Baz, A., Alhakami, H., Agrawal, A., Khan, R.A.: A multi-perspective benchmarking framework for estimating usable-security of hospital management system software based on fuzzy logic, ANP and TOPSIS methods. KSII Trans. Internet Inf. Syst. **15**(1), 240–263 (2021)
5. Khatari, M., Zaidan, A.A., Zaidan, B.B., Albahri, O.S., Alsalem, M.A.: Multi-criteria evaluation and benchmarking for active queue management methods: open issues, challenges and recommended pathway solutions. Int. J. Inf. Technol. Decis. Mak. **18**(4), 1187–1242 (2019)
6. Hendawi, A., et al.: Benchmarking large-scale data management for Internet of Things. J. Supercomput. **75**(12), 8207–8230 (2019). https://doi.org/10.1007/s11227-019-02984-6
7. Cornwell, N., Bilson, C., Gepp, A., Stern, S.E., Vanstone, B.J.: The role of data analytics within operational risk management: a systematic review from the financial services and energy sectors. J. Oper. Res. Soc. **74**(1), 374–402 (2023)
8. Huynh, T.L.D., Shahbaz, M., Nasir, M.A., Ullah, S.: Financial modelling, risk management of energy instruments and the role of cryptocurrencies. Ann. Oper. Res. **313**(1), 47–75 (2022). https://doi.org/10.1007/s10479-020-03680-y
9. Rabin, K., Tiwari, A.K., Hammoudeh, S., Albulscu, C.T.: Financial modeling, risk management of energy and environmental instruments and derivatives: past, present, and future. Ann. Oper. Res. **313**(1), 1–7 (2022)
10. Pekár, J., Pcolár, M.: Empirical distribution of daily stock returns of selected developing and emerging markets with application to financial risk management. Central Eur. J. Oper. Res. **30**(2), 699–731 (2022)

11. Soto-Beltran, L.L., Robayo-Pinzon, O.J., Rojas-Berrio, S.P.: Effects of perceived risk on intention to use biometrics in financial products: evidence from a developing country. Int. J. Bus. Inf. Syst. **39**(2), 170–192 (2022)
12. Kwateng, K.O., Amanor, C., Tetteh, F.K.: Enterprise risk management and information technology security in the financial sector. Inf. Comput. Secur. **30**(3), 422–451 (2022)
13. Christensen, J.H.E., Lopez, J.A., Mussche, P.L.: Extrapolating long-maturity bond yields for financial risk measurement. Manag. Sci. **68**(11), 8286–8300 (2022)
14. Pandey, K.K., Shukla, D.: Stratified linear systematic sampling based clustering approach for detection of financial risk group by mining of big data. Int. J. Syst. Assur. Eng. Manage. **13**(3), 1239–1253 (2022). https://doi.org/10.1007/s13198-021-01424-0
15. Albrecher, H., Azcue, P., Muler, N.: Optimal ratcheting of dividends in a Brownian risk model. SIAM J. Financ. Math. **13**(3), 657–701 (2022)

Research on Consumer Energy-Saving Awareness Based on Online Reviews of Energy-Efficient Home Appliances

Xi Luo and Jieping Liu(✉)

Department of Big Data Management and Application, Chengdu Neusoft University, Chengdu, Sichuan, China
nsu.liujieping@gmail.com

Abstract. Green consumption is a key development strategy outlined in China's "14th Five-Year Plan" and the 2035 Vision Outline. Research reports indicate that 80% of consumers are in favor of sustainable consumption practices. Despite this, several studies suggest that there exists a noticeable gap between intention and behavior in consumers' eco-friendly consumption patterns. Utilizing web crawler technology, this study collected data on 8,020 home appliances that were energy-efficient and screened 1.481 million customer opinions associated with these products on Jingdong platform. The study measured consumer energy-saving attention through word frequency statistics. The research concluded that consumers' attention towards energy-saving behaviors when utilizing energy-efficient home appliances was low, accounting for merely 4.279% of their energy consumption. This outcome highlights the observed gap between consumers' environmental intentions and their actual energy-saving behaviors. Besides, based on descriptive analysis and Kruskal-Wallis test results, the study discovered significant differences in consumers' levels of energy-saving attention towards various categories and energy efficiency levels of eco-friendly home appliances. As a solution, the study recommends that businesses can guide customers to improve their energy-saving attention through effective customer service guidance, offers such as gift coupons, augmented yet meaningful experience cards, or even through setting up a dedicated energy-saving comment section. The proposal aims to amplify other consumers' comprehension of energy-saving information and usage experience of ecological products.

Keywords: Energy-Saving Attention · Green Consumption · Energy-Saving Home Appliances · Online Comments

1 Introduction

In China, as the economy rapidly grows, the concepts of green environmental protection and low-carbon development have been widely disseminated among the citizens, leading to greater recognition of green consumption. The "Survey Report on the Present Status of Public Green Consumption in China" revealed that 83.34% of respondents

J. C. Hung et al. (Eds.): FC 2023, LNEE 1132, pp. 100–107, 2024.
https://doi.org/10.1007/978-981-99-9538-7_14

supported green consumption behavior, with 46.75% expressing a strong inclination towards it. Studies carried out by Alibaba's New Service Research Center indicated that over 60% of participants were conscious of green consumption, with post-90s and post-2000s generations exhibiting considerably greater awareness than their older counterparts, amounting to 70% and 79%, correspondingly. "Developing green consumption" was included as one of the critical strategies for China's future development in the 14th Five-Year Plan and the 2035 Vision Goal Outline, indicating the significance of green consumption in the country's prospects.

Ajzen proposed the TPB model in 1985, based on the TRA theory, which suggests that behavioral attitudes influence behavioral intentions that further reflect in the behavior, a viewpoint substantiated by a substantial corpus of empirical studies. Despite the model's popularity, there exists abundant research indicating that the impact of attitudes or intentions on behavior is nonlinear and can differ, leading to the "gap between intention and behavior" or "inconsistency between words and deeds." Consequently, even though 80% of customers endorse green consumption, research suggests that their attitudes towards this practice may be affected by variables such as purchasing power and information transparency, creating a discrepancy between their attitudes and actual green consumption behavior.

One of the crucial green consumption behaviors is consumers buying and being mindful of energy-saving home appliances that they can recognize in their daily consumption; furthermore, online comments provide consumers' subjective assessment of their experience after buying the product. Therefore, the research on energy-saving home appliances analyzes the frequency of keywords associated with energy-saving consumption in online comments about energy-efficient home appliances to understand how customers perceive them, and proposes measures to promote such behavior from the perspective of online comments.

2 Correlational Research

Presently, research on green consumption focuses on exploring driving forces, fostering green consumption values, and examining relevant institutional mechanisms. Various approaches, including the TPB model, regression analysis, and SEM, have been employed by scholars to investigate the factors that influence eco-friendly behavior. Studies contend that numerous variables, such as age, education, income level, occupation [1, 2], green awareness and environmental consciousness [3, 4], media promotion [5, 6], gamification and intelligence [7, 8], and packaging design, have an impact on eco-friendly behavior, according to individual consumer characteristics.

Nonetheless, a linear relationship between driving factors and eco-friendly behavior does not exist. Scholars have found that a hole exists between factors that drive ecological behavior, consumer attitudes towards eco-friendly behavior, and their willingness to act on such attitudes. As an illustration, William et al. discovered that 30% of UK consumers expressed significant concerns about environmental matters, yet they struggled to translate it into their purchasing behavior. Similarly, according to Essiz Oguzhan [9] et al., although consumers have positive perspectives towards green consumption, they usually find it challenging to transform these values into green purchasing behavior, known as the "Green Gap." Several scholars have explored the causes that underlie this gap. According to Chen Kai et al. [10], the gap arises from various factors such as perceived efficacy and expectations of others' consumption behavior, group, and individual past consumption behavior, balancing green and non-green product attributes, green product prices, availability factors, and situational factors of green consumption behavior. Li Chuang et al. [11] also found that trust and information reduce this gap, while behavioral costs such as extra expenses, time, and energy increase it. In a similar vein, Li Pengxiang [12] research identified several causes behind this gap, including self-control, purchasing behavior inertia, social reference group norms, balancing green and non-green product attributes, high prices and low availability of green products, and a lack of information, transparency, and trust.

In summary, extensive research has shown that there is a fundamental discordance between consumers' attitudes and intentions towards green consumption practices and their actual behavior. This discrepancy highlights the need for more in-depth research into how to bridge this gap. Unfortunately, several studies rely too heavily on the TPB model, and their investigation methods are restricted by SEM tools. As a result, we must expand our research beyond the traditionally limited methods. Therefore, this article ("this study" can be more specific) advocates for analyzing consumer usage comments on products, seeing as they could accurately reflect their actual consumption patterns, including their perception of energy-saving protocols. By examining customers' online reviews of energy-saving products, we have a more natural avenue to better comprehend their energy-saving proclivities and investigate whether this gap exists.

3 Data Collection and Pre-processing

3.1 Data Collection

The data utilized for this study originates from JD.com, the largest home appliance e-commerce platform in China. This study selected four of the most commonly purchased household appliances from this platform - refrigerators, air conditioners, washing machines, and televisions. A web crawler program was employed to gather a total of 8,916 products and 1.842 million corresponding comments in regards to the designated four household appliances. The data retrieved includes essential product attributes such as the product ID, product title, and product parameters. Moreover, this data includes factors such as the total number of comments, hot review tags, comment content, and score for each product. Furthermore, the product parameters include energy efficiency levels that conform to the "Management Measures for Energy Efficiency Labeling" in China.

3.2 Data Pre-processing

Initially, this study utilized regular expressions to extract the energy efficiency levels of the 8,916 products from their respective product parameters. The results identified that 896 items did not indicate their energy consumption level, although 8,020 products possessed a relevant energy consumption rating. Following this, the comments related to those 8,020 products were subject to a screening process. Comments with a text length of fewer than five words or default replies like "This user did not fill in the evaluation content" were eliminated. Consequently, this process yielded 8,020 products and their 1.481 million associated, valid online reviews.

4 Data Empirical Analysis

4.1 Measurement of Consumer Energy Efficiency Concerns

Consumer online reviews provide subjective evaluations of the consumer's experience of a product after actual use. From these reviews, it can be inferred that if consumers are mindful of saving energy, they will express this in their comments. As such, a method to quantify consumers' energy conservation concerns can be achieved by analyzing the frequency of energy conservation keywords in their reviews. The frequency with which these keywords appear can be used to determine the level of consumers' attention towards energy conservation information.

Additionally, this study has identified nine relevant energy conservation keywords. These keywords include energy consumption, energy efficiency, energy conservation, power-saving, low carbon, green consumption, green lifestyle, environmental protection, eco-friendliness, as well as trade-in.

4.2 Overall Energy Conservation Concern Analysis

The general level of energy conservation concern is calculated as the ratio of comments that contain energy conservation keywords to the total number of "relevant" comments. That is, comments that contain one or more energy conservation keywords. According to our statistics, amongst the 1.481 million reviews analyzed, 63,377 showcased energy conservation keywords, and thus the overall level of energy conservation concern was established at 4.279%. Among the reviews that contain energy conservation keywords, they are distributed as shown in table 1:

Table 1. The distribution of all energy-saving keywords with a frequency greater than 0

count	mean	std	min	25%	50%	75%	max
63377	1.574	0.893	1.0	1.0	1.0	2.0	11.0

It is apparent that consumers display limited awareness of energy-saving efforts, with only 4.279% showcasing concern. Even within the consumers focusing on promoting energy-saving efforts, the average value is just 1.574. This study aims to delve deeper into consumers' energy-saving awareness in different product categories and energy efficiency levels, and thus utilized differential analysis via ANOVA, t-tests, non-parametric testing, and descriptive analysis. Typically, ANOVA necessitates normality data to yield accurate outcomes. To this end, this study performed normality tests on both frequency data of all energy-saving words and frequency data of energy-saving words greater than zero separately, employing tools such as histograms, Q-Q plots, and Kolmogorov-Smirnov tests. However, the results showed non-normal distribution data. Therefore, this study utilized descriptive analysis and the non-parametric Kruskal-Wallis test to conduct differential analysis on energy-saving awareness in different product categories and energy efficiency levels.

4.3 Analysis of Differences in Energy-Saving Attention Among Different Product Categories

Table 2 provides differential analysis of energy-saving attention for different product categories based on descriptive analysis methods.

Table 2. The distribution of energy-saving attention and energy-saving keywords (with a frequency greater than 0) for different product categories

	energy-saving attention	count	mean	std	min	25%	50%	75%	max
Refrigerator	4.028%	18169	1.335	0.619	1	1	1	2	8
Air condition	8.111%	26715	2.039	1.055	1	1	2	3	11
Washing machine	1.406%	6254	1.236	0.528	1	1	1	1	6
Television	4.787%	12239	1.088	0.371	1	1	1	1	7

Additionally, this paper employed the Kruskal-Wallis test to examine the frequency data of energy-saving keywords for different product categories, with and without restrictions of frequency greater than zero. Results demonstrated that the Kruskal-Wallis statistics and p-values were 21,642.563, 0, and 9,086.129, 0, respectively. Since the p-values were both less than 0, the null hypothesis was rejected, indicating significant differences among different product categories. As per descriptive analysis, consumers exhibit the highest energy-saving awareness for air conditioners and the lowest for washing machines.

4.4 Analysis of Differences in Energy Efficiency Level and Energy Efficiency Concern

Table 3 showcases the differences in energy efficiency concerns for different energy efficiency levels, based on descriptive analysis methods.

The Kruskal-Wallis test was also conducted on the frequency data of all energy-saving keywords and frequency data of energy-saving keywords with frequency greater than zero for different energy efficiency levels. The Kruskal-Wallis statistics and p-values were 2,984.956, 0 and 1,720.614, 0, respectively. As all p-values were less than 0, the null hypothesis was rejected, signifying significant differences between different energy efficiency levels. As per descriptive analysis, consumers showcase increased

Table 3. Distribution of Energy Efficiency Concern and Keyword Frequency (greater than 0) for Different Energy Efficiency Levels

	Energy Efficiency Concern	count	mean	std	min	25%	50%	75%	max
Level One Energy Efficiency	5.037%	35055	1.711	0.975	1	1	1	2	11
Level Two Energy Efficiency	2.753%	6181	1.337	0.659	1	1	1	2	6
Level Three Energy Efficiency	3.646%	14316	1.433	0.794	1	1	1	2	9
Level Four Energy Efficiency	4.203%	4765	1.115	0.429	1	1	1	1	6
Level Five Energy Efficiency	5.609%	3060	1.867	0.834	1	1	2	2	9

energy efficiency concerns when energy-saving appliances are rated level one or five. Moreover, amongst the five energy efficiency levels, ratings one and three are more preferred by consumers.

5 Conclusion

This study focuses on analyzing online reviews of energy-efficient appliances to examine the energy-saving awareness of consumers. To this end, we obtained a dataset of 8,020 energy-efficient appliance products and 1.481 million corresponding online reviews using web crawling technology. By conducting statistical analysis on the occurrence frequency of nine energy-saving keywords in the online reviews, we measured the level of energy-saving awareness among consumers. The data was further subjected to descriptive analysis and Kruskal-Wallis test to examine the differences in energy-saving awareness across different product categories and different levels of energy efficiency. The findings of our study indicate that:

(1) Despite the fact that various surveys suggest that 80% of consumers support environmentally-friendly consumption, the actual energy-saving awareness exhibited by consumers when utilizing energy-efficient appliances is meager, at 4.279%. This outcome is indicative of the "intention-behavior" gap that exists among consumers when purchasing energy-efficient appliances.

(2) Results from the statistical tests demonstrate noteworthy disparities in consumers' energy-saving awareness levels that vary according to the energy efficiency level and

appliance category. Consumer awareness is at its highest when using air conditioners, whereas washing machines have the lowest awareness levels. Additionally, when energy-efficient appliances are classified into either level 1 or level 5 efficiency categories, consumers exhibit a greater level of energy-saving awareness.

A multitude of research studies suggest that online reviews have a substantial impact on consumers' behavior regarding consumption. Hence, directing consumers towards eco-friendly consumption through online reviews is currently a challenging issue that scholars are investigating. In the context of online reviews of energy-efficient appliances, boosting users' knowledge of energy conservation, and motivating them to actively offer feedback and comments on energy-saving information while utilizing the products, can foster the promotion of energy-efficient appliances among other consumers and encourage green consumption. As such, companies can use various methods such as customer service guidance, offering coupons or attaching experience description cards to the product to encourage users to share their energy-saving perceptions during actual usage. Furthermore, establishing a specific area for energy-saving experience reviews can invite consumers to provide feedback on energy-saving perceptions, increase their awareness of energy conservation, and make the product's energy-saving information and user experience easier for others to comprehend.

Acknowledgements. This work was supported by Sichuan Research Center of Electronic Commerce and Modern Logistics, Research on Green purchase intention and behavior of E-commerce consumers under the background of "14th Five-Year Plan" (Research Project Number : DSWL21-37).

References

1. Ya, P.: The influencing factors of Chinese consumers' Green consumption willingness and the construction of green consumption mechanism. J. Commer. Econ. **15**, 57–59 (2019)
2. Xin, D.: Consumer innovation and green consumption behavior: theoretical mechanism and empirical research. J. Commer. Econ. **19**, 56–58 (2020)
3. Subburaj, A., Sangeeta, M., Sonia, M.: How green consumption value affects green consumer behaviour: the mediating role of consumer attitudes towards sustainable food logistics practices. Vision J. Bus. Perspect. **25**(1), 65–76 (2021)
4. Wang Zhaohua, L., Bin, W.B., Guanhua, M.: The impact of energy-saving information exposure on green consumption behavior——an empirical study of large-scale text data from e-commerce data platforms. Chin. J. Manage. Sci. **30**(01), 241–251 (2022)
5. Li, Z., Jianxin, S., Ling, Z.: Research on the relationship between media propaganda and green consumer behavior-based on social cognitive theory. J. Shandong Technol. Bus. Univ. **35**(05), 79–91 (2021)
6. Xie, S., Madni, G.R.: Impact of social media on young generation's green consumption behavior through subjective norms and perceived green value. Sustainability **15**(4), 3739 (2023). https://doi.org/10.3390/su15043739
7. Du, S., Xu, J., Zhang, D., Yang, X.: How gamification drives e-commerce users' green consumption behaviors: a netnography study of ant-forest. Nankai Bus. Rev. **25**(02), 191–204 (2022)

8. Wang, J., Zhao, J.: The implementation effect of information embedded regulatory tools on promoting online green consumption behavior in digital era: green purchase scenario simulation and regulatory tools design experiments. J. Manage. World **38**(04), 142–162 (2022)
9. Essiz, O., Yurteri, S., Mandrik, C., Senyuz, A.: Exploring the value-action gap in green consumption: roles of risk aversion, subjective knowledge, and gender differences. J. Glob. Mark. **36**(1), 67–92 (2023). https://doi.org/10.1080/08911762.2022.2116376
10. Kai, C., Qian, P.: Research on the influencing factors of Green consumption attitude-behavior Gap. Enterp. Econ. **08**, 25–30 (2014)
11. Chuang, L., Ying, S.: How to improve the consistency of intention and behavior in the context of green consumption? J. Arid Land Resourc. Environ. **34**(08), 19–26 (2020)
12. Pengxiang, L.: Analysis on the influencing factors of attitude-behavior Gap in Green consumption. Contemp. Econ. **09**, 82–85 (2021)

Research on Radar Target Recognition Based on Deep Learning

Wenchao Ren[1], Zeyuan Liu[2], Rongxin Lv[3](✉), and Yilin Ge[4]

[1] School of Mechanical and Electrical Engineering, Shenyang Aerospace University, Shenyang, China
[2] School of Aerospace, Shenyang Aerospace University, Shenyang, China
[3] School of Creation, Shenyang Aerospace University, Shenyang, China
13840230825@163.com
[4] International Institute of Engineers, Shenyang Aerospace University, Shenyang, China

Abstract. Traditional shallow learning radar target image recognition ways usually depend on complex artificial feature extraction process, which needs a lot of human expertise. to effectively extract the characteristic information of enemy combat units, and realize the type recognition of the UAV radar, which is very important to ensure the UAV's combat capability and combat victory. Aiming at the problem that the shallow learning method of radar target image recognition is difficult to extract advanced features and the small number of enemy unit radar image samples available in practical engineering has a great effect on the diagnosis accuracy of the deep neural network model, this paper uses the transfer method in deep learning. The learning way is to fine-tune the pre-trained hotspot deep CNN convolutional network (GoogleNet) respectively, which are used for the recognition and classification of radar target images. The noise reduction radar target figure is used as input to train a constructed model, Then, the trained model can realize radar target image recognition and classification. The model are validated using the MSTAR dataset gived by DARPA/AFRL, and the results show that model achieve high diagnostic accuracy, proving the effectiveness of the proposed way in radar image recognition of enemy combat units.

Keywords: UAV · Radar Image · Target Recognition · Deep Learning · Transfer Learning

1 Introduction

With the rapid iteration about artificial intelligence, deep learning, as a hot pattern recognition method, has been applied in various fields on a large scale. Zhang Xiaohong et al. [1] explained how deep learning is applied in the field of vehicle navigation, and expounded the latest trend of vehicle motion constraints based on deep learning. Zhang Linlin et al. [2] applied deep learning to the field of language recognition to effectively improve the recognition accuracy of short texts in similar languages. Peng Tongxin et al. [3] applied deep learning to the field of vehicle flow prediction to enhance the overall prediction performance of the model and enhance the stability of the model.

J. C. Hung et al. (Eds.): FC 2023, LNEE 1132, pp. 108–114, 2024.
https://doi.org/10.1007/978-981-99-9538-7_15

Because of the high efficiency, plasticity and universality of deep learning, scholars have used a variety of deep neural network models to the field of UAV radar target image recognition. Yan Yuan [4] proposed a set of comprehensive solutions to the problem of radar small-sample target recognition by uniting meta-learning and transfer learning, aiming to put forward appropriate model learning and classification ways according to different practical application scenarios, so as to enhance the efficiency and accuracy of radar small-sample target recognition.

Transfer learning is a study to solve the domain adaptation direction, that is, for the source domain with a large amount of label data and available parameter Settings, the target domain with a small quantity of labels is processed by transferring the learned knowledge [5]. In transfer learning, the pre-trained model can be applied to different tasks by fine-tuning it and training it with a small quantity of samples. Moreover, transfer learning can be realized on ordinary personal computer, which does not require too much hardware, and it usually takes only a short time to complete the model training. Therefore, scholars have great interest in transfer learning, and it has been widely used in UAV small target detection [6], emotion recognition [7], image classification [8] and other fields. There are also some researches on UAV radar target recognition. Wang Guoshuai [9] introduced the Deep adaptation network (DAN) method into HRRP target recognition, and added an adaptive layer to the original CNN model. Experimental consequences show that contrasted with the conventional transfer learning way and DAN method, the proposed method can improve the recognition rate of target domain data by about 15% under the influence of sea clutter, which significantly enhances the generalization performance and robustness of the model.

Experiments are performed in this paper depended on the MSTAR dataset gived by DARPA/AFRL. This dataset was collected by an X-band synthetic aperture radar (SAR) sensor used by Sandia National Laboratory [10–12].

2 Google Net Model

In a general way, the most direct method to enhance network performance is to add the depth and width of the network, but perpetually adding will bring a lot of trouble: a sea of parameters, if the training data set is restricted, it is prone to overfit.

The larger the network, the more parameters, the greater the computational complexity, hard to apply; The deeper the network is, the gradient dispersion trouble is easy to occur, and it is hard to optimize the model. Researchers want to lower parameters while adding network depth and width. To lower parameters, it is natural to turn full connections into sparse connections. Nevertheless, in implementation aspect, the actual computation quantity will not be improved qualitatively after full connection turns into sparse connection, because most of the hardware is optimized for dense matrix computation. Although sparse matrix has less data, it is hard to lower the computation time consumed. In this need and situation, GoogLeNet (see Fig. 2) believes that the most basic approach is to use sparse joins instead of full joins and convolution operations. For the sake of keeping the sparsity of neural network structure and making full use of the high computational performance of dense matrices, GoogleNet proposed a modular structure named Inception to finish this purpose. Inception is a network-in-network structure. Based on this structure, the width and depth of the entire Network can be enlarged, and the performance can be improved by 2 to 3 times. A typical Google Net deep learning Network is mainly composed of Inception modules.

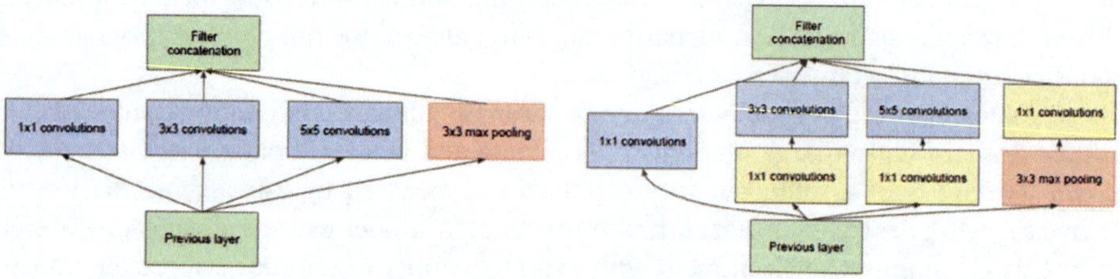

(a) Inception module, naïve version (b) Inception module with dimension reductions

Fig. 1. Inception structure diagram

Inception is a net within-networks (see Fig. 1), which contains a total of 4 subnets. This structure stacks the convolution ($1 \times 1, 3 \times 3, 5 \times 5$) and pooling operations (3×3) generally applied in CNN (the dimensions of convolution and pooling are the identical, and the channels are increased together). On the one hand, the width of the network is increased. On the other hand, it also adds the adaptability of the network to the size.

2.1 1 × 1conv Branch

1×1conv branch is the leftmost branch in the figure above. The network is widened by 1×1 convolution for BatchNorm final activation.

2.2 1 × 1conv → 3 × 3conv Branch

In this step, the convolution kernel is 3×3 in size, but before 3×3 convolution, the feature map will first go through 1×1 convolution layer parameter reduction (1×1 convolution will significantly reduce network parameters, see Fig. 1).

Similarly, input a set of 192 features with 32×32 size and output 256 sets of feature data. The first figure is directly realized by 3×3 convolution, which needs $192 \times 256 \times 3 \times 3 \times 32 \times 32 = 452984832$ multiplication. In the second picture, 1×1 convolution was used to reduce to 96 features, and then 3×3 convolution was used to recover 256 groups of features, which required $192 \times 96 \times 1 \times 1 \times 32 \times 32 + 96 \times 256 \times 3 \times 3 \times 3 \times 32 \times 32 = 245366784$ multiplications. The method of dimension reduction by 1×1 convolution saved half of the computation. Some people may ask, isn't the number of features reduced when 1×1 convolution is reduced to 96 features, which will affect the effect of the final training? The answer is no. As long as the quantity of features in the final output remains the identical (256 groups), the dimensionality reduction in the middle is analogous to the impact of compression and does not affect the final training consequence.

2.3 1 x 1conv → 5 x 5conv Branch

In InceptionV1, a convolution kernel with kernel = 5 is used for feature extraction. In V2, 5 × 5 is replaced by two 3 × 3 convolution kernels, because the two are equivalent and the number of convolution parameters of 3 × 3 is about 1/3 of the convolution operation of 5 × 5. So the two 3 × 3 convolution operations in this code are actually the 5 × 5 convolution operations on the right.

2.4 3 x 3pooling → 1 x 1conv

Use a pooled kernel here, but it's more like convolution. The previous pooling operation stride is the same size as the convolution kernel, and it is not filled, so the size of H × W feature map becomes H/s × W/s after the pooling layer. Here, the pooling operation stride = 1, padding = 1, kernel = 3, in fact, the size of feature map will not change after the pooling operation, only the pooling layer is used to extract feature expression different from the convolution operation.

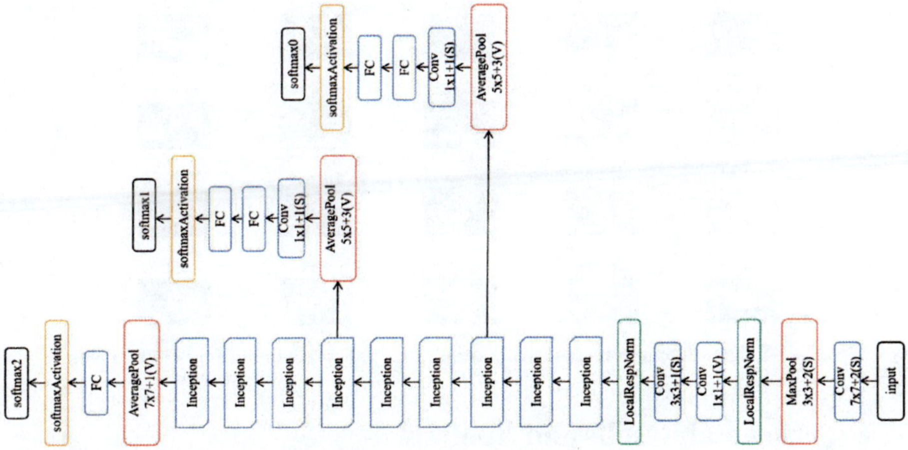

Fig. 2. Google Net network model

Since the purpose of this experiment is to identify the target of UAV radar gray image, the last three layers of Google Net are substituted by the fully connected layer, Softmax layer and classification output layer with the same number of nodes as the number of image categories, and the process is shown in Fig. 3.

The parameters in the pre-training network Google Net have also been fully trained on ImageNet applying millions of figures to extract a large number of features required for image classification and recognition. Therefore, the target recognition of radar gray image can be realized by training its replacement layer with a small dataset.

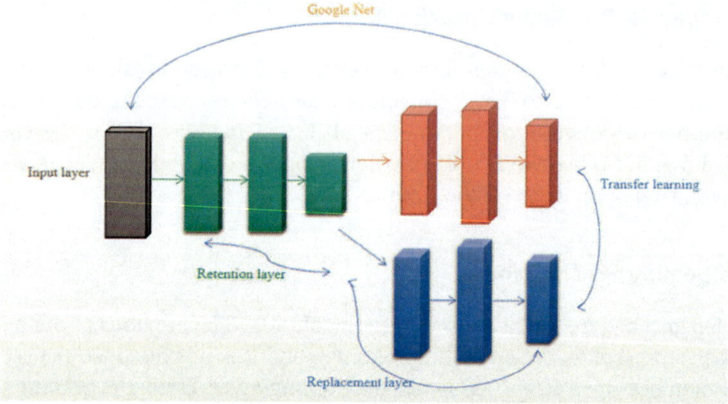

Fig. 3. Migration Learning on Google Net

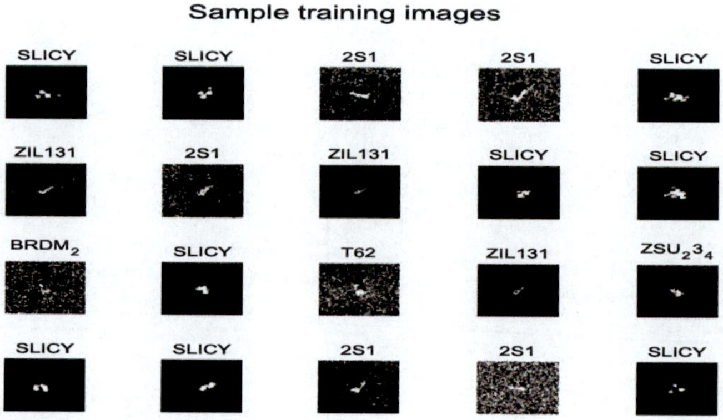

Fig. 4. Target recognition results of partial radar grayscale images by Google Net

3 Experimental Results and Analysis

The last three layers of GoogleNet were substituted with a fully connected layer with eight neurons, a Softmax layer, and a categorical output layer. The sample set constructed in front was imported into MATLAB as the input of the GoogleNet transfer learning model constructed, and the minimum batch size was set as 30, the maximum training times as 15, the learning rate of the retention layer as 0.0001, and the learning rate of the replacement layer as 20. With the progress of training, the recognition results of the GoogleNet transfer learning model on some radar gray image targets are shown in Fig. 4.

The same group of experiments were repeated for 10 times, and the final results were shown in Table 1.

As shown in Table 1, the trained GoogleNet transfer learning model achieves an average recognition accuracy of 98.06%. The network instance confusion matrix was shown at Table 2, the recognition rate is 99.45%.

Table 1. 10 test results

	1	2	3	4	5	6	7	8	9	10	Average
Accuracy(%)	97.63	98.21	98.32	98.17	98.66	97.33	98.68	98.01	97.61	98.01	98.06
Training time/s	2004	2021	2033	2000	1986	1978	2032	2025	2027	1998	2010.4

Table 2. Network instance confusion matrix (recognition rate 99.45%)

Class	SLICY	BTR60	2S1	BRDM2	D7	T62	ZIL131	ZSU23/4
SLICY	99.30	0.00	0.60	0.00	0.10	0.00	0.00	0.00
BTR60	0.40	97.10	1.00	0.20	0.00	1.10	0.20	0.00
2S1	0.00	0.00	99.90	0.00	0.00	0.00	0.10	0.00
BRDM2	0.00	0.00	0.00	100.00	0.00	0.00	0.00	0.00
D7	0.00	0.00	0.00	0.00	100.00	0.00	0.00	0.00
T62	0.00	0.00	0.00	0.00	0.00	100.00	0.00	0.00
ZIL131	0.00	0.00	0.00	0.00	0.00	0.00	100.00	0.00
ZSU23/4	0.00	0.00	0.20	0.00	0.00	0.50	0.00	99.30

4 Conclusion

In this paper, a way of UAV radar image target recognition based on CNN and transfer learning is proposed, which solves the difficulty of deep feature extraction in the shallow learning method of radar target recognition and the problem that the accuracy of recognition is greatly impacted by the small number of available radar image samples in practical engineering. A new model is constructed by fine-tuning the pre-trained CNN model GoogleNet, which is used for target radar image recognition. The model was verified by using the MSTAR dataset gived by DARPA/AFRL. In view of the difficulty in obtaining samples in reality, the identification accuracy of each armored vehicle was 98.06% with a small sample of 100.

Acknowledgment. This research was supported by The 2022 Provincial College Student Innovation and Entrepreneurship Training Program (Project Number: S202210143018).

References

1. Zhang, X., Zhou, Y., Zhu, F., Hu, H.: A new model of vehicle motion constraint based on parameter self-learning and its inertia estimation error suppression analysis. J. Surveying Mapp. **51**(07), 1249–1258 (2022)
2. Zhang, L., Yang, Y., Chen, Z., Pan, Y., Li, Yu.: Language recognition method of similar language short text based on deep learning. Comput. Appl. Softw. **37**(02), 124–129+176 (2020)

3. Peng, T., Han, Y., Wang, C., Zhang, Z.: Mixed deep learning model for short-term subway traffic prediction. Comput. Eng. **13**(5), 297–305 (2022). https://doi.org/10.19678/j.iSSN.1000-3428.0061309
4. Yan, Y., Sun, J., Sun, J., Yu, J.: Syst. Eng. Electron. **43**(03), 684–692 (2021)
5. Deng, J., Zhang, Z., Eyben, F., et al.: Autoencoder-based unsupervised domain adaptation for speech emotion recognition. IEEE Sig. Process. Lett. **21**(9), 1068–1072 (2014)
6. Hao, X., Yang, X., Zhao, H., Li, T., Li, J.: Radar target imaging algorithm based on RCS static broadband data. Terahertz Sci. Electron. Inf. J. China **20**(03), 208–212 (2022)
7. Hung, J.C., Lin, K., Lai, N.: Recognizing learning emotion based on convolutional neural networks and transfer learning. Appl. Soft Comput. J. **84**, 1–19 (2019)
8. Khan, S., Islam, N., Jan, Z., et al.: A novel deep learning based framework for the detection and classification of breast cancer using transfer learning. Pattern Recogn. Lett. Recogn. Lett. **125**, 1–6 (2019)
9. Wang, G.: Radar target recognition based on migration study. Electron. Res. Inst. China Electron. Technol. Group Company (2021). https://doi.org/10.27728/dcnki.GDZKX.2021.000029
10. Koch, G., Zemel, R., Salakhutdinov, R.: Siamese neural networks for one-shot image recognition. In: Proceedings of the 32nd International Conference on Machine Learning (2015)
11. Vinyals, O., Blundell, C., Lillcrap, T., et al.: Matching networks for one-shot learing. In: Proceedings of the Conference on Naural Information Processing Systems, pp. 3630–3638 (2016)
12. Snell, J., Swersky, K., Zemel, R.S.: Prototypical networks for few-shot learning. In: Proceedings of the Advances in Neural Information Processing Systems, pp. 4077–4087 (2017)

Research on the Application of Computer Digital Technology in Textile Art Design

Changlan Jia(✉)

International Fashion Institute, Shandong Vocational College of Light Industry, Zibo, China
jiawing0286@163.com

Abstract. With China's scientific and technological level continuously improving in **recent** years, computer digital technology has become widely used in a variety of industries. People's lives are enriched by textile art. Using computer digital technology can make textile art patterns and patterns more novel and diverse, laying a solid foundation for the advancement of textile art design level. When compared to traditional textile art, the use of computer digital technology can increase work efficiency, save time, and improve the overall benefits of businesses. The application of computer digital technology has had a significant impact on textile art design, resulting in a significant change in the textile art design process. It has challenged people's preconceived notions and ways of thinking, infusing new life into the design community.

Keywords: Computer digital technology · Textile art design · Use strategy · Analyses

1 Introduction

It has become popular in all aspects of society due to the rapid development of computer technology. Computers use numbers as the basis for information processing, so they can also be used as computer digital technology. Computer digital technology is widely used in a variety of industries and has had a significant impact on people's daily lives. The use of computer digital technology in textile art design primarily employs computers for color separation, color matching, ink jet and wax spraying, which diversifies textile product styles and meets people's aesthetic needs [1, 2].

2 The Relationship Between Computer Digital Technology and Textile Art Design

No matter what kind of art design is not an independent individual, the latest technology from other disciplines will be fully absorbed in the process of art design to improve their own development level [3, 4]. Because of the advancement of electronic information technology, the use of computer digital technology in textile art design provides technical support for the development of textile art while also innovating traditional

J. C. Hung et al. (Eds.): FC 2023, LNEE 1132, pp. 115–121, 2024.
https://doi.org/10.1007/978-981-99-9538-7_16

design concepts and design ideas (As show in Fig. 1). Science and technology, as one of China's main productive forces, can only form a new type of productive force by combining textile art design with computer digital technology, thereby promoting the rapid development of social economy [5, 6].

The continuous advancement of computer digital technology has a positive impact on society, economy, and other aspects, as well as improving people's lifestyle and aesthetic concept, and people have higher expectations for textile art design. Computer digital technology is integrated into textile art design, allowing the two to be organically combined to fully meet people's aesthetic needs while also giving textile art design more creativity and vitality [7, 8]. It is also an effective measure to promote the development of computer digital technology in order to make it live and artistic. The application of computer digital technology to textile art design will have a positive impact on people's daily lives and provide people with quality services in an artistic form, which is also the main future development trend of computer digital technology [9, 10].

Fig. 1. Computer digital technology promotes innovative development of textile art design

3 The Influence of Computer Digital Technology on Textile Art Design

The rational application of computer digital technology in textile art design overcomes the limitations of traditional design methods, making textile art design more concise and flexible while saving textile production enterprises a significant amount of time, energy, and money. The use of computer digital technology in textile art design has a far-reaching and positive impact on textile art, and it is also a significant innovation in the textile development process, completely changing people's traditional design concepts and ideas and injecting new life and motivation into the textile art design industry [11–13].

3.1 Innovation of Pattern Drawing Methods

If the traditional pattern design method is used, such as drawing regular patterns and straight lines, traditional tools such as compasses and rulers must be used, which is difficult and complicated. The use of computer digital technology can make drawing textile patterns easier. For example, when drawing free-form curves, the radian of the curve can be adjusted to a proper range using computer digital technology, and the curves will be smoother and more natural, providing people with a refreshing visual experience. To accomplish this, we can employ software such as Photo Shop, Free Hand, and others.

Fig. 2. Innovative development of computer digital technology in pattern design

These softwares can draw the figures and lines in the pattern precisely. It is simple to draw according to the pattern that the designer wishes to express during the drawing process. If there are any errors or parts that need to be adjusted, the Undo function in the software can be used to return and redraw and optimize the pattern. Another example is that when designing patterns with computer digital technology, designers can use computer digital technology to fill the patterns into set size pictures, which is also impossible in traditional textile art design (As show in Fig. 2).

In contrast to the traditional method of textile pattern design, the process of returning and splicing requires using the cutting mode as the basis and splicing the design manuscript by cutting, resulting in the phenomenon of splicing dislocation. After the application of computer digital technology in textile art design, seamless splicing can be easily completed, and the pattern can be endlessly circulated by using the continuous drying function, resulting in a more reasonable pattern layout and the prevention of waterway. When scanning the pattern, computer digital technology can perform useful operations like returning, cutting, and splicing.

3.2 Color Transformation is More Convenient

If the textile pattern is designed using the traditional design method, the design draft must be copied again, and the color area is used as the basis for color filling again. It must be redrawn five times if there are five color drafts. This method will waste a lot of drawing materials as well as a lot of valuable time and energy. Furthermore, the use of computer digital technology for textile pattern art design, as well as related software, can easily complete color transformation (As show in Fig. 3).

Fig. 3. Computer digital technology makes color transformation more convenient

To effectively replace colors, for example, you can use vector software that includes straws and paint bucket tools. Another example is that the brightness, color saturation, and contrast of the pattern can be effectively adjusted using the image software's tone adjustment and color balance functions, resulting in a variety of finished patterns. Designers can also easily change colors by using the color matching function in design software and the computer's automatic configuration function.

3.3 Faster Pattern Circulation

The main channels of pattern circulation in the process of designing textile patterns by traditional methods are books, photos, pattern drawing manuscripts, copying, and so on, and their transmission methods are also relatively traditional, and the speed of transmission cannot be guaranteed. The use of computer digital technology in textile art design can hasten pattern circulation.

The channels and methods of pattern circulation are becoming more diverse as network technology advances, and rapid pattern circulation can be realized by using websites, information exchange platforms, and other means. When users require patterns, they can download them quickly via the network, greatly reducing pattern circulation time. Designers are more convenient and faster in the process of design exchange due to the accelerated circulation of patterns, and provide diversified access channels for textile production enterprises to obtain samples. Time is money, and this can save a lot of time while also increasing the economic benefits for businesses.

4 Computer Digital Technology in Textile Art Design Application Strategy

The use of computer digital technology in the art design of preventive products has overcome the limitation of pattern color in traditional design methods, allowing the design connotation to be presented more intuitively and completely. The combination of computer digital technology and textile art design can reflect the design theme that designers want to express, sublimate the design content theme, and fully meet the consumption needs and aesthetics of consumers. Following that, we will examine the design strategy of textile wall-hanging decoration and bedding.

4.1 Decorative Wall-Hung Textiles Art Design and Application Strategy

Decorative wall-hung textiles are typically hung on interior walls with strong ornamental and decorative properties. Designers should first thoroughly investigate the needs of consumers and the market in order to design textile works of art that are more in line with consumers' aesthetic needs during the artistic design process of this textile using computer digital technology (As show in Fig. 4).

For example, in the wall-hung decorative textile "Above the Starry Sky," the designer first used computer digital technology to print Van Gogh's famous painting "Starry Sky" on the wool fabric, and then, based on the artistic conception of the painting, the wool felt material with similar colors was printed on the wool fabric by needle punching, which enhanced the overall unevenness and artistic sense of the wall-hung decorative textile. The designer also fully integrates the composition characteristics and uses wool felt to create the shape of a windmill on it, making the overall content of the artwork more substantial and complete.

Fig. 4. Computer digital technology promotes the development of decorative wall hanging textiles

4.2 Bedding Textile Art Design and Application Strategy

Bedding textiles are necessities in people's daily lives, and designers should use computer digital technology to consider the practical needs of consumers during the artistic design process. Bed sheets, pillows, bedding, and other bedding products, for example, are closely related to the overall interior decoration style. Designers should therefore unify the overall design style of bed products so that some rows of bed products have the same textile art style, forming a good visual aesthetic feeling indoors and creating a comfortable rest environment for people (As show in Fig. 5).

Fig. 5. Computer digital technology promotes the development of home textile design

5 Conclusion

Using computer digital technology to carry out artistic textile design can overcome all of the drawbacks of traditional design methods and save a significant amount of time, energy, and design materials. Designers should consider the practicality and aesthetics of textiles during the design process in order to fully meet the needs of consumers and create massive overall benefits for textile production enterprises.

Acknowledgments. This work was supported by 2020ZJJGLX062, 2020ZJJGLX063, 2020JY21.

References

1. He, M.: (2017) Research on the application of digital technology in textile art design. Think Tank Era **10**, 190–191 (2017)
2. Ma, S.Q.: (2014) Application of digital technology in textile pattern design. J. Brand Res. **1**, 8 (2014)
3. Zhou, M., Ge, X.F.: (2018) Application of digital printing design technology in textile pattern design. Art Educ. **1**, 221–222 (2018)
4. Ren, X.L.: (2006) The influence of digital technology on textile art design. J. Shandong University Art Design **1**, 80 (2006)
5. Zhu, J.L.: (2021) Application of graphic design technology in textile pattern design. Chem. Fiber Textile Technol. **50**(02), 13–14 (2021)
6. Mao, Z.: (2015) Application and analysis of computer technology in national textile design. China National Exhibition **10**, 168–169 (2015)
7. Shu, W.: (2021) Research progress of computer-aided tie-dyeing pattern design. Textile Auxiliaries **38**(12), 14–16 (2021)

8. Yu, M.G.: (2018) Application of computer aided technology in tie-dyeing pattern design. Textile Auxiliaries **35**(7), 45–48 (2018)
9. Lu, L.: (2017) Analysis and research of tie-dyeing pattern design based on computer technology. J. Pu'er Univ. **33**(6), 68–69 (2017)
10. Yao, J.: (2017) Application of computer aided design in tie-dyeing technology teaching. Light and Textile Indus. Technol. **46**(3), 69–70 (2017)
11. Bao, Z. H., Xu, H., Wang, Z. H. (2010) "Application of computer color matching technology in textile pattern design", Journal of South-central Minzu University(Natural Science Edition), 2010, 29(1): 107–110
12. Huang, M.J.: (2021) Application of computer graphic design software in textile creative design. Chem. Fiber Textile Technol. **50**(1), 29–30 (2021)
13. Chen, C.: (2020) Research on the application of PS graphic design in clothing and textile design. Textile Reports **39**(8), 77–79 (2020)

Check for updates

Application of Multimodal Information Technology Based on BIM Technology in Intelligent Construction

Ping Fu(✉)

College of Architecture and Engineering, Chongqing Vocational and Technical University of Mechatronics, Bishan, Chongqing 402760, China
17783436533@163.com

Abstract. With the development of the times, information technology is gradually becoming popular, and multimodal information technology is gradually being used in intelligent construction. Through multimodal information technology, the quality and efficiency of building management can be improved, and the cost of building intelligence can be saved, thus saving resource utilization efficiency. However, there are still some shortcomings in the use of multimodal information technology in intelligent construction. This article studied the application of multimodal information technology based on Building Information Modeling (BIM) technology in intelligent construction, aiming to continue improving the quality of intelligent construction through BIM technology. This article tested the cost reduction of intelligent construction using multimodal information technology after using BIM technology through experiments. The data showed a minimum reduction of 10% and a maximum reduction of 19%. This proved that BIM technology had good results in intelligent construction of multimodal information technology.

Keywords: Intelligent Construction · Multimodal · Building Information Modeling Technology · Information Technology

1 Introduction

With the rapid development of the economy, building informatization has become an inevitable trend. There are so many construction projects nowadays. If information technology is not used for management, it would bring a great burden to management personnel, so studying how to improve the quality of information intelligent construction is of great significance.

Many scholars are studying architecture, and Ingle P V believed that high-rise construction sites, especially those located in urban areas with limited space, faced difficulties in timely delivery of materials. Management technology could further benefit from state-of-the-art equipment such as radio frequency identification tags and ubiquitous sensor networks, which achieved significant results in automated logistics management of

J. C. Hung et al. (Eds.): FC 2023, LNEE 1132, pp. 122–130, 2024.
https://doi.org/10.1007/978-981-99-9538-7_17

construction sites [1]. Alaloul W S believed that engineering and architecture were lagging behind the trend in implementing artificial intelligence solutions. However, through applications and algorithms, it could help bridge the technological gap [2]. Khanzadi M believed that there were problems in research on building technology in the context of construction projects, and research on the synergistic effects and environmental impacts of new technologies should be strengthened to initiate the process of successfully entering the fourth industrial revolution [3]. Although scholars conducted a lot of research on architecture, there were still many shortcomings.

The development and application of multimodal information technology have brought good news to intelligent construction. Through the intelligent management of multimodal information technology, buildings can be built according to people's needs, which can greatly improve people's satisfaction with buildings. This article studied the application of multimodal information technology based on BIM technology in intelligent construction. This article experimentally tested the reduction in construction costs of multimodal information technology using BIM technology after intelligent construction, and found that the reduction was still significant. This proved that BIM technology could play a good role in multimodal information technology intelligent construction.

2 Use of Multimodal Information Technology in Intelligent Construction Using BIM Technology

2.1 Basic Concepts of Intelligent Construction

Generally speaking, intelligent construction refers to optimizing the system, service internal structure, management content, and structure of a building according to the needs of users during the process of building design, so as to meet their various needs and provide users with a more comfortable and safe living and working environment [4, 5]. In an intelligent building environment, users can choose intelligent tools based on their actual needs, so as to achieve good operation of intelligent building functions. Intelligent buildings utilize advanced scientific and technological advancements, such as communication technology, information technology, and control technology. The core content of intelligent buildings is low-carbon environmental protection, energy conservation and emission reduction, thus further creating a good living environment for users [6, 7]. Intelligent buildings have numerous advantages, such as environmental friendliness, energy efficiency, and practicality. It can better achieve peaceful development of the environment and humanity, which is consistent with the concept of sustainable development. It belongs to the focus of energy structure adjustment and is also a hot project in the current construction industry [8, 9].

2.2 Problems in Intelligent Construction

Theoretical research is still relatively lacking: Currently, there are still many problems in intelligent buildings, mainly manifested in the research on the design theory of intelligent buildings. The current design theory of intelligent buildings cannot match the relevant technical application requirements of intelligent building design, and the corresponding theoretical structure is also incomplete. Therefore, in the actual design and development process of intelligent buildings, it is necessary to conduct in-depth research on the corresponding architectural design theories [10, 11].

Imperfect intelligent product system: There is currently no dedicated intelligent construction product system in the construction market. From this, it can be seen that there are still significant shortcomings in the intelligent design of buildings. To change this situation, it is necessary to increase the design and research of intelligent application systems in building design management to ensure the smooth development of building intelligence. At the same time, in the process of building intelligence development, it is also necessary to increase the development of products related to intelligent buildings to meet market needs and accelerate the development of the industry [12, 13].

2.3 Application Advantages of Multimodal Information Technology in the Field of Architecture

Advantages in architectural design: Compared with traditional architectural design, the application of multimodal information technology frees architectural designers from the inconvenience and low design efficiency caused by manual drawing. The use of computer-aided design software based on multimodal information technology can assist designers in online design and modification. In this software, relevant graphics, images, text and other information can be edited in real-time, thus greatly reducing the workload of designers and improving work efficiency [14, 15]. With the continuous development of big data technology, the application of multimodal information technology in the field of architectural design is becoming increasingly intelligent. Design software can provide targeted design solutions to designers according to their requirements, thus using database comparison methods to identify problems and provide correction suggestions.

Advantages in construction management: Construction management is a complex task that involves many tasks. Therefore, it is necessary to use project management methods to supervise and manage the construction process, which is also a necessary condition to ensure the smooth progress of construction projects [16, 17]. In the past, construction project management work mainly relied on a large amount of manpower to collect, organize, classify, and query information, which not only restricted the management efficiency of construction projects, but also caused an increase in construction costs. With the application of multimodal information technology, paperless network office has been achieved. By utilizing the 0A system, the sharing of construction management information can be achieved, and previously complex management work can also be carried out on network platforms, thus greatly improving the efficiency of construction management. For example, in the process of building construction management, warehouse managers can record the usage of relevant materials online, so as to timely grasp the changes in material costs and achieve effective control of building construction costs [18, 19].

2.4 Key Technologies of Multimodal Information Technology

To achieve intelligent management of buildings, it is necessary to rely on advanced information technology and combine it with the needs of intelligent construction management to carry out comprehensive design, so as to meet specific building intelligent management goals. When designing an intelligent integrated system for buildings, the key technologies of multimodal information technology are shown in Fig. 1:

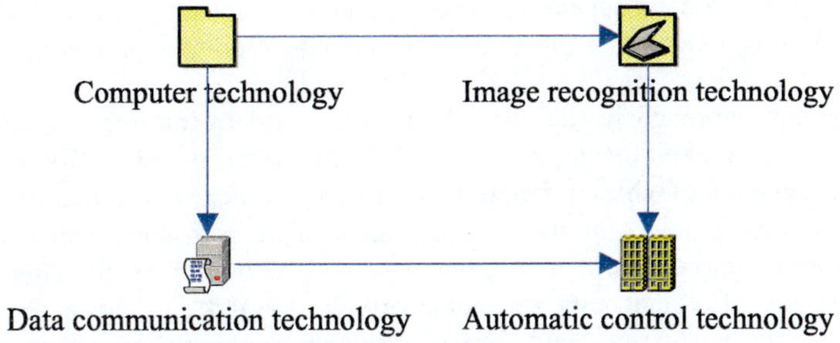

Fig. 1. The key technology of multimodal information technology

The specific analysis is as follows:

(1) Computer technology: The software of any system needs to operate on computers as media. Therefore, in order to improve the application effect of building intelligent integrated systems, advanced computer management technology needs to be adopted to improve the performance of computer media, including stability, security, service sustainability, fault tolerance, and efficiency.
(2) Image recognition technology: When integrating various subsystems, the integration objects include image acquisition subsystems such as vehicle access monitoring systems and video capture monitoring systems. Therefore, in order to convert the collected data into readable digital data, effective image recognition techniques must be used to complete the recognition, collection, and storage of input image data, thereby achieving the conversion of target data.
(3) Data communication technology: When designing an intelligent integrated system for buildings, it uses a centralized data storage management mode. It is produced in real-time by various subsystems based on the operation of their own equipment. However, if the centralized and aggregated storage of these data is to be achieved, specialized data communication technologies need to be used to ensure that the data produced by each subsystem can be continuously, stably, reliably, and accurately reported to the integrated system data center, thereby achieving data storage.
(4) Automation control technology: The building intelligent integrated system, named "intelligence", is able to automatically analyze the collected and reported data based on pre-set rules. It makes corresponding processing based on the analysis results, without the need for manual intervention, thus improving the management efficiency and quality of the building. Therefore, in order to make the system more valuable, it is necessary to carry out comprehensive automation control.

2.5 Development Trends of Intelligent Construction

With the rapid development of intelligent buildings, the corresponding equipment level would also improve, thereby promoting the development of related industries and bringing greater opportunities to them. To meet the requirements of intelligent buildings, related equipment must constantly undergo innovation in order to form a trend of common development. Generally speaking, in order for intelligent buildings to achieve better development, various advanced equipment must be applied to the building. In order for related enterprises to stand out among many enterprises, they must continuously carry out reform and innovation, thereby promoting the better development of intelligent buildings [20].

Intelligent comprehensive buildings can utilize modern information technology to unify the connection and management of building clusters, thereby fully realizing systematic management of building functions. In urban complexes, information technology has been used to achieve automatic management of modern complexes; system concatenation and comprehensive management should be implemented. This starts with the construction of a smart city and aims to build a smart city cluster. By doing so, various resources can be reasonably utilized, thus improving the effectiveness of social governance and reducing governance costs.

2.6 Overview of BIM Technology

BIM technology, as a way of expressing information based on building 3D models, provides new ideas for information sharing in various stages of the construction process. Based on technology, design results can be directly utilized. Based on models, documents or charts can be automatically generated, and data results can be visually observed and analyzed. Effective cooperation with various parties involved in the project can be achieved, thus achieving refined management of engineering projects. The method used in BIM technology mainly utilizes design models to automatically generate acceptance plans, thus enabling every construction quality inspector to understand the distribution of acceptance plans in advance; non destructive handover was carried out for each testing unit; By comparing and analyzing the acceptance results, erroneous acceptance was avoided. The characteristics of BIM technology are shown in Fig. 2:

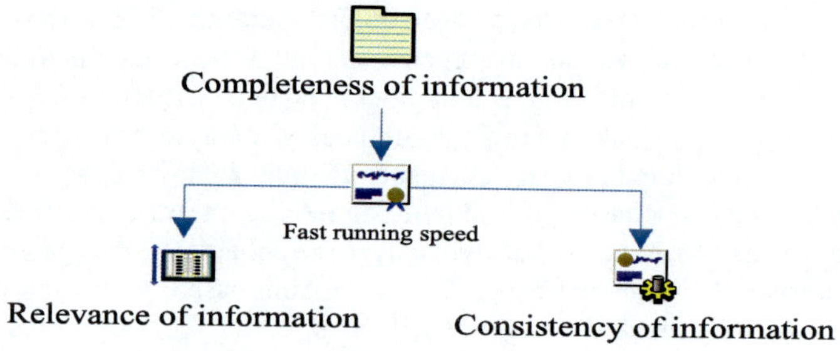

Fig. 2. Characteristics of BIM technology

2.7 Significance of BIM Technology in Intelligent Construction

The application of BIM technology can effectively solve the "bottleneck" problem in the current intelligent building information management: At present, there are many problems in the intelligent building information management. The root cause is that each construction unit has established their own independent information system, and information sharing between each unit is impossible, thus forming a "information silo". Through BIM technology, a single source of engineering data can be established. In engineering, all participants can use BIM as the same source of information, thus enabling precise information exchange and sharing among all participants, and ensuring the correctness and consistency of project information.

In the design of various specialties, information sharing has been achieved: in the project information management system of various specialties in construction engineering, the required project parameters and related information can be directly obtained from the BIM information model without the need for repeated input of data. In a specific field, when some parameters or information in a certain field change, the parameters or information in other fields in that field would also change, thus achieving collaborative design in multiple fields.

BIM technology is used to achieve information management throughout the enterprise lifecycle: BIM technology is built on top of the entire lifecycle of construction projects. It can carry out project design, construction, on-site management, and operational management, thereby achieving dynamic, integrated, and visual project information management throughout the entire process, thus meeting the management and monitoring of construction quality, safety, cost, and progress of engineering projects.

2.8 Specific Use of BIM Technology in Intelligent Construction

BIM technology can help calculate the construction period of intelligent construction. When using BIM technology in intelligent construction, the first step is to establish an objective function, and the calculation method of the objective function is shown in Formula 1:

$$A = B + C \tag{1}$$

In Formula 1, A represents the total cost, B represents the direct cost, and C represents the indirect cost. The relationship between direct costs and duration costs under normal construction period is shown in Formula 2:

$$T = \sum\nolimits_{i=1}^{n} B \tag{2}$$

In Formula 2, T represents the duration cost, and the relationship between indirect cost and duration cost is shown in Formula 3:

$$T = \frac{C}{A \sum_{i=1}^{n} B} \tag{3}$$

3 Simulation Experiment of BIM Technology and Multimodal Information Technology in Intelligent Construction

This article investigated ten intelligent buildings using multimodal information technology under BIM technology, and investigated people's satisfaction with these buildings. If there was a high level of satisfaction, it indicated that the use of multimodal information technology in intelligent construction under BIM technology was successful. The satisfaction of ten buildings constructed using multimodal information technology under BIM technology is shown in Fig. 3:

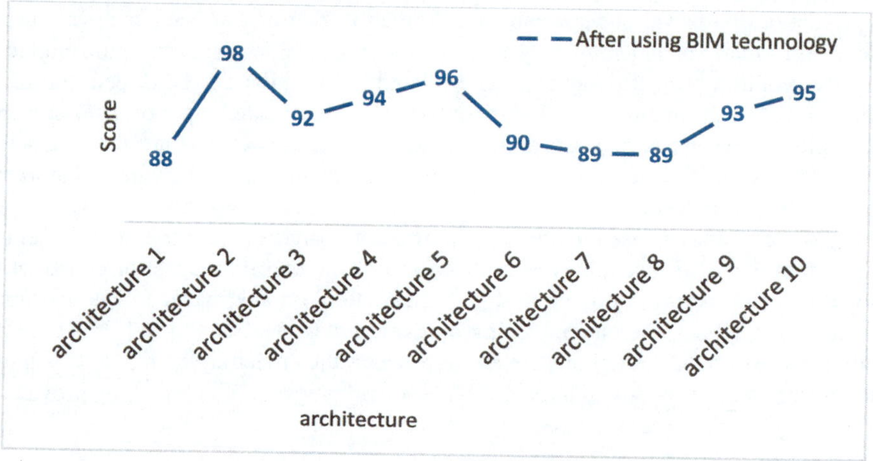

Fig. 3. Satisfaction with intelligently constructed buildings

From the experimental results in Fig. 3, it could be seen that people's satisfaction with intelligent construction using multimodal information technology under BIM technology was the lowest at 88 points and the highest at 98 points. This indicated that people highly recognized the intelligent construction of buildings using multimodal information technology under BIM technology, proving that multimodal information technology under BIM technology had good results in intelligent construction.

This article also tested how the intelligent construction of multimodal information technology under BIM technology could reduce construction costs. The construction cost reduction of intelligent construction using multimodal information technology under BIM technology is shown in Fig. 4:

From the experimental results in Fig. 4, it could be seen that the cost reduction of intelligent construction using multimodal information technology using BIM technology was at least 10% and up to 19%. The cost reduction was still quite significant. Through experimental data, it could be known that the intelligent construction of multimodal information technology under BIM technology could play a role in reducing construction costs. Therefore, it proved that BIM technology and multimodal information technology had a high degree of compatibility for intelligent construction.

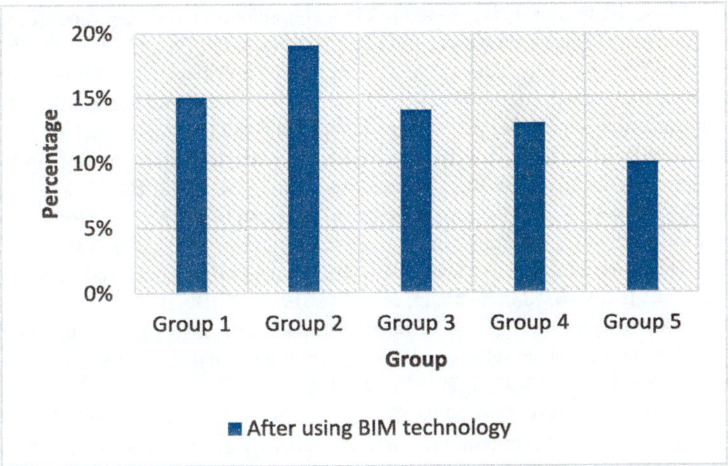

Fig. 4. Construction cost reduction of intelligent construction using multi-modal information technology under BIM technology

4 Conclusions

Intelligent construction is a hot topic nowadays. Good intelligent construction has the function of reducing construction costs and meeting people's various construction needs. The combination of intelligent construction and multimodal information technology can achieve good results. This is because information technology can optimize the accuracy of building construction, and improve efficiency, so as to reduce errors. Combining BIM technology is also a good choice. Nowadays, the use of BIM technology in building construction is also gradually increasing. This article studied the application of BIM technology and multimodal information technology in intelligent construction, which aimed to combine intelligent construction and BIM technology with multimodal information technology. The experimental test showed that the cost required for construction decreased after using two technologies, indicating that multimodal information technology and BIM technology achieved good results in intelligent construction. Due to space limitations, the experiments conducted in this article were still insufficient. It would be improved in the future. In the end, it is hoped that intelligent construction would continue to do better.

References

1. Ingle, P.V., Mahesh, G.: Construction project performance areas for Indian construction projects. Int. J. Constr. Manag. **22**(8), 1443–1454 (2022)
2. Alaloul, W.S., Alzubi, K.M., Malkawi, A.B.: Productivity monitoring in building construction projects: a systematic review. Eng. Constr. Archit. Manag. **29**(7), 2760–2785 (2022)
3. Khanzadi, M., Sheikhkhoshkar, M., Banihashemi, S.: BIM applications toward key performance indicators of construction projects in Iran. Int. J. Constr. Manag. **20**(4), 305–320 (2020)

4. Tabei, S.M.A., Bagherpour, M., Mahmoudi, A.: Application of fuzzy modelling to predict construction projects cash flow. Periodica Polytechnica Civil Eng. **63**(2), 647–659 (2019)
5. Zaman, U., Florez-Perez, L., Khwaja, M.G.: Exploring the critical nexus between authoritarian leadership, project team member's silence and multi-dimensional success in a state-owned mega construction project. Int. J. Project Manag. **39**(8), 873–886 (2021)
6. Ismail, M., Fathi, M.S.: Leadership in construction: Leadership styles practiced in construction project–a review. J. Adv. Res. Bus. Manag. Stud. **13**(1), 24–30 (2018)
7. Prasad, K.V., Vasugi, V., Venkatesan, R.: Critical causes of time overrun in Indian construction projects and mitigation measures. Int. J. Constr. Educ. Res. **15**(3), 216–238 (2019)
8. Khlaifat, D.M., Alyagoub, R.E., Sweis, R.J.: Factors leading to construction projects' failure in Jordon. Int. J. Constr. Manag. **19**(1), 65–78 (2019)
9. Keshk, A.M., Maarouf, I., Annany, Y.: Special studies in management of construction project risks, risk concept, plan building, risk quantitative and qualitative analysis, risk response strategies. Alex. Eng. J. **57**(4), 3179–3187 (2018)
10. Mathar, H., Assaf, S., Hassanain, M.A.: Critical success factors for large building construction projects: perception of consultants and contractors. Built Environ. Project Asset Manag. **10**(3), 349–367 (2020)
11. Gurgun, A.P., Koc, K.: Administrative risks challenging the adoption of smart contracts in construction projects. Eng. Constr. Archit. Manag. **29**(2), 989–1015 (2022)
12. Mahmoudi, A., Abbasi, M., Deng, X.: A novel model for risk management of outsourced construction projects using decision-making methods: a case study. Grey Syst. Theory Appl. **10**(2), 97–123 (2020)
13. Fischer, A.: Artificial intelligence - intelligent construction - particles for a better life. ZKG Int. **72**(5), 1 (2019)
14. Li, J.L., Chen, L.M., Xu, H.: Intelligent construction, operation, and maintenance of a large wastewater-treatment plant based on BIM. Adv. Civil Eng. **2021**(5), 1–11 (2021)
15. Amirtash, P., Parchami Jalal, M., Jelodar, M.B.: Integration of project management services for international engineering, procurement and construction projects. Built Environ. Project Asset Manag. **11**(2), 330–349 (2021)
16. Khoso, A.R., Md Yusof, A., Chai, C.: Robust contractor evaluation criteria classification for modern technology public construction projects. J. Publ. Procurement **21**(1), 53–74 (2021)
17. Kermanshachi, S., Safapour, E.: Identification and quantification of project complexity from perspective of primary stakeholders in US construction projects. J. Civ. Eng. Manag. **25**(4), 380–398 (2019)
18. Hire, S., Sandbhor, S., Ruikar, K., et al.: BIM usage benefits and challenges for site safety application in Indian construction sector. Asian J. Civil Eng. **22**(7), 1249–1267 (2021)
19. Sang, L., Yu, M., Lin, H.: Big data, technology capability and construction project quality: a cross-level investigation. Eng. Constr. Archit. Manag. **28**(3), 706–727 (2021)
20. Garcia-Nieves, J.D., Ponz-Tienda, J.L., Salcedo-Bernal, A.: The multimode resource-constrained project scheduling problem for repetitive activities in construction projects. Comput.-Aided Civil Infrastruct. Eng. **33**(8), 655–671 (2018)

Construction of Financial Market Risk Early Warning Model Based on Artificial Intelligence Technology

Sijia Li(✉)

School of Management and Engineering, Nanjing University, Nanjing 210009, China
lsjnju@126.com

Abstract. Since the 1980s, world economic integration, the financial industry has become the core of social development, and the accompanying financial risks have become a relatively difficult problem for countries. The complexity and diversity of financial business and the rapid increase of uncertainty factors have led to a significant increase in risks and may lead to financial crisis when accumulated to a certain extent. Once a financial crisis breaks out, it will lead to social and even political crisis in addition to economic recession. Its main purpose is to provide investors with timely information about financial market risks, so that they can make better decisions and thus increase profits.

Keywords: Financial market · Artificial intelligence technology · Risk warning

1 Introduction

As an emerging developing country, China's economy has been basically geared towards good development, and all industries have shown a vigorous and upward trend. However, there are still potential financial risks that cannot be ignored: imperfect financial laws and regulations, lack of strong supervision and management, and market access mechanisms that need to be discussed. After China's entry into WTO, the financial supervision has been gradually relaxed, and the competition among global industries has led to the further expansion of the scope of China's implied financial risks [1]. In addition, due to the large scale growth of China's shadow banks at this stage, and there is a foam in the real estate industry, various signs show that China's finance is facing financial risks that cannot be underestimated. Therefore, it is urgent for China suitable for its own situation, so as to monitor various financial risks in a timely manner.

The existing domestic and foreign literature on financial risk focuses more on some developed countries and countries that have had financial crises. Most of the financial risk prevention strategies are proposed for countries that have experienced financial crisis. The research conclusions have their own shortcomings: the subjectivity is too strong in the empowerment of the indicators of the analytic hierarchy process. so we can not directly use foreign pre-warning models when using financial risk pre-warning models for China. The some of them are still introducing foreign theories and models [2]. The detection process is shown in Fig. 1 below.

J. C. Hung et al. (Eds.): FC 2023, LNEE 1132, pp. 131–136, 2024.
https://doi.org/10.1007/978-981-99-9538-7_18

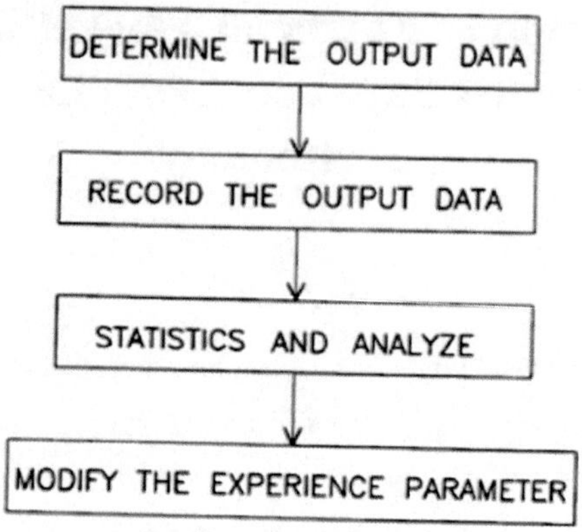

Fig. 1. Detection Process

Chinese scholars still rely on foreign indicator selection systems, and have not conducted research based on China's specific national conditions. In different periods and countries, the influencing factors of financial risks have different characteristics. In recent years, China is an emerging market that has been in the process of reform. Financial risk has its unique characteristics and representativeness. The theoretical analysis of financial risk in China will further enrich the theoretical system of financial risk.

2 Related Work

2.1 Overview of Domestic Research

There were few empirical studies. Most of the methods used were signal methods, supplemented by multivariate statistics and information systems. Chen Songlin (1997y24) combined the internal and external generation mechanism of financial risk, analyzed the systematic risk factors and non systematic risk factors respectively, and built a more perfect financial risk monitoring indicator system. He believed that because some financial risk played a catalytic role, it promoted the continuous spread of financial risk, and finally led to the financial crisis. Based on this, he added other risks that might trigger the crisis to the model, and built A relatively complete risk measurement model [3]. Cao Wenlian and Xu Xiaobo drew on the achievements of foreign research, combined with the actual domestic conditions, and theoretically combined with macro, meso and micro level research, analyzed and constructed the early warning indicator system of financial risk in China. Liu Zhiqiang re established a set of early warning indicator system after learning from foreign research methods on early warning indicators. The system is composed of two parts: one is about the asset quality, profitability, credit growth level and interest rate of domestic financial institutions; Second, indicators such as foreign debt investment and exchange rate [4]. At the same time, he proposed that this set, but there is no necessary explanation on the determination of indicator threshold in his research. The service supply chain is shown in Fig. 2.

He Xiaobo and Zhang Yuhong used KLR signal light display method to build input, calculation and output modules, cluster analysis method, subjective analysis method

Fig. 2. Service supply chain

(analytic hierarchy process) and objective analysis method (entropy method) to build a risk early warning system for commercial banks, and conducted empirical analysis. This method intuitively shows the risk status of banking and financial institutions. Feng Yun and Wu Chongfeng proposed a multi time scale early warning process. By classifying and supplementing the early warning process, they classified it into three levels: long-term, medium-term and short-term. They proposed a method to add multi time scale and supplement indicator sets with changes in financial risk. This method makes the system sensitive to changes in market insight. Zhang Ying sorted out the early warning methods for the currency crisis, predicted the financial crisis in six aspects, and established an early warning indicator system.

Whose research objects are mostly developed countries, does not conform to the special situation of China. However, domestic research on it has just begun, and most of the existing research methods and models abroad are used to study the economic situation of a certain period in China. The existing early warning model research is slightly outdated, and it is urgent to build a risk early warning model suitable for domestic situations.

2.2 Concept Analysis of "Artificial Intelligence"

Since the birth of the name "artificial intelligence", until now, the field of artificial intelligence has become a hot field of extensive research. Researchers still have not formed a clear and unified view on what is "artificial intelligence". As for the definition and connotation of AI, Russell's representative view is that AI is divided into four types of systems: "thinking rationally acting rationally like human beings". These four definitions point to two dimensions respectively. One is the internal standard of AI, which focuses on thinking or behavior, The second is the realization of artificial intelligence, which is to imitate human beings or to learn independently.

Searle classified AI in terms of the extent to which the development of AI can be achieved, namely weak AI and strong AI. Weak AI means that no matter how intelligent AI is, its essence is still the simulation of human behavior. AI itself does not have the ability to think about behavior and act accordingly; Strong AI means that AI can think and act independently, and its behavior is no different from human behavior. We can think of it as a universal intelligent AI with human intelligence level or even beyond human intelligence level. Searle believes that people can create weak AI, but opposes strong AI. The different dimensions and approaches make the development of AI form different schools and research directions.

3 The Fragility of the Financial System Itself

The financial institutions and the internal fragility caused by the irrational economic activities.

① The fragility of financial markets.

The formation, accumulation and gradual extreme deterioration of financial system risks caused by the sharp fluctuation of asset prices in financial markets (capital markets and foreign exchange markets, etc.). For example, when the interest rate of a country's money market rises, the inflow of foreign funds will probably lead to the rise of asset prices in the market, which will lead to foam. The influx of capital will also lead to the rise of the value of the local currency. If the funds are withdrawn, the asset quality will shrink rapidly and substantially, leading to hidden risks and financial crisis. In addition, more and more homogeneous financial markets will also lead to financial risks. Homogeneous risk refers to the financial risk generated when the financial market gradually moves towards comprehensive operation and internationalization, which makes the thinking mode, professional technology and financial supervision standards among financial markets tend to be the same, and ultimately leads to the consistency of expectations and behaviors. Consistent actions have the same direction of action and cannot offset each other, resulting in sharp fluctuations in asset prices and rising financial risks.

② The Lending Characteristics of Financial Institutions and the Internal Vulnerability of Their Structures

The vulnerability of financial institutions due to their high leverage characteristics, coupled with the problem of principal-agent of financial institutions, has strengthened the risk of financial system. On the one hand, due to the excessive debt scale and the high leverage ratio of the balance sheet of financial institutions, a small decline in asset prices is likely to cause cost losses of financial institutions. On the other hand, the asymmetry between its earnings and the risks and responsibilities it bears, the lack of supervision and the government's assistance have caused financial institutions to ignore the interests of shareholders and depositors, leading them to pursue high profits in high-risk businesses and increasing the risk of investors. Based on the "Financial Vulnerability Hypothesis" put forward by Mins in 1997, he divided enterprises into three types, namely, risk aversion, investment and Ponzi enterprises, according to their increasing risks, and analyzed that commercial banks must experience cyclical crisis and bankruptcy wave due to their inherent characteristics. The risk of a single bank will spread to the entire banking system, and eventually the financial crisis will break out.

③ Irrational Behavior of Financial Market Subjects

In most cases, financial market economic entities are rational in their activities, but a large number of irrational behaviors will eventually lead to the outbreak of financial crisis. The irrational behavior in the financial crisis is divided into three stages: craziness, fear and collapse, describing the different psychological processes of investors. However, there are widespread irrational behaviors. Investors' common irrational expectations have led to "irrational prosperity" and the existence of asset price foam. In the case of relatively flat supply and demand, asset prices must decline sharply after the surge.

4 Model Building

In the 1970s, the American scholar Sadi (a professor at the University of Pittsburgh in the United States) proposed the AHP method. According to the nature of different problems and the overall goal to be achieved, the method decomposed the problem into multiple influencing factors according to the complexity of the problem, and arranged them hierarchically according to the subordination relationship between the factors, thus forming a multi-level structural model, Finally, the comparison method is used to establish the importance order and corresponding weight value of each level and each factor. Since the 1980s, the Analytic Hierarchy Process (AHP) has been introduced into China. Because of its characteristics of combining qualitative and quantitative methods to solve different decision-making factors, it has been widely used in many fields of economic activities in China, such as scientific research evaluation and economic management.

Centering on the overall status of financial risk of AI technology, the hierarchy diagram, as shown in Fig. 3.

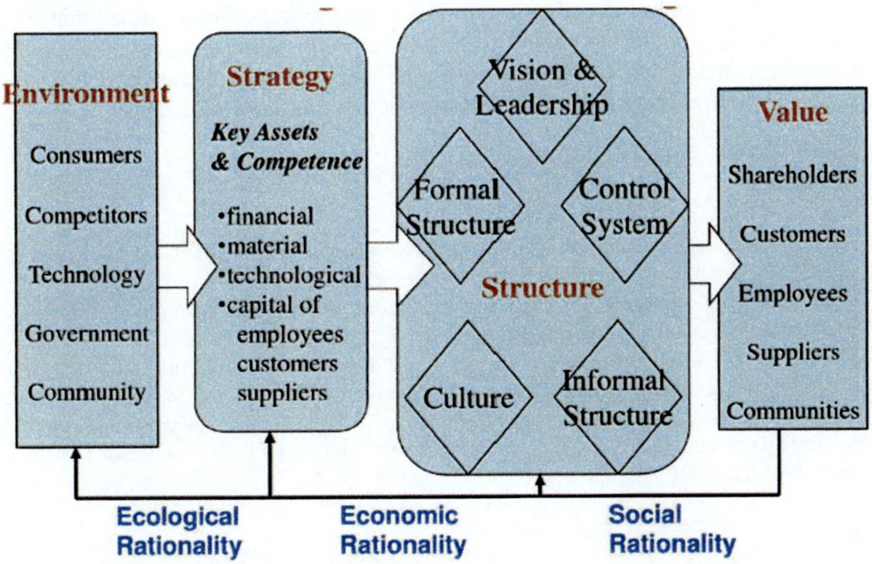

Fig. 3. Hierarchy Chart of Financial Risk Early Warning Indicators

Regression analysis is an effective method to establish risk early warning models. However, because there is usually a linear relationship between economic variables, when using regression methods to establish early warning models, multiple collinearities are usually encountered. Therefore, using the principal component method to solve the collinearity between independent variables becomes a means to deal with the problem of collinearity. This paper constructs the principal component regression model. When collinearity occurs, the regression model constructed by the least squares method will increase the variance of parameters, making the regression equation very unstable, and the regression equation and regression coefficient will not pass the significance test. This paper selects the principal component regression method when studying the risk early warning model, uses the 14 indicators of the signal display model to take the logarithmic variable as the independent variable, and uses the CVaR of the logarithmic return rate of the Shanghai Stock Exchange Index as the substitute variable of financial risk to establish the principal component regression model.

5 Conclusion

At the present stage, China's financial industry has a low level of development and its regulatory system needs to be improved urgently. Therefore, maintaining financial stability and preventing financial risks have become an urgent task. Especially today, when the global economy is gradually becoming synchronized, we should effectively control our own financial risks to prevent the spread of international financial risks. In order to prevent and control financial risks in China, we need to further understand financial risks and their supervision, and strengthen the prediction and control of financial risks in the financial system. On the other hand, on the basis of building a complete and effective early warning indicator system, it is urgent to build the measures that match it. The measures should include efficient laws and regulations, reasonable operation mechanism, complete organizational, we should comprehensively consider the impact of political, cultural and social non economic factors on China's finance.

References

1. Hong, W.: Research on financial crisis early warning mechanism of listed companies based on logistic model and factor analysis. Ind. Econ. Rev. (2017)
2. Zhu, M.R., Fang, W., Song, Y.: The research of sme financial crisis warning model based on neural network. DEStech Trans. Econ. Bus. Manag. (2016). (iceme-ebm)
3. Xiao, Q., Wang, Y.: Dynamic early warning of cash flow risk of listed companies on the growth enterprise market based on state space model. Pioneering Sci. Technol. Mon. (2019)
4. Ma, F., Zhou, Y., Mo, X., et al.: The establishment of a financial crisis early warning system for domestic listed companies based on two neural network models in the context of COVID-19. Math. Probl. Eng. **2020** (2020)

Design of Charge Point Based on Rectangular Self Convolution Window Algorithm Data Measurement

Wanzhong Li(✉)

Shandong Anlan Power Technology Co, LTD., Zibo 255000, Shandong, China
shandonganlan@163.com

Abstract. The charge point design based on the rectangular self convolution window algorithm data measurement is to use the rectangular self convolution window to measure the electric field strength in a certain area. This method can simultaneously measure electric field strength, voltage and current. The main advantage of this method is that it can be applied to any type of charging equipment at any location without changing its form or structure. In addition, the method has no effect on the charging system itself; Therefore, it can be used to test different types of charging systems. In this paper, the main circuit units of electric vehicle charging pile are designed, including rectifier unit, power conversion unit and human-computer interaction unit, and the rectifier unit and power conversion unit are modeled. For the rectifier unit, VIENNA rectifier is used to replace the common power frequency rectifier to provide DC with good power quality. Then DC-DC full bridge power conversion is performed on the output DC current, and the output is adjusted by phase shifting control. Finally, the human-computer interaction unit is designed, and the work flow chart of the human-computer interaction system is given.

Keywords: Data measurement · Rectangle self convolution window algorithm · Charging pile design

1 Introduction

In recent years, the global carbon emissions have experienced a certain degree of control, but also a trend of gradual rebound. The resulting air pollution problems need to be solved urgently. Environmental protection and energy conservation have gradually become the focus of the world. The emission of pollutants from automobiles has always accounted for about 50% of the total urban air pollution emissions. Pollution gases, including carbon monoxide, hydrocarbons, nitrogen oxides, etc., not only pollute the air around us, pose a threat to our physical health, but also cause irreparable damage to the atmosphere. This urgent problem needs our attention [1].

In order to deal with this global problem, the automobile manufacturing industry and automobile related industries are constantly transforming to green, focusing on the low pollution pollution of automobiles, and looking for clean and low pollution

J. C. Hung et al. (Eds.): FC 2023, LNEE 1132, pp. 137–143, 2024.
https://doi.org/10.1007/978-981-99-9538-7_19

energy as a substitute for oil. Therefore, new energy vehicles have emerged, which are characterized by low energy consumption and low pollutant emissions [2]. They are of great significance in alleviating ecological pollution and energy shortage.

Major countries in the world have their own long-term development plans for electric vehicles. For example, the French government once spent 2 billion francs to replace 10% of the internal combustion engine vehicles of domestic government departments with pure electric vehicles, and improve the charging stations and infrastructure in Paris. The British government invested a total of 230 million pounds in the construction and development of electric vehicles between 2011 and 2014. Compared with developed countries, China's electric vehicles started late, but the country put the development of electric vehicles in an important position.

2 Related Work

2.1 Development Status of Electric Vehicles and Their Charging Systems

As pure electric vehicles still have considerable shortcomings in their own technology, hybrid vehicles can promote low-carbon travel, and at the same time, they have solved the problems of insufficient endurance and long waiting time for charging. Although they also have high costs, they have solved the problem of empty window. In developed countries, This type of vehicle has a greater probability of taking the lead in replacing the internal combustion engine vehicle.

In the process of electric vehicle research, three representative technologies are developing vigorously [2]. One is the motor drive technology. The motor and its driving technology have always been the core technology of the entire system of electric vehicles. The control mode of this motor is very different from the old motor control. It requires efficient and reliable control of new motors, such as switched reluctance motor, DC brushless permanent magnet motor and other control technologies. Second, it is the intelligent system of electric vehicles -- the on-board energy regulation system. This system models a certain type of electric vehicle and its battery, and then collects relevant data during the operation and charging of the vehicle. Finally, through strong analysis and processing capabilities, it makes reasonable configuration in many aspects for electric vehicles, and keeps all components on the vehicle in the best working state [3]. Third, it is the vehicle battery improvement technology. The capacity of the vehicle battery pack limits the endurance of the electric vehicle. The high weight of the vehicle body also makes this problem more prominent, and there may be potential safety hazards. Therefore, it is necessary to reduce the battery volume, increase the battery capacity, improve the safety and stability of the battery, reduce the cost of the battery, and improve the battery's ability to use in harsh environments based on the increasingly changing weather environment. The key is to improve the energy density of the battery in order to improve the vehicle's endurance. Therefore, lithium-ion batteries are gradually being applied to automobiles. The theoretical value of their battery energy density is higher than that of lithium iron phosphate batteries and lithium manganate batteries currently used. However, the technology is not mature and is still in the process of gradual experiments [4].

Electric vehicles are developing and making progress in the continuous research, and their supporting charging equipment is also making continuous progress. In the course of the development of charging equipment, a lot of research has been carried out on charging piles, such as research on charging technology, research on charging planning of charging piles, research on improving battery equipment, etc.

2.2 Current Status of Design and Development of Charging Piles in China

Due to the late start of the research on electric vehicles in China, the construction and deployment of charging facilities in China are also relatively late, In order to improve the competitiveness of the state-owned automobile industry, break through the level of automobile research and development, China has invested a large amount of research and development funds, launched major state-owned projects such as the "863 Plan", the special research plan for electric vehicles, energy conservation and electric vehicles, and has made relatively rich achievements. China has made significant research on automotive batteries, improving the stability and durability of power cells and fuel cells, and reducing the manufacturing cost of batteries; The vehicle mounted motor has been improved to improve the energy efficiency of the motor, which has reached more than 94%.

The development time of domestic electric vehicle charging facilities is only ten years, but it has also made great progress [7]. At the moment when the national government continues to strengthen policy support, and the country promotes charging facilities and clean energy, the development of charging facilities has sprung up all over the country.

At present, charging pile and charging station are still the focus of domestic promotion, and charging pile is one of the most vigorously promoted. By the end of 2016, China has built more than 80000 public charging piles and more than 50000 private charging piles, and the number of charging piles is growing at a rapid growth rate. The charging pile has small floor area and flexible application. The charging station can meet multiple different charging demands of a variety of vehicles, and its charging efficiency is higher than that of the charging pile. Slow charging is still the mainstream of charging pile, which is easy to realize, easy to maintain and relatively safe. Combined with two different charging modes, the charging station mode is adopted in places with dense charging, which can meet the requirements of both charging speed and charging types. In areas with relatively less charging demand, flexible charging piles should be set to meet the regional demand for electric vehicle charging.

At the moment of low-carbon development, in combination with China's economic situation and the automobile industry, China has implemented policies on the one hand, and on the other hand, promoted the operation of charging piles to be economical and reasonable. However, the operation mode and charging pile itself still have huge development space and potential, and the urban construction planning of charging pile and charging station can be further studied. Therefore, there is still a long way to go to bring electric vehicles and charging facilities to the market and consumer groups.

3 Analysis and Design of Charging Pile System Research Scheme

3.1 Research on Charging Technology

Charging technology generally refers to wireless charging and conductive charging (contact charging). Wireless charging generally requires three core technologies: radio wave, electromagnetic induction and magnetic resonance. Foreign technologies for wireless charging are at a mature stage, while the core of domestic research on wireless charging technology is to realize wireless charging by induction. This charging method uses high-frequency coils. A charging device is installed at the primary side, and a receiving device is installed at the secondary side. When the charging device is working, high-frequency AC power is generated and transmitted to the secondary side. Then, it is rectified by the vehicle mounted charger at the secondary side, and DC power is transmitted to the battery.

Conductive charging (contact charging) is to connect the charging equipment and the receiving equipment directly through the conductor to form a loop. In actual places, contact plugs are usually used to connect the charging equipment of charging piles with the vehicle power battery, and conduct electricity through metal contact to provide electric energy. This charging mode is widely used in real life, and the technology is very mature, simple and easy to implement. However, direct charging by plugging and unplugging the connector for many times will gradually cause loss to the interface over time. The charging equipment will need more complex maintenance work, and many safety protection measures need to be designed for the loss. Therefore, compared with wireless charging, it has certain disadvantages.

The charging process of fast charging technology is generally short, which can quickly increase the battery charge to about 80% in a short time. High current is selected for charging, and the charging power is 15 KW–100 KW. Such high power and high current require very high hardware, and more comprehensive protection measures are also required.

Fast charging technology is relatively widely used in charging stations. Generally, three-phase four wire power grid voltage is used for power supply. In order to obtain better grid power, hardware should be reasonably designed to suppress the harmonic component of the grid, and also prevent the harmonic generated by the charging device from feeding back to the grid, which also has great requirements on the cost of hardware. In addition, the battery to be considered will generate higher temperature when charged under high current. Therefore, in order to prolong the battery life and prevent battery damage, reasonable control strategies and protection circuits should be designed.

3.2 Design Requirements of Charging Pile

The popularization of electric vehicles is accelerating year by year, and the in-depth development of its related charging facilities is gradually in progress. When the popularization of charging piles as the mainstream charging facilities of electric vehicles is in full swing, in order to achieve the application of charging piles.

The design requirements of charging devices are as follows:

(1) The charging pile shall be able to charge reliably, complete its main charging functions, and ensure personal safety during charging.
(2) The charging pile shall be provided with relatively complete protective measures and a stable and safe charging working environment.
(3) The interactive system of the charging pile shall be complete and convenient, and the user can conduct charging related operations intuitively and reliably according to the preset operation information.
(4) After the charging process is completed, the user should be able to query the relevant information of the charging process and receive the print slip from the printer.
(5) The charging pile shall have the ability to save and display data. During the charging process, it shall display and save data such as temperature, electric energy and quantity, and user amount, so as to facilitate users to query relevant historical information.
(6) The charging pile shall communicate reliably with the system background to ensure that the staff can control the charging pile system at any time, adjust the system and correct system errors in time.
(7) Reasonable pricing method shall be adopted for pricing and charging according to the national electricity price standard.

4 Design of Charging Pile Based on Data Measurement of Rectangular Self Convolution Window Algorithm

4.1 Rectangular Self Convolution Window Algorithm

The stronger the ability of window function to suppress spectrum leakage, the higher the accuracy of harmonic detection using FFT in asynchronous sampling. Cosine combined window is widely used in windowed interpolation FFT power harmonic analysis because of its superior sidelobe performance and strong ability to suppress spectrum leakage. The frequency domain performance of the cosine window is closely related to the number of cosine combination terms and coefficients.

Because the signals in the power grid are characterized by large noise and strong fluctuation, and the detection accuracy of amplitude and phase is required to be high, this chapter focuses on the construction of a fast window function with low side lobe peak value and side lobe attenuation. Based on the understanding, its construction characteristics are analyzed. Combined with the self convolution function, the cosine self convolution window function is constructed. Taking the two term, three term, four term and five term cosine composite window as an example, the first to third order self convolution windows of common cosine windows are constructed through time-domain self convolution, and the main lobe and side lobe characteristics of cosine self convolution windows are analyzed. Compare the sidelobe performance of various cosine windows, and select the four fifth order Nuttall windows with the best sidelobe performance among the four cosine windows and the five fastest sidelobe decay (MSD) windows with the best sidelobe performance among the five cosine windows to respectively construct their cosine self convolution windows to perform weighted FFT operations on signals.

$$KeyScore(S_i) = \frac{\sum_{i=1}^{n} tf_i}{\log(|length(S_i)|)} \tag{1}$$

4.2 Structural Framework of Charging Pile

The charging pile designed in this paper selects the following technical parameters according to the national technical standards to directly charge the electric vehicle.

Because of the design limitations, the power of the charging pile is set lower to make the design more complete and feasible. Here, the maximum output power is 15kW, which can quickly complete the task of DC charging the electric vehicle.

According to the design requirements of the charging pile described above, this paper chooses the staged constant current charging mode. The overall system framework of the DC charging pile should include the main charging system, printer, data acquisition module, alarm module, etc. Its main framework is shown in Fig. 1.

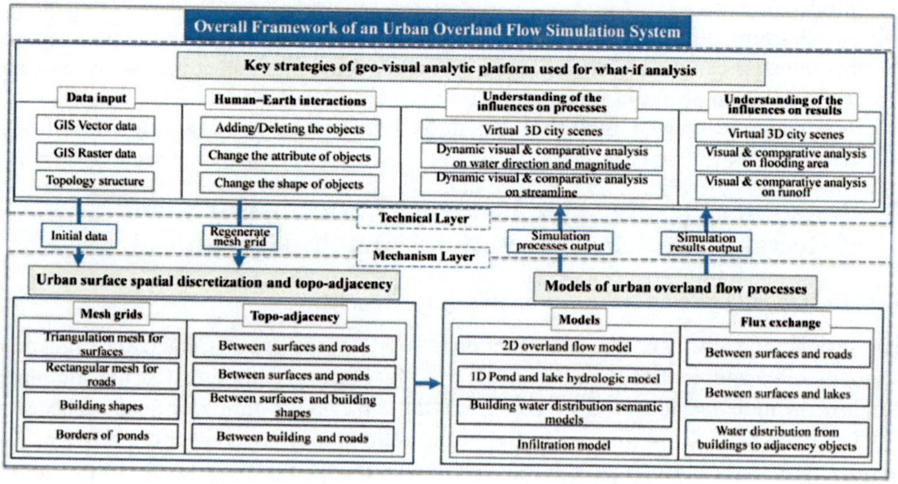

Fig. 1. Overall framework of charging pile system

According to the information provided in the figure, the three-phase alternating current provided by the power grid is connected with the electric energy meter through the AC contactor, and then enters the main charging circuit. After rectification, high-frequency inverter and secondary rectification, the lithium iron phosphate battery pack is charged regularly under the control of the main control module. The electricity meter is connected with the AC contactor for charging. Once the system is abnormal, it can directly feed back to the alarm locking module, so that the system is locked for processing by the staff. Card readers and printers are devices that provide convenient means for consumers to use. Card readers and IC cards constitute a means for consumers to swipe their cards. Printers enable consumers to print vouchers after consumption, so that consumers have consumption history as a basis.

5 Conclusion

The circuit parameters involved in this paper are used in the simulation design to ensure that the experimental circuit is true and feasible, and can be used in the future. This design has a certain reference value for the actual construction of the circuit, and puts forward a more advanced idea for the development direction of the electric vehicle charging pile. It is hoped that this charging system can be applied in the future work, and on the basis of improving efficiency, the cost of the circuit will be greatly reduced to ensure the commercial operation of the charging pile. The circuit designed in this paper has many shortcomings. For example, the design is more complex, there are many parameters to be controlled, the system is more difficult to achieve, and more switching devices are used, which will increase the cost, and the process of controlling switching devices is also more complex. Therefore, there is still much work to be done in the system simplification and practical trial. I hope that we can make efforts by ourselves and others in the future, The charging system is further optimized to make it practical.

References

1. Meng, X.H., Cao, S., Zhang, Y.H., et al.: Design of temperature measurement and control system of chemical instrument based on internet of things (2021)
2. Fei, W., Jiacheng, C., Wanliang, W.: A dynamic multi-objective evolutionary algorithm based on prediction. J. Comput. Des. Eng. (2022)
3. Qi, W., Yu, J., Liang, Q., et al.: Design of graph neural network social recommendation algorithm based on coupling influence. Int. J. Pattern Recogn. Artif. Intell. (2022)
4. Li, Y.S., Yi, K., Zhao, X., et al.: Design of target tracking filter for multi-rate system based on radar measurement (2022)

Design and Safety Research of University Management System Based on NFC Technology

Li Xuanyi(✉)

Jilin Communications Polytechnic, Changchun 130012, China
nihaoxiatian628@163.com

Abstract. The role of system design in university management is very important, but there is a problem with the unreasonable design. The B/C structure does not solve the security problems in university management, and the data integrity rate is low. Therefore, this paper proposes an NFC technology to construct a management system model. First, the information management knowledge is used to classify the management data, design according to the system development standards, and realize the standardized processing of the system data. Then, NFC technology is used to form an information management set and optimize the system design. MATLAB simulation shows that under specific management requirements, the management data analysis accuracy and system design time of NFC technology are better than those of the B/C structure.

Keywords: managing data knowledge · system design · NFC technology · Authoring results

1 Introduction

System design is a crucial evaluation content in university management, which is of great significance to university management [1]. However, in the actual development process, there are little problems in the system design, which has a certain impact on the management of universities [2]. Some scholars believe that the application of intelligent algorithms to university management can effectively carry out design and safety data analysis and provide corresponding support for development [3]. On this basis, this paper proposes NFC technology to optimize the system design process and verify the effectiveness of the model.

2 Related Concepts

2.1 Mathematical Description of NFC Technology

NFC technology uses management data knowledge to optimize the design and according to the multi-dimensional indicators in university management [4], discover the dangerous information in university management and integrate the case data to finally judge the feasibility of the system design [5]. NFC technology combines information management knowledge, uses system mining and intelligent algorithms, optimizes system design results, and can improve the intelligent system design rate.

J. C. Hung et al. (Eds.): FC 2023, LNEE 1132, pp. 144–149, 2024.
https://doi.org/10.1007/978-981-99-9538-7_20

Hypothesis 1: The system design is $\sum x_i$, the design set is x_i, the design is y_i, and the judicial function of the system design is $f(x_i)$ as shown in Eq. (1).

$$f(x_i) = \underline{\underline{\sum x_i | y_i \emptyset \xi}} \tag{1}$$

2.2 Selection of System Architecture

Hypothesis 2: The system design function is $F(x_i)$ and the design weight coefficient is q_i, then the university management method selection is shown in Eq. (2).

$$F(x_i) = \frac{y_i \prod q_i \cdot \xi}{z_i} \tag{2}$$

2.3 Handling of System Design Schemes

Before NFC technical analysis, it is necessary to conduct a standard analysis of the design in the system design and map the design to the development table to determine the abnormal semantic design. First, the design is comprehensively selected, and the threshold and weight of the design are set to ensure the accuracy of NFC technology. The design is a semantic transformation design that needs to be standardized. If the design is in a non-standard distribution, the accuracy of the overall development is reduced. In order to improve the accuracy of NFC technology and improve the level of system design, the NFC technology analysis should be selected, and the specific method selection is shown in Fig. 1.

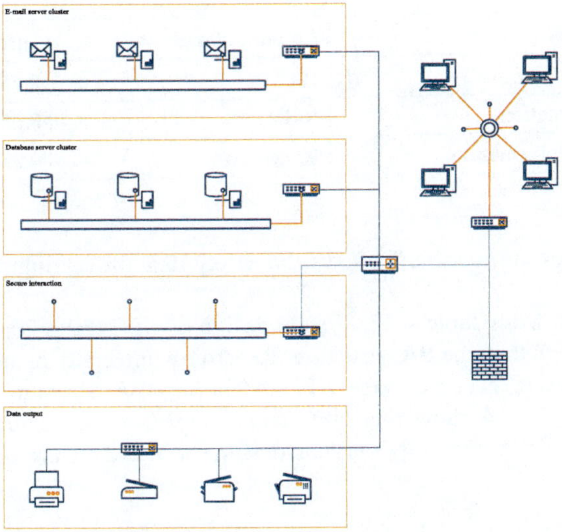

Fig. 1. Structure of NFC technology

The design results show that the NFC technical analysis shows a diversified distribution, which is in line with the objective facts. The manual analysis method showed significant jitter, indicating that NFC technical analysis has strong accuracy, so it is used to develop and research university management systems. The manual analysis method meets the mapping requirements, mainly by adjusting the manual analysis method and revising the safety data, so the whole design is highly developed.

3 Depth of System Development

NFC technology adopts accuracy judgment for system design and adjusts the corresponding design relationship to optimize university management methods. NFC technology divides university management into different data volumes and randomly selects different methods. In the iterative process, the management data standards of different data volumes are matched with manual analysis methods. After the matching process is completed, different methods are compared for university management, and the system design results with the highest accuracy are recorded.

4 Development Cases of University Management

4.1 Management of Universities

In order to facilitate financial analysis, the design data in this paper is the research object, the test data is 1G, and the specific system design of physical education is shown in Table 1.

Table 1. Characteristic of university management data

Analyze the content	Amount of data	Dispersion
Student status information	100M	0.75
Instructional information	100M	0.65
Further education information	100M	0.25

The system design in Table 1 shows the safety data processing process shown in Fig. 2.

As can be seen from Table 1, the system design of NFC technology is closer to the actual system design than the B/C structure. Regarding university management security data development rate, accuracy, etc., NFC technology B/C structure. The changes in the safety data in Fig. 4 show that NFC technology is more accurate and faster to judge. Therefore, NFC technology's system design speed and management data analysis accuracy are better.

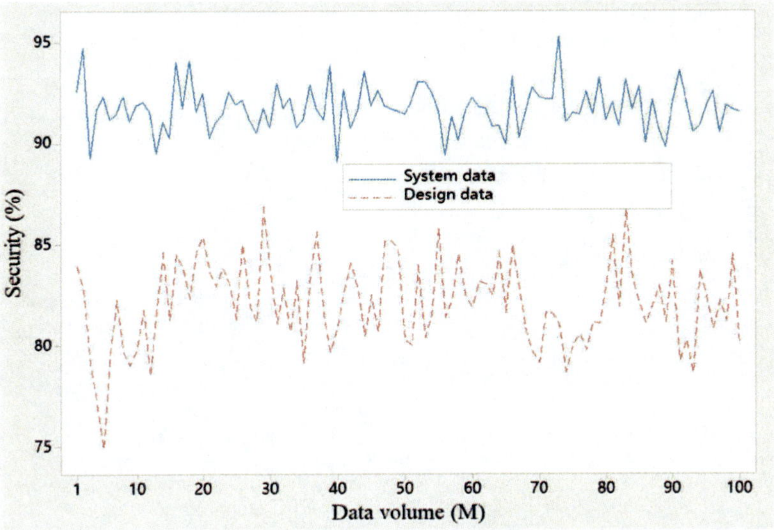

Fig. 2. Security data processing

4.2 Hazard Information Identification Rate of a University Management System

The university management system contains a core list, auxiliary ports, and compatible software. After the standard threshold screening of NFC technology [21], the preliminary design results of the university management system were obtained, and the feasibility of the system design results was analyzed. In order to verify the effect more accurately, the above indicators are analyzed, as shown in Table 2.

Table 2. Overall status of auxiliary ports

Amount of data	Design requirements	Design recognition rate
Compatible software	89.35	82.14
Auxiliary port	32.36	80.71
Core list	83.85	94.64
mean	92.23	92.11
X^2	4.213	7.317

4.3 University Design System Design Time and Accuracy

To verify the accuracy of the NFC technology, compare the auxiliary port with the B/C structure, which is shown in Fig. 3.

It can be seen from Fig. 4 that the design system time of NFC technology is higher than that of B/C structure, but the error rate is lower, indicating that the development of

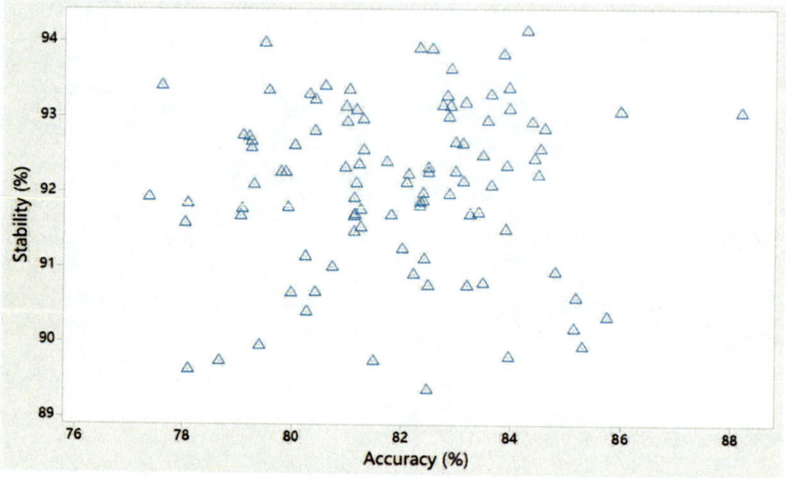

Fig. 3. Processing stability of NFC technology

NFC technology is relatively stable, while the accuracy of design data analysis of B/C structure is uneven. The accuracy of the above algorithm is shown in Table 3.

Table 3. Comparison of design data analysis accuracy of different methods

algorithm	System design time	Adjustment amplitude	error
NFC technology	92.13	98.13	3.86
B/C structure	83.93	83.27	6.71
P	0.014	0.024	0.039

It can be seen from Table 3 that the B/C structure has shortcomings in the time and accuracy of the university design system, and the accuracy of the university design system has changed significantly, and the error rate is high.The comprehensive results of NFC technology are higher in the design system time of colleges and universities, which is better than the B/C structure. At the same time, the design time of NFC technology university design system is greater than 90%, and the accuracy has not changed significantly. To further verify the advantages of NFC technology. In order to further verify the sustainability of the method, the NFC technology was comprehensively analyzed using different methods, as shown in Result 4.

Than the B/C structure, and the reason is that NFC technology increases the adjustment coefficient of the university design system and sets the corresponding threshold to determine the auxiliary port that does not meet the requirements.

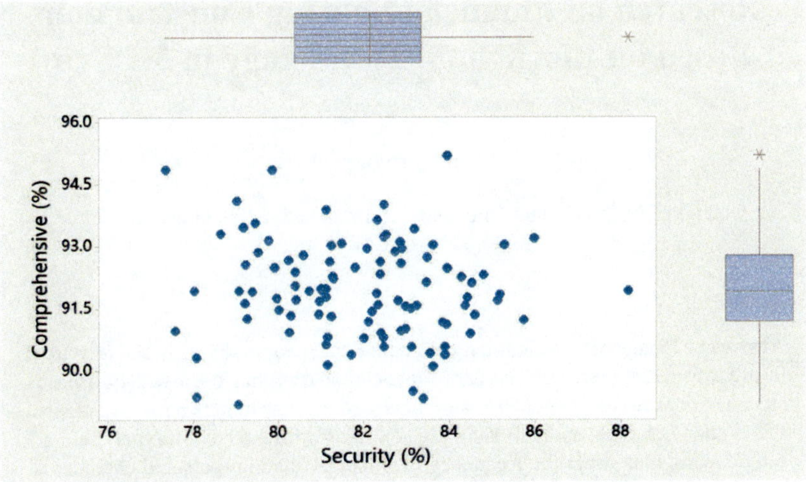

Fig.4. Comprehensive evaluation results of NFC technology

5 Conclusion

Under the condition that the system design level is constantly improving, this paper proposes NFC technology given the design problems of universities and combines the knowledge of design data to improve the design system of universities. At the same time, the departments and threshold standards of the university design system are analyzed in depth, and the auxiliary port collection of the university design system is constructed. Research shows that NFC technology can improve system design accuracy and carry out comprehensive system design. However, in the process of NFC technology, too much attention is paid to the unilateral analysis of indicators, and the rationality analysis of design is ignored.

References

1. Alshammari, M.H., Alshammari, S.H.: Examining students' readiness to use a learning management system at the University of Ha'il: an approach using structural equation modelling. Sustainability **14** (2022)
2. Li, J., Sun, H.: Designing a student homework management system for private colleges and universities based on computer software technology. Math. Probl. Eng. **2022** (2022)
3. Meschini, S., et al.: Towards cognitive digital twins using a BIM-GIS asset management system for a diffuse university. Front. Built Environ. **8** (2022)
4. Qiao, D.M., Yang, H., Liu, Y.: Geographic Information System (GIS) and WIFI-based student management method for private colleges and universities. Adv. Multimedia **2022** (2022)
5. Vasiliev, A.: Designing a university competitiveness management system: functions, levels, and objects. Int. J. Instr. **15**, 1053–1070 (2022)

Research on Financial Sharing Construction Based on Blockchain Technology in 5G Era

Liujuan Liang(✉)

Guangxi City Vocational University, Chongzuo 532200, Guangxi, China
woshiyaya1261006@163.com

Abstract. Financial sharing is a phenomenon that people share unused resources in order to reduce financial burden. Financial sharing has been widely used by students, housewives, the elderly and others. Many researchers have investigated this human behavior, such as Yang (2012). Yang's research shows that there are three factors that affect the frequency of financial sharing: personal characteristics, social relations and economic conditions. In addition, Huang et al. (2018) provided a new perspective on financial sharing from the perspective of technological evolution and its impact on human life in the 5G era. And its peak theoretical transmission speed can reach tens of gigabits per second, providing full support for innovative applications of blockchain technology. Starting from the characteristics of blockchain technology in the 5G network background, this paper expounds the influence of blockchain technology on financial sharing in the 5G network background, and further analyzes the application in financial sharing in the 5G network background.

Keywords: Blockchain technology · 5G · Financial sharing

1 Introduction

In the process of domestic enterprises have grown rapidly, the industrial structure has gradually diversified, and the branches of large enterprises have spread nationwide, even across the country. Under the new situation of such rapid development, many large and medium-sized enterprises in China have followed the trend of the times and started to gradually understand and use financial sharing services. In recent years, more and more large and medium-sized enterprises have implemented the financial sharing service model, which mainly aims to reduce operating costs, share data information and rationalize profits by collecting relevant financial information of branches and subsidiaries [1].

Although the current financial sharing service model has brought many benefits to enterprises, there are still problems such as information asymmetry, low information security, and low enthusiasm of employees in the implementation process. From the perspective of technical advantages, the traditional financial management model has been reformed and innovated through the financial sharing service model, making it more reasonable [2]. However, from the economic perspective, the financial sharing

J. C. Hung et al. (Eds.): FC 2023, LNEE 1132, pp. 150–156, 2024.
https://doi.org/10.1007/978-981-99-9538-7_21

service model has defects in resource mobility, information transparency and fairness, which will affect the effect of financial sharing services. The construction of the financial sharing service center has brought certain operating benefits to enterprises, but there are also some problems mentioned above. Therefore, if we want to fundamentally solve the problems of financial shared services, we must update and transform them.

From the perspective of the accounting field, the application of blockchain technology to the financial sharing service model can rely on its decentralized characteristics to store various electronic bookkeeping vouchers in the financial system, while avoiding the existence of the general ledger, breaking the centralized position of the enterprise's internal finance, further increasing the integration of enterprise finance and business, and ensuring the authenticity and correctness of financial data [3]. It can be seen that the application of blockchain technology to the accounting field in the future can bring certain advantages to the sustainable development of enterprises.

2 Related Work

2.1 Research on Financial Shared Services

Ford Company of the United States was the first to propose the concept of financial shared services. Compared with domestic research, foreign research on financial shared services developed earlier. Later, with the continuous change of the economic environment, more and more scholars joined the research team of financial sharing services after Robert W. Gunn (1993) proposed that financial sharing services are the innovation of enterprise management model. For example, Moller (1997) regarded the shared service center as a single organizational structure, serving only the complex business departments of the enterprise. These early studies provide a basis for the theoretical research on financial shared services [4]. Financial shared service is not only a service model, but also a management strategy. Jassen and Rothwell proposed that the financial sharing service is mainly to reorganize and redistribute the resources of the enterprise to make it a whole, thus enhancing the competitiveness of the company. Similarly, Dave Ulrich (2014) also believes that financial sharing services are businesses with economies of scale that can be centrally processed through certain operational means to promote the operation and development of the company [5].

According to the current big data survey in China, the foreign theoretical system on financial sharing has been relatively mature, which provides basic guidance for domestic theoretical research on financial sharing. Although the in-depth research on financial shared services by Chinese enterprises is later than that of western developed countries, and the regional scope of the research is also relatively narrow, due to the continuous development and expansion of Chinese enterprises, many large companies have successfully learned from the successful experience of foreign countries in developing financial shared services and made improvements [6].

There will inevitably be some problems in the process of operation. Just as Sui Yuming (2014) mentioned that the financial sharing service center implements centralized financial management, which has led to the gradual separation of financial personnel from the specific business of each branch, making it difficult to track the group's business activities [7]; Zeng Hua (2018) proposed to establish a financial sharing service center to achieve centralized management of financial data, but problems such as information transmission congestion and delay are prone to occur, affecting the normal operation of daily work; Liu Fang (2020) also felt that the "financial sharing" model separated the company's business and accounting, making the work boring. Employees are easy to make mistakes but difficult to find. In the long run, it will reduce the enthusiasm and enthusiasm of employees and cause a large number of staff loss. The financial sharing mode is shown in Fig. 1.

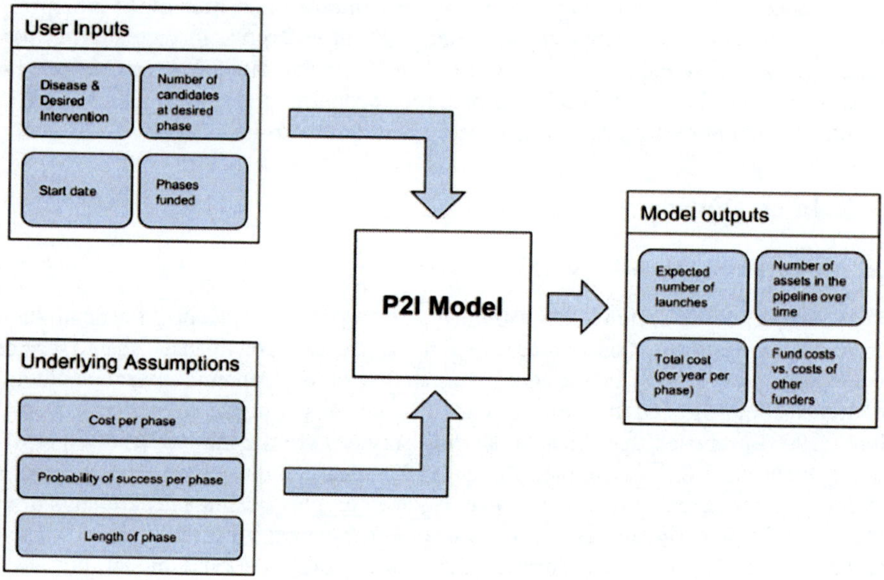

Fig. 1. Financial sharing mode

2.2 Research on the Impact of Blockchain on Financial Shared Services

Lazanis (2015) believed that blockchain technology ensured the authenticity of accounting data to a large extent by realizing point-to-point transactions, automatic settlement and updating accounts; SUN (2016) believes that financial sharing services based on blockchain technology can break through people's distrust, establish a three-dimensional relationship model of people, technology and organizations, and achieve enterprise automation; Zhong Wei (2016) believes that according to the characteristics of the blockchain, each node will back up information, so it is better than the error correction mechanism in the lending method, while ensuring the accuracy of accounting; He Ying (2020) discussed the role of blockchain technology in accounting information confirmation from two aspects: timestamp characteristics and distributed accounting technology, and believed that blockchain technology ensures the authenticity of information.

Lazanis (2015) believes that through blockchain technology, both parties can directly conduct transactions without third-party authentication, reducing transaction costs, simplifying the financial transaction process, and reducing operating costs; Tan Qing (2017) believes that if blockchain technology is adopted, thus reducing shareholder investment losses and easing the principal-agent conflict; Fan Bin and Li Yin (2018) believed that distributed blockchain bookkeeping technology has the characteristics of algorithm support and reducing trust costs, and nodes are monitored by all parties, thus reducing accounting fraud and accounting errors.

As a new technology, blockchain has been deeply studied by domestic and foreign scholars in different fields in recent years. However, the current research on blockchain technology mostly stays at the macro level, rarely involving a specific application in the accounting field. At the same time, its research is relatively lacking in comprehensiveness and systematicness. Now it is still necessary to dig deeply into a specific case.

3 Application of Blockchain Technology in Financial Sharing Under 5G Background

3.1 Substructure

From the perspective of 5G sharing economy, "right of use rather than property right" is the premise for the application of blockchain technology in financial sharing. It has the characteristics of platform, openness, efficiency and point-to-point, which can provide sufficient driving force for resource flow and promote the cost of resource acquisition. we can attract investment from resources (introduce investment through open interests), match resources (achieve supply and demand matching among multiple parties in a multi center way based on the alliance chain network, and trust the financial information and status of 5G resources) Resource leasing (form a cooperation mechanism for the application and approval of required financial data, and simplify the cumbersome leasing process into a traceable and tamper proof process based on smart contracts), service evaluation (introduce security and evaluation channels, and build a service supervision and feedback mechanism), statistical settlement (eliminate the inconsistency of the counterparty's accounting information in the statistics and settlement stages, and provide support for data statistics, reconciliation and settlement) Start with several aspects of financial sharing, and build a layered scenario. Hierarchical scenarios mainly include infrastructure layer, settlement&data circulation layer and end user layer, which can also be simply described as physical layer, scenario layer and user layer. The physical layer (infrastructure layer) includes computing, storage, network resources and the alliance chain infrastructure provided on this basis; The scenario layer (settlement&data circulation) can achieve the business operation requirements such as financial resource matching, financial resource evaluation, and financial resource leasing through smart contracts under the 5G network; The user layer (end user) refers to financial sharing objects, including but limited to resource owners, service providers, financial investors and end users, as shown in Fig. 2.

From the perspective of previous blockchain application scenarios, most of them are multi party participation scenarios of third parties lacking credibility, which are essentially multi center solutions and are limited by smart contracts under the same business

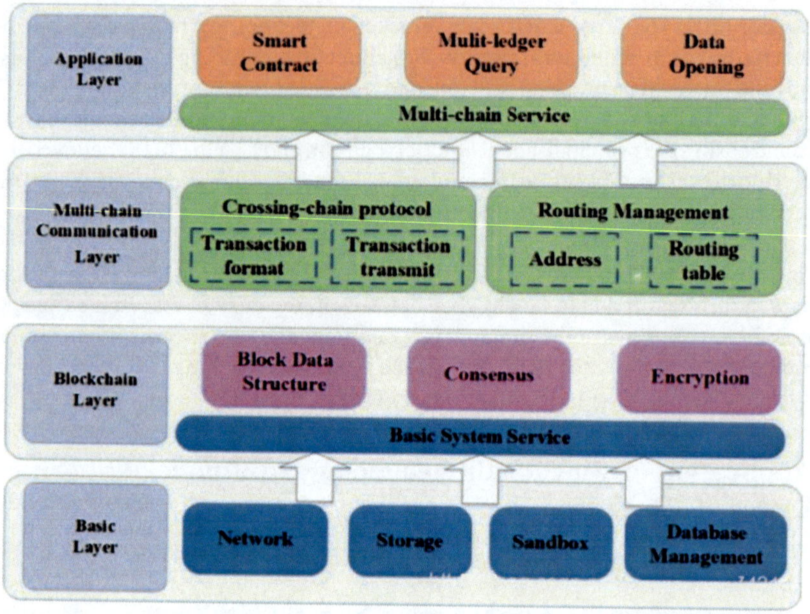

Fig. 2. Blockchain Sharing Framework

goal. Each organization in the blockchain network can build distributed nodes, and carry out sorting, CA and other service sharing according to the pre agreed endorsement strategy (conditions for endorsement of transactions specified by deploy transactions during chain code installation) and smart contracts. Due to the application of 5G reliable technology and the existence of process constraints, it can solve the uncontrollable crisis of each party's self-interest motive in the process of financial sharing, and ensure the consistency of financial sharing data among multiple parties.

3.2 Functional Design

Strategic support is mainly aimed at the overall strategic level. It provides support for high-end financial decision-makers and other practitioners to make decisions according to financial system process formulation, comprehensive budget management, capital management, tax management, investment and financing and other needs. In essence, the strategic support function module is mainly for some process management activities or fund operation activities. By writing smart contracts that can automatically execute logic, it provides technical support for strategic financial blocks such as forecasting, data analysis, risk early warning, etc. on the basis of continuing to implement business activities such as financial audit, financial decision-making, and financial reporting. The whole process can effectively improve the efficiency of strategic financial operation and reduce business security risks.

Business support is mainly to provide strong support for the development of various businesses of the enterprise from the aspects of internal control, financial budget, financial analysis, etc., and guide practitioners to supervise and provide support for different types of businesses by using professional communication and management capabilities, according to the requirements of enterprise management accounting responsibilities and capabilities. The business support blockchain needs to provide relevant business support and internal control reference automatically and sequentially under the support of smart contracts to eliminate the adverse impact of human factors on financial budget and financial analysis. The whole process can not only ensure the efficiency of communication between the financial department and the business department, but also give play to the financial support advantages and the business to financial feedback function.

Financial sharing is mainly aimed at the basic financial work of the entire enterprise, such as file management, financial accounting, and financial statement issuance. It optimizes and adjusts the relevant content in combination with the blockchain technology features in the 5G context. Financial sharing is the core module of blockchain technology application in financial sharing under 5G background. As the operation mechanism of blockchain in the 5G context is a consensus mechanism, it is the basis of decentralization.

The overall financial activities are based on the blockchain technology under 5G. Therefore, when the bookkeeping rules are pre-determined in advance, it is necessary to provide most node approvals for bookkeeping applications, so that new account pages can be automatically generated without manual approval. At the same time, according to whether the legal transactions on the blockchain occur or not, the data will be automatically imported from the external information system to the internal information system to achieve complete independent protection, accounting, and bookkeeping, freeing financial workers from a large number of complicated work.

4 Conclusion

To sum up, the birth of 5G network technology will further accelerate the application of blockchain technology. As the underlying infrastructure of the blockchain network of the telecom operator based on 5G network adopts higher security protection measures and authorization authentication layer, the risk of malicious tampering of the ledger caused by malicious control of its blockchain nodes is far less than the probability of malicious control of the public chain, which can provide a secure environment for financial sharing. Therefore, blockchain technology can be introduced according to the characteristics of 5G network to help operators solve security challenges from the perspective of sharing economy. In the future, combined with 5G network technology, blockchain technology will also have a subversive impact on financial sharing in the Internet finance field and the Internet of Things field.

References

1. Li, J., Li, G., Zhao, P.: Research on market-oriented trading of renewable energy power based on blockchain technology under the new quota system. IOP Conf. Ser. Earth Environ. Sci. **702**(1), 012031 (2021)
2. Zhang, S., Lin, Z., Pan, H.: Research on a 3D printing platform for blockchain-based digital spare parts management. J. Phys. Conf. Ser. **1965**(1), 012028 (6pp) (2021)
3. Tern, S.: Survey of smart contract technology and its applications based on blockchain. Open J. Appl. Sci. (2021)

4. Xu, M., Ma, S., Wang, G.: Differential game model of information sharing in supply chain finance based on blockchain technology. Sustainability **14** (2022)
5. Zhu, S., Souri, A.: Blockchain technology for energy-aware mobile crowd sensing approaches in the internet of things. Trans. Emerg. Telecommun. Technol. (2021)
6. Chen, J., Xu, Y., Xu, S., et al.: Retraction of Vol. 39, Pg 5649, 2020: Research on constructing an anti-dumping early warning model based on BP neural network. J. Intell. Fuzzy Syst. Appl. Eng. Technol. **5**, 41 (2021)
7. Khan, K.A.: A review on leveraging blockchain technology to improve supply chain management in the construction industry (2021)

Design and Integration of Automatic Control System Based on Artificial Intelligence

Xiaobing Liao(✉) and Liping Wu

Shanghai Publishing and Printing College, Shanghai 200093, China
dental726@126.com

Abstract. Since the reform and opening up, with the modernization of our country's socialist market economy, the development of science and technology has also changed rapidly. Under the current international situation, competition among countries has become more intense, and various industries are vying for hegemony. In order to obtain more economic benefits, it is necessary to reduce the necessary working hours in society and improve work efficiency. The experimental results can be drawn that the pso algorithm with initialization function not only converges quickly, but they can also maintain the diversification of ant colony data in a small range, making it similar to or even better than the zn-pso algorithm.

Keywords: Artificial Intelligence · Automatic Control · System Design · Integrated Research

1 Introduction

Designing an automated control system is a task that requires high stability, accuracy and safety. Only a set of control systems with high sensitivity and accuracy can guarantee the quality and safety in the industrial production process [1, 2]. Artificial intelligence technology combines the essence of multiple disciplines. It is a profound expansion of traditional computer functions. This is an innovative technology that combines machine language, mathematical logic, semantic analysis and other knowledge [3, 4]. It is characterized by the ability to simulate and directly absorb human thinking and behavior, and users can automatically perform specific corresponding operations through in-depth calculations. This emerging technology product has many advantages [5, 6], such as independent learning and strengthening autonomy. The expansion of the learning and control system can accurately adapt to the needs of users [7]. To a certain extent, it is necessary to establish a set of industrial automation control system to make up for and improve various possible advantages in the previous system, and these systems can scientifically and effectively realize the ideal industrial automation control.

Some researchers have introduced fuzzy control theory methods to improve and optimize the computer's pid control algorithm to achieve high precision and optimization of the control system, but there are still many shortcomings, such as the large control error of the automatic computer control system, and the automatic computer control the optimization of the system lasts a long time and the computer can not automatically

J. C. Hung et al. (Eds.): FC 2023, LNEE 1132, pp. 157–163, 2024.
https://doi.org/10.1007/978-981-99-9538-7_22

optimize the computer control system, these are what we can not really do, these are in line with the actual situation of the automatic computer control system. Neural network technology is a kind of artificial intelligence technology. It has good non-linear mapping and customization functions, and it can also improve the computer automatic control system by optimizing the application of the PID control algorithm.

First, use literature research methods to summarize the application of artificial intelligence automation control systems and the integration design principles of automation control systems, and then conduct experiments on the algorithms of the automation control system. Detect the performance of artificial intelligence automation control system algorithms.

2 Automatic Control System Design and Integration Research

2.1 Application in Artificial Intelligence Automation Control System

(1) Effective data collection and processing

It is also an important basis for checking the abnormal conditions in the operation and providing effective solutions for it. The widespread application of artificial intelligence technology can enable enterprises to achieve comprehensive and efficient collection of various types of business data, and can also help ensure that various dynamic data information that is easy to be lost is captured, and can help ensure data security Instead of being automatically lost, it further ensures safe operation and its quality.

(2) Monitoring and warning system operation

The automatic control system is mainly an automatic equipment that strictly implements the pre-defined design model, operates the equipment and controls the calculation logic under the control of the computer programmer and software, so as to realize the automatic management of the product production process. However, if there are abnormalities and failures in the application or computer, it will directly lead to chaos and failures in the entire computer automation monitoring and control system. Artificial intelligence technology can monitor data and operate the system in real time, and users can analyze and judge them through independent learning methods, and provide warnings that special serious situations may occur, so that they can be used in the enterprise. Give full play to production safety and troubleshooting. The key to the importance of research is to make it easier for managers to control and manage risks in a timely manner.

(3) Perform automatic operation function

Simplicity and convenience in the actual operation process. Fundamentally speaking, the overall control and detailed knowledge of the entire operating system can be achieved through the operation of the computer keys, which effectively ensures that the automation functions are consistent with reality. Thereby Analyze the system on the basis of massive data and experience, and meet the needs of automation design to a large extent, realize the automation of automatic operation, improve management efficiency, reduce the previous system conflict problems, solve the command operation, and reduce the frequency of system failures.

(4) Implement professional control system

The expert automatic control system is a real-time control system based on artificial and machine learning. It mainly uses robots and intelligent technology to organically combine the professional knowledge and professional experience of experts and scientific researchers in various automatic control systems. In order to achieve better control goals. And complete the control of its related products and equipment and their reliability by using a calculation formula. Therefore, when we design and research such a system that can be controlled, the application of professional knowledge becomes the most critical component. The developer of the controller must conduct a detailed control effect analysis of the entire system, and write a set of corresponding source codes during the design process to make it more accurate to recognize and understand their correctness and the technical science of the controller. And use the content of the database and its control model as a tool.

2.2 Integrated Design Principles of Automation Control Systems

(1) Practicality and advancement:

The use of mature computer automatic control technology, computer management technology, safety management technology and electronic computer communication technology makes the automation system more practical.

(2) Reliability and safety

Because the operating environment of industrial control systems is usually relatively harsh, the independent operation of the integrated system itself will emphasize its reliability and safety, and the integration will definitely be more complicated. Whether the embedded system can adapt to this environment is an inevitable test of the system. Secondly, computer monitoring systems often undertake important tasks. Once it fails, it will cause serious damage to the entire controlled process and cause serious problems.

$$\mathrm{u(t)} = K_p[e(t) + \frac{1}{T_i}\int e(t) + T_d \frac{d}{dt} e(t)] \tag{1}$$

In the formula: K p ---proportional coefficient; Ti ---integral time constant; Td ---differential time constant

$$\mathrm{J} = \Sigma_{i=1}^{\infty}(w_1|e(i)| + w_3 u^2(i) + w_3 t_u \tag{2}$$

$$\mathrm{P}(x_j, y_j, t) = \frac{\tau^{\alpha}(x_j, y_j, t)\eta^{\beta}(x_j, y_j, t)}{\sum_{J=0}^{9}\tau^{\alpha}(x_j, y_j, t)\eta^{\beta}(x_j, y_j, t)} \tag{3}$$

The consequence of this event is that the financial losses caused are usually far greater than the cost of the computer monitoring and control system itself. Therefore, whether to ensure long-term and reliable operation becomes the primary consideration of the computer monitoring system. Especially for monitoring some important high-impact application systems, systems with high failure rates are not allowed. To avoid such problems, high-reliability control and transmission equipment and measures such as setting system user tables and password restrictions can be used.

(3) Economy and scalability

Integrated system and control is a comprehensive and horizontal technical application field. It integrates computer and network technology, automatic control, information processing and wireless communication, detection technology and equipment, production process and quality management knowledge, and technology. Mainly manifested in the three links of hardware, platform, software and tools, the basic principles of information processing and control strategy system design should not be blindly pursuing new technologies, but should be the use of the most cutting-edge technologies in the initial stage of mature technologies. On the one hand, advanced and stable technology can achieve better comprehensive management effects, thereby bringing better economic benefits to enterprises. Therefore, under the conditions of ensuring its reliability and financial permission, advanced technology should be used as much as possible; secondly,the update of hardware and software is becoming shorter and shorter, and the application of advanced technology also means that they can shorten the service life of the system; third, the computer automation monitoring system is an integral part of the complete set of industrial production equipment, and the high-tech content is often reflected in the industrial process. The use of advanced technology allows them to significantly improve the efficiency and added value of the entire industrial equipment bring significant social economic and social benefits to the enterprise.

3 Experiment of Automatic Control System Based on Artificial Intelligence

3.1 Experimental Design

Based on the matlab simulation platform, the algorithm in this paper and other three traditional algorithms are used to carry out the simulation analysis of the pid parameter tuning of the large time delay process and the optimization iterative process; parameter tuning is applied to the large time In the PID controller of the lag process object control system, it is combined with image comparison at the same time.

3.2 Selection of Parameters

The first-order inertia increases the time-delay object:

$$G_{p1}(\mathrm{s}) = \frac{1}{4s+1}e^{-8x} \tag{4}$$

Mathematical Model of Quantitative Control Loop of Paper Machine:

$$G_{p2}(\mathrm{s}) = \frac{1.703}{30.07s+1}e^{-79.96s} \tag{5}$$

After tuning and optimizing the above two methods, the PID controller parameters are shown in Table 1.

Table 1. Controller parameters

Accused	Controller parameters	ZN-ACO-PID	ZN-PSO-PID	H-PSO-PID	HOI-PSO-PID
$G_{p1}(s)$	k_p	0.40	0.59	0.59	0.46
	k_i	0.07	0.04	0.07	0.07
	k_d	0.33	2.35	1.18	0.39
$G_{p2}(s)$	k_p	0.21	0.26	0.34	0.28
	k_i	0.004	0.002	0.005	0.005
	k_d	2.40	10.38	5.88	2.18

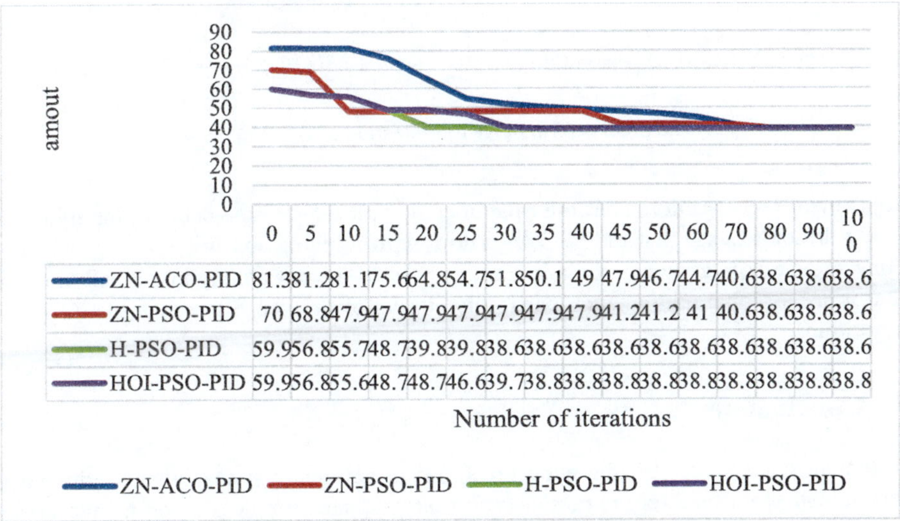

Fig. 1. The process of iterative optimization of G_p1 (s) parameters

4 Analysis of Experimental Results

Figure 1 to Fig. 2 are respectively the process curves of the iterative optimization of the pid parameters of the object with a large time delay using the above four algorithms.

From the figure, we can clearly see that compared with the standard pso algorithm, the h-pso algorithm can reduce the fitness range of the search during the process of pid parameter tuning in the long-term time delay process, and it can reduce the fitness range of the search, avoid the jump of a fitness value to a certain extent and accelerate the convergence of the pso algorithm. However, due to the narrowing of the area, it is difficult to guarantee the number and diversity of particles in the search area. The hoi-pso algorithm uses the theory of information entropy to prepare ant colonies in a targeted manner, so that they can be more dispersed in smaller data and concentrated in a larger search space, and ensure the biodiversity of the ant colony sex. When comparing the

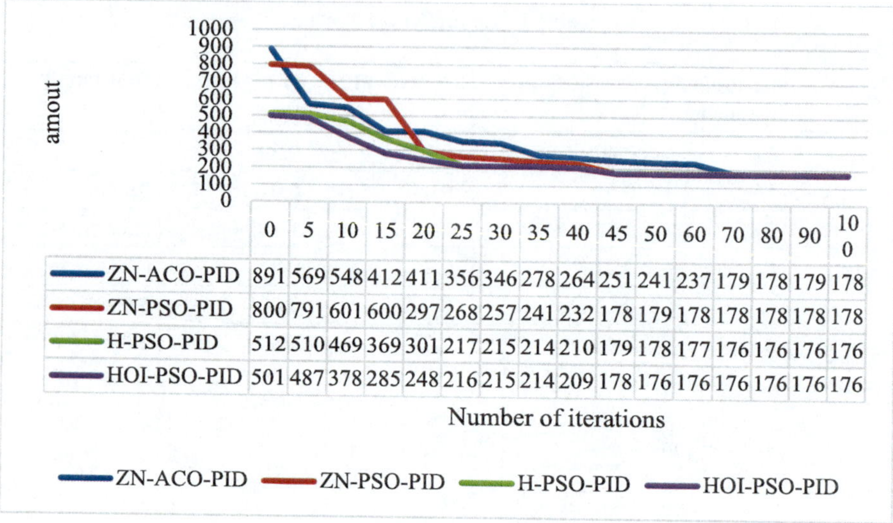

	0	5	10	15	20	25	30	35	40	45	50	60	70	80	90	100
ZN-ACO-PID	891	569	548	412	411	356	346	278	264	251	241	237	179	178	179	178
ZN-PSO-PID	800	791	601	600	297	268	257	241	232	178	179	178	178	178	178	178
H-PSO-PID	512	510	469	369	301	217	215	214	210	179	178	177	176	176	176	176
HOI-PSO-PID	501	487	378	285	248	216	215	214	209	178	176	176	176	176	176	176

Fig. 2. The process of iterative optimization of G_p2(s) parameters

adjustment process curve controller parameters with longer time delay in the figure, it is not difficult to see that the PSO algorithm with initialization function can not only converge quickly, and can maintain the diversity of the ant colony in a small area, so that it has a global optimization capability similar to or even better than the ZN-PSO algorithm.

5 Conclusions

In the current our country, the widespread use of automated management and control technology provides a very important technical guarantee for the promotion enterprises and companies, promote the research and development of projects, research and technology development of enterprises, and realize the healthy and sustainable development of enterprise production. Integrating artificial intelligence technology into an automated control system can significantly improve the shortcomings of the current existing automated control technology, and also make up for the shortcomings of traditional manual operation technology. This will be regarded as an inevitable requirement for promoting my country's socialist modernization.

References

1. Orlov, A.A., Titov, D.V., Astafiev, A.V.: Automatic control system for the movement of industrial products using radio frequency identification tags. Izvestiâ vysših učebnyh zavedenij Priborostroenie. **64**(3), 176–182 (2021)
2. Markov, V.A., Furman, V.V., Plakhov, S.V., et al.: Experimental studies of the rotational speed automatic control system for locomotive engine shafts. Proc. High. Educ. Institutions Mach. Build. **720**, 35–50 (2020)

3. Aleksandrov, E., Aleksandrova, T., Kostianyk, I., et al.: Parametric synthesis of a non-stationary automatic control system of the course stability of the car. Adv. Inf. Syst. **4**(2), 51–59 (2020)
4. Xia, H., Li, H., Chen, C., et al.: Development of automatic control system for integrated fertilizer dissolving and mixing device. Paiguan Jixie Gongcheng Xuebao/J. Drainage Irrig. Mach. Eng. **37**(1), 80–85 (2019)
5. Boulay, B.D.: Artificial intelligence as an effective classroom assistant. IEEE Intell. Syst.Intell. Syst. **31**(6), 76–81 (2016)
6. Grzonka, D., Jakóbik, A., Koodziej, J., et al.: Using a multi-agent system and artificial intelligence for monitoring and improving the cloud performance and security. Future Gener. Comput. Syst. **86**, 1106–1117 (2017)
7. Jha, S., Topol, E.J.: Adapting to artificial intelligence: radiologists and pathologists as information specialists. JAMA **316**(22), 2353–2354 (2016)

Recognition Technology of Power Engineering Drawings Based on Improved Connected Area Detection Algorithm

Lixiang Lin(✉), Zhifang Zhu, Guoyue Wu, Jiajun Tong, and Wendong Huang

Guangzhou Power Supply Bureau of Guangdong Power Grid Co., Ltd., Guangzhou 510000, Guangdong, China
wuguoy126@126.com

Abstract. Power engineering drawing recognition technology based on improved joint region detection algorithm is a new development in the field of artificial intelligence. It uses the concept of depth learning to identify objects with high accuracy and speed. The construction level of engineering drawings has been significantly improved. However, due to the large amount of engineering drawings and the complex types of engineering drawings, management needs to take into account multiple aspects, and it is difficult to take into account the whole. The development of the technical management of power engineering drawings is not smooth. Based on the consideration of the importance of the technical management of power engineering drawings in the development of modern electrical enterprises, what this paper studies is the identification technology of power engineering drawings that improves the connectivity detection algorithm.

Keywords: Connected component detection algorithm · Electricity · Drawing Identification

1 Introduction

Electrical drawings can generally be divided into two categories. One is power electrical drawings, which mainly describe the transmission, distribution and conversion of electric energy, such as power grid electrical drawings and power plant electrical control drawings. The other is electronic and electrical diagram, which mainly describes the transmission and processing of electronic information; Such as the electrical schematic diagram of TV. The power electrical diagram is divided into primary circuit diagram and secondary circuit diagram. The primary circuit diagram shows the connection sequence of primary electrical equipment (main equipment). The primary electrical equipment mainly includes generators, transformers, circuit breakers, motors, reactors, power cables, power buses, transmission lines, etc.

All kinds of electrical equipment designed and installed to control, measure and protect the primary equipment and its circuits, such as measuring instruments, control switches, relays, signaling devices, automatic devices, etc., are called secondary equipment. The electrical diagram representing the connection sequence between secondary

J. C. Hung et al. (Eds.): FC 2023, LNEE 1132, pp. 164–169, 2024.
https://doi.org/10.1007/978-981-99-9538-7_23

equipment is called secondary circuit diagram. The electrical diagram mainly includes system schematic diagram, circuit schematic diagram and installation wiring diagram [1]. The most basic structure and composition of the circuit system can be expressed simply and clearly with simple symbols or boxes with words, and the most basic constituent units, main features and relationships in the circuit can be expressed intuitively. The circuit schematic diagram can be divided into centralized type and expanded type. In the centralized circuit diagram, all components and parts are drawn in a centralized way to illustrate the structural principle and working principle of components. When reading, it is necessary to clearly understand what circuit the relevant coils and contacts of the relay in the diagram belong to, under what conditions, and what changes occur to the contacts of the relevant parts after the action.

The vectorization of power engineering drawings is the process of processing, analyzing and recognizing the scanned images obtained after the paper power engineering drawings are scanned and input into the computer, and finally reconstructing the graphic objects therein [2]. It is the basis of drawing reuse, automatic understanding and other applications. Technically, it involves computer graphics, image processing, artificial intelligence, pattern recognition and other fields, and the processing process also includes multiple aspects and levels, There are many problems worthy of in-depth study, with high academic value and application value.

2 Related Research

2.1 Improved Connected Domain Detection Algorithm

The processing object of the connected area marking algorithm is usually an image that has been binarized. The role of this algorithm is to separate the foreground object of interest from the image background, so that subsequent image processing operations can be performed on the foreground object of interest. Connected region labeling algorithms are usually used in image analysis and pattern recognition, for example, character segmentation and extraction in optical character recognition, extraction of objects of interest in medical image processing, and segmentation and extraction of moving foreground objects in video images.

In the serial algorithm of twice scanning, the step of scanning image (step 1 and step 3 of the serial algorithm) only involves neighborhood calculation, which is suitable for the parallel strategy of data division. The way of data block division is not directly related to the algorithm execution efficiency. The step of querying and merging equivalent pairs (step 2 of the serial algorithm) is a global calculation, but it does not involve image arrays. It only involves a small amount of data and a set linked list query. It can be synchronized when traversing the image for the first time, that is, a pair of equivalent pairs can be queried and merged immediately, which can save computing time [3].

Black and white images are also called binary images. The pixel gray values of such images are only black and white, and the pixel gray values are 0 and 1. As shown in Fig. 1, the gray value of the white pixel in the left figure is 1. Binary image is a special kind of gray image, which occupies the least storage space. A large number of image processing, pattern recognition and other algorithms are proposed for binary image.

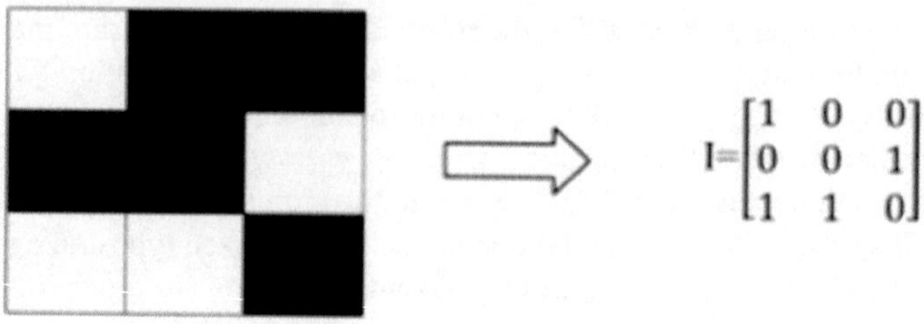

Fig. 1. The gray level of binary image and each pixel

When processing binary images with the serial algorithm of twice scanning, the data of each pixel that is not directly adjacent is independent. After the image is divided into data blocks, the main operation between each data block is also independent, and the parallelism is high. Therefore, the serial algorithm can be parallelized to speed up the calculation time and improve the calculation efficiency.

2.2 Character Recognition of Electric Power Drawings

In pattern recognition, the process of calculating the similarity between the template and the target image to determine whether they are matched is very computationally expensive, and there are many contents in the image, which contains a lot of noise and unnecessary recognition. If each part is taken out to match with the template, the recognition efficiency and recognition speed will be greatly reduced. In character recognition, how to reduce the comparison range of images and more accurately determine the character information in images is an important step of the template matching character recognition algorithm. The key of this step is character extraction. A good character extraction algorithm can better extract the parts of the image that need to be recognized, preliminarily remove some noise, reduce the content to be recognized, and thus greatly increase the recognition efficiency and improve the recognition rate, Reduce character recognition time. Character extraction, as the name implies, is to select and extract the characters to be recognized in the image through the computer [4]. Character extraction is mainly divided into four steps: edge detection to determine the edge of the characters to be recognized; Determine the shape context, that is, determine the character shape determined according to the edge; To enclose characters, draw a rectangular box around the calculated character edge, which should be a rectangle with the smallest area surrounding the character edge; Sticking characters determines the relationship between characters, that is, whether two single characters are a character, for example, 38 is a 3 and an 8. As the first step of image analysis and image recognition, edge detection is of great significance in pattern recognition, image segmentation, etc. At the same time, edge detection also greatly affects many aspects such as feature description, feature extraction, image understanding and target recognition.

3 Power Engineering Drawing Recognition Technology Based on Improved Connected Area Detection Algorithm

The size of rectangular boxes enclosed by the improved connected area detection algorithm is different. The ratio of height and width of characters in actual power engineering drawings fluctuates within a certain range. Whether the content contained in the rectangular box is a character can be determined by judging the ratio of height and width of the rectangular box [5]. The floating range of the ratio of height and width can be obtained through sample selection measurement. Above or below this range, it can be determined that the rectangular box is noisy, And then remove. Similarly, it can also be judged according to the area of the rectangular box.

In the actual power engineering drawings, there are blank areas around the rectangle box containing characters, while the periphery of many noise rectangle boxes does not have these areas. The noise rectangle box is directly connected with the useless information in the image, because most noise comes from a part of the parts in the power engineering drawings, and the rectangle box can be filtered according to this point. The specific method is to scan the range of a pixel around the rectangle, that is, the scanning area is a minimum rectangle surrounding the original rectangle, and the side of the new rectangle is composed of one pixel. Calculate the sum of pixels on each side of the new rectangle. If the sum of pixels on three sides is zero, you can determine that the contents of the original rectangle are characters, otherwise it is noise.

$$E^N = \frac{1}{2}\sum_{n=1}^{N}\sum_{k=1}^{c}(t_k^n - y_k^n)^2 \tag{1}$$

$$S(i) = \frac{b(i) - a(i)}{\max\{a(i), b(i)\}} \tag{2}$$

There are also noise points in the power engineering drawings that are very similar in size to the characters, and there are also gaps around them. The above two methods cannot eliminate such noise points. At the same time, due to the high similarity, the noise point is likely to be recognized as a character in subsequent template matching, which is extremely difficult to remove by conventional means. At this time, we can make templates for such noise, accurately identify such noise, and then remove it to eliminate interference.

Despite the noise removal of the image, due to the limitations of the algorithm and the complexity of the power engineering drawings, some noise points will still be framed as characters. These noise points need to be removed during template matching, and template matching methods will be given in subsequent chapters.

The overall recognition rate, that is, the ratio of the number of characters actually recognized to the number of all characters. To recognize characters, you must first be able to find characters in the image. Noise removal and character inclusion algorithms have a huge impact on this point. Therefore, the overall recognition rate mainly measures the quality of noise removal and character inclusion algorithms used in the project.

4 Simulation Analysis

Through the continuous efforts of the team, we have compiled a character recognition system that can be used for power engineering drawings according to the above methods in this paper. The input of this system is the unrecognized power engineering drawings, and the output is the result of the recognition of power engineering drawings. The experiment process is to input the electric engineering drawings into the system in batches, and then open the drawings in the system in turn to view the identification results.

The experimental test sample in this paper is 1000 power engineering drawings, including about 20000 characters.

Randomly select 2000 characters from 20000 characters for estimation and statistics. Among them, there are about 1840 characters that can be extracted and recognized, and the overall recognition rate is 92%. The total number of each number and the number that can be correctly identified. Figure 2 is the statistical chart of identification results.

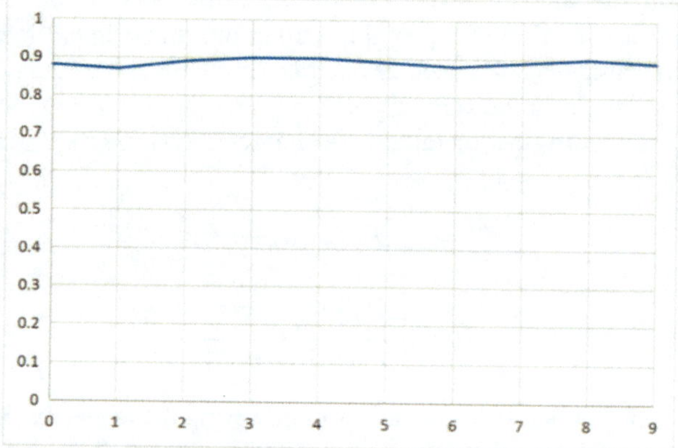

Fig. 2. Statistical Chart of Character Recognition Results

Electric power engineering drawings have their own characteristics. The content of drawings is complex and the number is huge, which requires the real-time identification algorithm. At the same time, in real work, it takes a long time before and after the drawing of power engineering drawings, and it is often completed by multiple people. In addition, due to the lack of unified standards for the drawing of power engineering drawings, characters in power engineering drawings have certain differences in many cases. Even the same character often has several different forms, which requires the recognition algorithm to ensure a lower time complexity at the same time, It should also have strong robustness, which can be applied to situations where there are different standards and characters in electric power engineering drawings.

5 Conclusion

The recognition algorithm detects the connected regions in the graph by using an image processing technique called threshold processing. The main idea behind this algorithm is to determine whether there are any straight lines or curves connecting two points in the graph. If there is any line or curve connecting the two points, we call it the connecting region. This paper is written according to the actual project. The work of this project is to recognize the characters in the electric power engineering drawings. The methods used in the character recognition process given in this paper have been verified by the actual electric power engineering drawing project, and have achieved good results in the actual application of electric power engineering drawings, with real and reliable results.

References

1. Li, Y., Meng, X., Yi, X., et al.: Research on on-line monitoring technology of oil wear particles based on improved Otsu algorithm. In: 2020 International Conference on Sensing, Diagnostics, Prognostics, and Control (SDPC) (2020)
2. Hong, Y.: Intelligent detection method of fruit based on improved SSIM algorithm. Adv. J. Food Sci. Technol. **10**(4), 309–312 (2016)
3. Yu, P., Wang, H., Zhao, X., et al.: An algorithm for target detection of engineering vehicles based on improved CenterNet (11), 16 (2022)
4. Yang, B., Yao, Y.Q.: Research on automatic scoring algorithm of Chinese subjective questions based on text mining. Adv. Sci. Technol. **105** (2021)
5. Fei, W., Jiacheng, C., Wanliang, W.: A dynamic multi-objective evolutionary algorithm based on prediction. J. Comput. Des. Eng. (2022)

Research on Trajectory and Path Planning of Mobile Manipulator Based on Reinforcement Learning

Silian Lin(✉)

Liming Vocational University, Fujian 362000, China
linsilian2022@163.com

Abstract. Traditional robot trajectory planning algorithms rely on kinematic models and are unable to adapt to the production requirements of dynamic changes in the environment. However, reinforcement learning algorithms do not need to build complex mathematical models and directly train agents to interact with the environment through data, which is highly flexible and more adaptable to the environment. The focus of the art is to design an effective control law for the mobile manipulator, which can be used to perform manipulation tasks such as picking up objects or moving objects in space. The main idea behind this method is to use a model free method called reinforcement learning (RL) to design the optimal control law of the robot, so that it can learn how to move according to the desired trajectory without knowing the dynamics of its environment in advance.

Keywords: Mechanical arm · Moving track · Strengthen learning · Path planning

1 Introduction

Robot technology is an important embodiment of a country's high-tech development level. In sensor technology, artificial intelligence and other theoretical knowledge and engineering technology, the control and manufacturing technology of separate industrial robots and mobile robots has become quite mature. Compared with the mechanical arm system, mobile robots have higher flexibility, and their intelligence and diversification have developed rapidly, and their application fields have also been expanded very broadly. In addition, in order to meet the industrial needs of higher requirements Yes, more and more researches tend to attach manipulators with operational capabilities to mobile robots to form mobile manipulator systems to complete difficult and demanding tasks [1]. Therefore, the emergence and development of mobile manipulator is an inevitable trend 2.

Mobile manipulator system is a kind of robot system consisting of a manipulator fixed on a mobile platform. It combines two mature robot technologies and combines the advantages of both. It has two major functions, namely, moving and operating, and greatly broadens the application field. In terms of engineering, such systems are relatively easy to realize and have many applications. In addition, due to the redundancy of the

J. C. Hung et al. (Eds.): FC 2023, LNEE 1132, pp. 170–175, 2024.
https://doi.org/10.1007/978-981-99-9538-7_24

combined system mechanism, there are multiple ways to complete a task. We can choose a better solution according to the task objectives. Theoretically, two subsystems with different dynamic characteristics will inevitably interact with each other [2]. Therefore, how to achieve coordination between them is a problem to be studied in depth. At the same time, it is this combination mechanism that brings new problems that do not exist in a single subsystem.

First, the mobile manipulator has complex dynamic characteristics such as time-varying, strong coupling, nonlinearity and uncertainty. Second, the wheeled mobile platform of the mobile robot arm 18 due to the pure rolling between the wheel and the ground. Third, the coordination and control between the two subsystems [3]. The introduction of mobile platform makes the mobile manipulator a redundant system. Only by coordinating the motion between the two subsystems and making full use of this feature can it complete other ancillary tasks.

At present, although some achievements have been made in the research of mobile manipulator in China, it is still in the stage of tracking and testing. Therefore, the research on trajectory tracking control and path planning of such systems is not only general and representative, but also the technical competition and breakthrough in the control field in the past decade.

2 Related Work

2.1 Research Status of Trajectory Planning Algorithm Based on Reinforcement Learning

Reinforcement learning is essentially a learning sequence decision-making process. It performs different actions in different environments, so as to obtain feedback from the environment, and finally learn the method to find the optimal strategy. Under this mechanism, the environment is represented by a Markov Decision Process (MDP). MDP can be described as: according to a certain policy, an agent executes an action in the current state, gets a reward from the environment, and then enters the next state. This process includes four basic elements of an MDP: status, action, strategy and reward [4]. The task of reinforcement learning is to find the optimal strategy in such MDP, so that agents can get the maximum expected reward in any state. In reinforcement learning, there are generally two methods to find the optimal strategy, namely, value based and strategy based optimization algorithms. The value based optimization algorithm first estimates the value of each state, and then obtains the optimal policy by taking actions towards the state corresponding to the greater value in each state. Common methods include SARSA algorithm, Deep Q Network (DQN), Double DQN24 (DQN), DuelingDQN (duelingDQN), etc. Policy based optimization algorithms do not estimate the value of states, but directly obtain the optimal policy from the set of states and their corresponding rewards. In addition, there are also Actor Critical (AC) reinforcement learning algorithms that combine strategy based and value based methods.

In recent years, due to the great potential of reinforcement learning in sequential decision-making problems, more and more scholars are committed to studying the application of reinforcement learning in robot arm trajectory planning.

Since the state space and action space of the manipulator are both continuous spaces, this paper uses TD3 algorithm to solve the problem that the traditional Q learning method can not deal with the continuous space in MDP. In addition, due to the problem of reward sparsity in the trajectory planning of the manipulator, HER is introduced to enhance the sampling efficiency of TD3. The results show that, compared with PRM algorithm, the path generated by this algorithm is shorter and smoother.

2.2 Research Direction of Mobile Manipulator Control

Due to the introduction of mobile platform and the influence of uncertain parameters, external disturbances and unmodeled dynamics in its mathematical modeling, the control problem of mobile manipulator systems becomes difficult, and its research progress has always been the focus of the control community.

At present, there are three main types of mobile manipulator research, namely, motion planning of mobile manipulator, coordinated control and trajectory tracking of mobile manipulator, and force control of mobile manipulator. The details are as follows:

The combination of two subsystems with different dynamic characteristics makes the mobile manipulator system redundant, that is, when completing a task, there are many ways to achieve it: the mobile platform or manipulator can be moved independently, and the mobile platform and manipulator can be moved simultaneously. The former reduces the difficulty of system control and planning, but does not give full play to the advantages of mobile manipulator; The latter makes the whole system more flexible and more efficient. Therefore, how to coordinate the two subsystems is the key to the motion planning of the mobile manipulator system.

The dynamic characteristics of the two subsystems of the mobile manipulator are different. The mobile platform is generally heavy in weight and slow in dynamic response. On the contrary, the manipulator is generally light in weight and fast in dynamic response. Some mobile platforms even introduce nonholonomic constraints. In addition, the influence of various uncertain factors should be considered.

3 Basics of Reinforcement Learning

3.1 Basic Definition

(1) Agent: an abstract concept, which can exist abstractly. For example, the chess players in chess games control the direction of all their own chess pieces in the field; It can also be the controlled object itself, such as a UAV that always needs to keep balance to control the movement of its own blades. In the problem of trajectory planning of manipulator, agent is the manipulator itself.
(2) State: describes the current environment perceived by the agent and the situation of the agent itself. In a fully observable environment, the information received by agents can fully represent the current environment. However, in some observable environments, the information that the agent can receive is not sufficient to fully represent the current environment. For example, in card games, the opponent's face cannot be directly observed. For completely observable problems, s is usually recorded as the state of the agent at time t. This paper focuses on the fully observable environment.

(3) Action: the main way for agents to interact with the environment, and also the way for state transfer. For example, in Go, the action is where to put the child, what to put out in card games, and how to move the joint point or where to move in the robot arm trajectory planning problem.

Figure 1 shows the relationship between environment and agent. Starting from the agent, take action a at time t to make the environment enter s + 1 from s, and obtain instant reward r + 1.

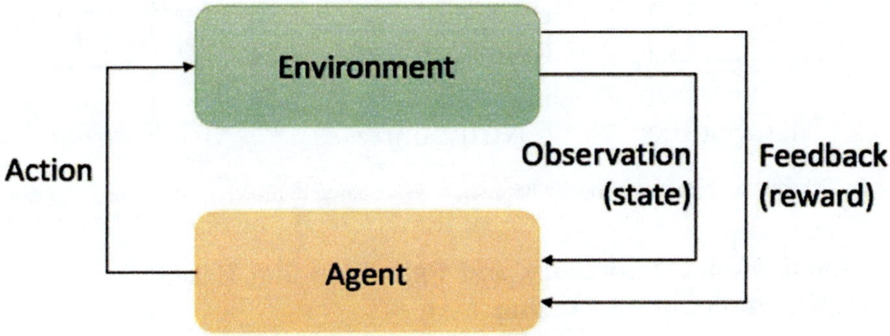

Fig. 1. Agent environment interaction diagram

(4) Markov Decision Process (MDP): It is a stochastic process in probability theory and mathematical statistics, without memory. This means that when all past states and current states are given, the next state of the process is only related to the current state, but independent of all past states. Their definitions are given below.

3.2 Strategy Iteration in Reinforcement Learning

The core of reinforcement learning, so as to maximize the expected cumulative rewards obtained by agents. Strategy iteration in reinforcement learning is usually implemented by generalized policy iteration (GPI). GPI is an iterative algorithm, which includes two steps: 1) strategy evaluation to evaluate the quality of the current strategy; 2) Policy promotion: improve the current policy according to the result of policy evaluation. By executing two steps alternately and iteratively, GPI can continuously optimize the strategy and finally converge to the optimum.

These algorithms have different application effects for different application environments.

Figure 2 shows a schematic diagram of a value based reinforcement learning algorithm. As shown in the figure, V (s) is V in each iteration. (s) And n is updated by greedy (V). The update of value function V (s) means that in all states, the agent will update the value under the old strategy according to the value obtained by the current strategy n. The renewal of the strategy is greedy to improve in the direction of increasing value.

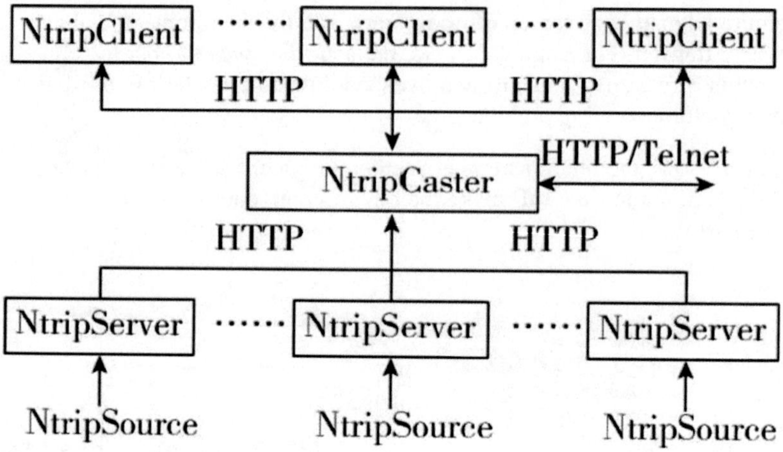

Fig. 2. Schematic diagram of strategy iteration of value based reinforcement learning algorithm

4 Mobile Robot Trajectory and Path Planning Based on Reinforcement Learning

(1) State space

The state space describes the current environment perceived by the agent and the situation of the agent itself. In the most general case, the state of the manipulator at a certain time is determined by the rotation state of all its joint points. By establishing a relative coordinate system at each joint point, the overall state of the manipulator can be described. In general application scenarios, each joint point of the mechanical arm can provide a rotational degree of freedom in a direction. Therefore, for joint point J, only one parameter is required θ, The rotation angle in the relative coordinate system represents the state of the joint point.

The motion state variable s of the manipulator at time t can be the combination of the state of each joint point of the manipulator, the end point position and the target position, namely:

$$S = \left[x_T, y_T, z_T, x_g, y_g, z_g, \theta_1, \theta_2, \ldots, \theta_N\right]^T \in N \tag{1}$$

(2) Action space

In addition to the state space N, the action set must be considered to realize the control of the manipulator under the MDP model. Each time the current state of the mechanical arm is detected, the system will select the best action to take in the next step from the action space according to the current state to achieve the best expected effect, as shown in Fig. 3. Therefore, there is a strong correlation between the state space and the action space.

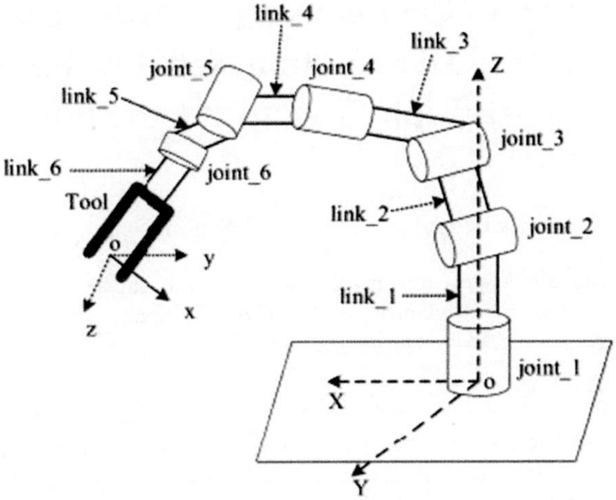

Fig. 3. Schematic diagram of mobile mechanical arm

5 Conclusion

The progress of social industrialization emphasizes the use of modern intelligent technology to liberate productivity. Manipulators have great help for intelligent assembly, flexible operation and productivity liberation, and their trajectory planning algorithms have an important impact on production efficiency. The traditional trajectory planning algorithm of manipulator often depends on the complex kinematics model built in advance, which can not adapt to the dynamic production requirements. Mobile manipulator is essentially a mobile robot system with a mechanical arm. This kind of system combines the advantages of mechanical arm and mobile robot and is widely used in civil and military fields.

References

1. Zhou, J., Luo, Q., Jiang, S., et al.: Open space path planning using inverse reinforcement learning, US20210294340A1[P] (2021)
2. Wang, Q., Shi, H.: Research on interactive recommendation system based on reinforcement learning. In: Society of Photo-Optical Instrumentation Engineers (SPIE) Conference Series. SPIE (2021)
3. Chen, J., Wang, Z.: A positioning method of mobile manipulator based on binocular vision (2022)
4. Hadjivelichkov, D., Vlachos, K., Kanoulas, D.: Improved reinforcement learning coordinated control of a mobile manipulator using joint clamping (2021)

Research on the Planning of Distributed PV Access to Distribution Network Based on Multiple Genetic Algorithms

Chengming Liu(✉) and Yunhui Fang

State Grid Hunan Electric Power Company Limited Economic and Technical Research Institute, Changsha 410007, Hunan, China
15609253474@163.com

Abstract. This article proposes a method for optimizing the routing and wire size of distributed photovoltaic access distribution networks using multiple genetic algorithms. This method can effectively integrate photovoltaic power into the existing power grid, while minimizing power loss and improving network reliability. This article provides a detailed description of the proposed method, including two stages: route selection and wire size determination, and provides evaluation results using a testing system.

Keywords: multiple genetic algorithms · distributed photovoltaic · distribution network planning

1 Introduction

In recent years, with the changes in energy consumption patterns and the enhancement of environmental protection awareness, distributed photovoltaic integrated power generation has become a form of high attention in new energy. However, due to its small grid connected capacity, uneven distribution, and large power fluctuations, distributed photovoltaic access has certain technical difficulties and planning strategy issues [1]. For the distribution network planning problem, traditional methods cannot effectively solve the distributed photovoltaic access problem, so a new planning method needs to be developed.

The distribution network planning method based on multiple genetic algorithms has become one of the current hotspots and difficulties in solving this problem. Multiple genetic algorithms can not only overcome the problems of large search space and slow convergence in traditional genetic algorithms, but also optimize between multiple objectives. In distribution network planning, multiple genetic algorithms can be applied to two stages: route selection and wire size optimization, in order to achieve minimum power loss and maximum operational reliability.

This method can effectively consider the integration of distributed photovoltaic access with existing power networks, while minimizing system load losses and improving the reliability of power supply. The contributions of this article include: designing a genetic algorithm for distributed photovoltaic access and applying.

J. C. Hung et al. (Eds.): FC 2023, LNEE 1132, pp. 176–181, 2024.
https://doi.org/10.1007/978-981-99-9538-7_25

2 Related Work

This paper takes the voltage limit, voltage fluctuation and harmonic pollution index as technical constraints, network active loss as the planning objectives of distributed PV access to distribution network. The realization of this goal requires the planning model as a foundation [2]. Therefore, based on a variety of genetic algorithms, this paper establishes a distributed photovoltaic access distribution network planning model with both security and economy, in order to realize the optimization of distributed photovoltaic access distribution network planning through a variety of genetic algorithms. The following is an introduction of the distributed PV access distribution network planning model based on multiple genetic algorithms. The distribution network planning model is shown in Fig. 1.

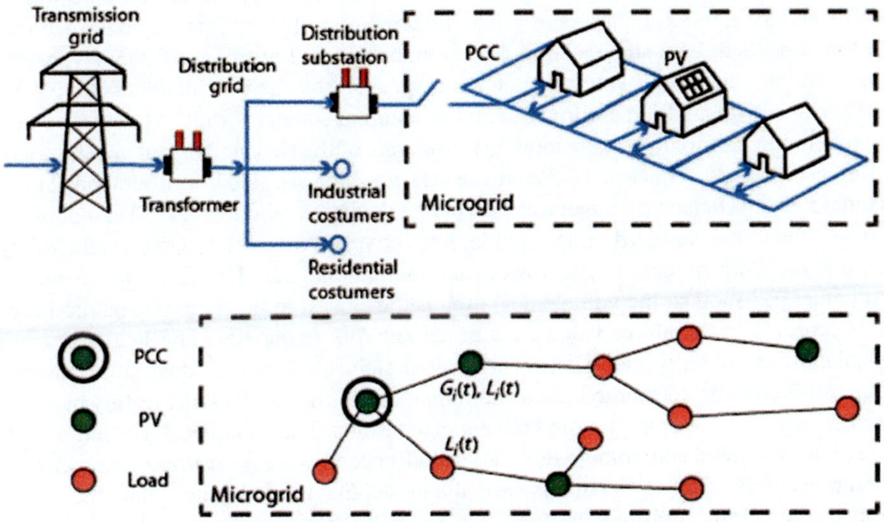

Fig. 1. Distribution Network Planning Model

The realization of the sub-target needs to be supported by a function, and the required function of the sub-target is the total access capacity function, the installed access number of distributed PV. At the same time, in order to improve the economy maintenance cost of distributed PV is also taken as a sub-goal. The achievement of economic subgoals also requires functions. The functions required for the economic sub-objective are operation and maintenance cost functions, investment costs per unit capacity, distributed PV service lifetime, and discount rate. The support function of this subtarget is the active network loss function, the node voltage, and the phase difference of either side voltage. The functions of these three sub-objectives can be combined into an whole objective function, namely the economic objective function.

The voltage over the limit constraint, the voltage fluctuation constraint and the harmonic pollution constraint. First, the voltage constraint. According to the "power quality power supply voltage limit" can be obtained, the voltage level is different, the voltage limit allowed by the distribution network is also different. Therefore, the voltage limit value can be taken as the voltage limit constraint condition. The voltage level, voltage frequency is different, the distribution network allowed voltage fluctuation limits are also different. Therefore, the constraint of voltage fluctuation is the voltage fluctuation value caused by distributed photovoltaic system. Third, the harmonic pollution constraints. Therefore, the harmonic current limit of the national standard allowing access to the distribution network can be taken as a constraint.

3 Solution

As a planning example, and uses the NPGA genetic algorithm to complete the model verification by comparing the planning results. The comparison between the planning results is shown in Table 1. Through the above two planning models, the optimization results and the operation parameters. The total amount of distributed pv obtained from the two models did not show a large deviation, namely, 3259 kW and 3285 kW, respectively, and the maximum voltage fluctuation specified by the two models is 0.56% and 1.1%, respectively, which are lower than the national standard limit. At the same time, the maximum harmonic voltage total distortion rate of the two models did not show a big difference, namely 1.1% and 1.73%, respectively, which are also lower than the national standard limit. The last parameter is the voltage deviation, and the 1.0582 standard unitary and the 1.063 standard unitary value, respectively. Although the maximum voltage deviation of both models is also lower than the national standard limit, but compared with other parameters, the voltage deviation is the closest to the national standard limit.

According to the above values, under the network parameters and load size of the calculation example, the voltage deviation is the closest limiting constraint. However, if the difference of distributed, then the indicators of the model built in this paper are smaller, which can ensure a more stable and safer distribution network operation state. The harmonic wave and voltage fluctuations will become the most important constraint conditions of distributed PV. In this case, the model that takes the more than the voltage limit as a constraint is difficult to play its role effectively and realize the effective control of the access capacity [3].

(3) Example planning and solution under different conditions

First, the influence of different load conditions on the case planning is discussed. Table 2 shows the operating parameters for the access capacity and the closest limits under different load conditions. According to the values shown in Table 2, the following conclusion can be drawn: Meanwhile, the data in Table 2 can also conclude that the voltage deviation has the greatest limit under this load condition. That is to say, the first limit constraint reached deviation. Combined with the above conclusions, the analysis shows that when the voltage deviation becomes the main constraint to ensure the operating state change at the same frequency with the load condition and increase with its increase. However, voltage fluctuation constraint conditions, and the harmonic and voltage fluctuation will replace the voltage deviation as the most important constraint condition. After the harmonic and voltage fluctuations become the most important constraints, as the load conditions, but will always be within the maximum limit range of the harmonic and voltage fluctuation constraints.

Table 1. Operation Parameters and Planning Results of the Distribution Network

model	parameter	Node 3	Node 4	Node 5	And or maximum
	Access capacity / kW	0.0542	0.0037	0.1320	0.3259
	The Node Electric JK / pu	1.0547	1.0532	1.0563	1.0582
Multi-port label and multi-constraint model	Total harmonic voltage distortion rate	0.44	0.03	1.07	1.1
	Electric bed fluctuations	0.28	0.01	0.39	0.56
	Line active power loss / kW	Line reactive power loss	70.28	line loss	116.28
	Access capacity / kW	0.2138	0.0458	0.0419	0.3285
	15 Electric JR / pu	1.063	1.0562	1.0551	1.063
Single-objective multi-constraint model	Total harmonic voltage distortion rate	1.73	0.37	0.34	L73
	voltage pulsation /%	1.1	0.09	0.12	1.1
	Active line loss of kW	Line reactive power loss	76.81	Meritorious loss	126.75

In addition to different load conditions, different power can also affect the case planning [4]. If the load conditions and access location are the same, and the power is different, then the distributed photovoltaic injected with capacitive, and its access capacity will be smaller. Accordingly, after the distributed photovoltaic injected with perceptual reactive power, its access capacity becomes larger. The connected distributed PV should be mainly operated per unit power.

Table 2. Access capacity and operating parameters closest to the limits under different load conditions

Access scheme/parameters		Photovoltaic number					The more limited the conditions
		DGI	DG2	DG3	DG4		
Way 1	on-position Access capacity	2 0.0526	3 0.1374	4 0.0551	0.0095	0.3401	node voltage
Way 2	----	2 0.0412	4 0.1819	6 0.0844	0.0796	0.3871	node voltage
Way 3	----	7 0.0303	9 0.1695	0.0067	0.0367	0.3653	node voltage
Way 4	----	9 0.0539	10 0.135	11 0.0683	0.0800	0.3372	node voltage
Way 5	----	12 0.0713	14 0.1410	16 0.0524	0.0782	0.3429	node voltage
Way 6	----	15 0.0522	16 0.1318	17 0.0545	0.0906	0.3219	node voltage

4 Conclusion

The research paper proposes a method for planning distributed photovoltaic access distribution networks using multiple genetic algorithms. The proposed method considers the integration of photovoltaic access power sources with existing power networks while minimizing power loss and improving. The method involves dividing the planning problem into two stages, namely, route selection and conductor sizing. In the first stage, the optimal routes for the distribution network are determined using multiple genetic algorithms. In the second stage, a different genetic algorithm is used to select the appropriate conductor sizes for each segment of the network. The proposed method was evaluated using a test system, and the results showed that it is effective in planning distribution networks with distributed photovoltaic access power sources.

References

1. Liu Scientific Research, Sheng, W., Ma, X.: Research on distributed photovoltaic access distribution network planning based on multiple group genetic algorithms. J. Solar Energy **42** (06), 146–155 (2021)
2. Martenxiao, M.X., Xingzhen, B.: Robust optimization planning method of distribution network platform area considering large-scale distributed photovoltaic access. Power Capacitors React. Power Compensation **42**(02), 183–190 (2021)

3. Chen, C., Yun, T., Wei, W.: Site selection method of distributed photovoltaic multi-point access to low-voltage distribution network. Autom. Instrum. **11**, 198–201 (2019)
4. Zhang, T., Gao, Z., Gu, J.: Research on site selection and capacity optimization of distributed photovoltaic access to distribution network. J. North China Electr. Power Univ. (Nat. Sci. Ed.) **46**(01), 60–66 (2019)

Research on Fault Diagnosis System of Mine Hoist Based on BP Algorithm

Dan Liu(✉) and Yingying Jiang

School of Electrical and Information Engineering, Heilongjiang University of Technology, Jixi 158100, Heilongjiang, China
Ld_1207@163.com

Abstract. This article introduces the research of a fault diagnosis system for mine hoists based on BP algorithm. The system aims to improve the reliability and safety of mine hoists. This article collected operational data of mine hoists and extracted fault data as the training set of the model. Then, a BP neural network model was established on the MATLAB software platform, and the model parameters were optimized and adjusted. Finally, use this model for real-time monitoring.

The contribution of this study is to propose a fault diagnosis system for mine hoists based on BP algorithm. This system can achieve real-time monitoring of mine hoist data, while greatly improving the quality of diagnosis system's judgment and prediction of mine hoist status. At the same time, the system has the function of data visualization, which enables users to understand the running status of the system more intuitively, and provides a more convenient and efficient operating platform for engineering technicians and managers.The controller is used to control all components to achieve stable operation; The sensor detects abnormal conditions in real time, such as overload or fault; It also provides information about the operation status and other related parameters. Finally, after receiving the data from the sensor, it will automatically adjust its parameters according to different operating modes.

Keywords: Mine hoist · BP algorithm · fault diagnosis

1 Introduction

On the one hand, it improves the automation level of the system and brings considerable economic benefits to the production. On the other hand, due to the sudden increase of factors affecting the operation of the system, it is more and more likely to cause failures or failures [1]. The occurrence of faults not only brings huge economic and property losses, but also sometimes endangers people's lives and causes catastrophic accidents. Therefore, the safety and reliability of modern equipment system operation has become an urgent problem that human beings must solve [2].

The history of fault diagnosis is closely related to the maintenance mode of equipment. In a quite long period after the Industrial Revolution, due to the production scale at that time, the technical level and complexity of the equipment were relatively low,

J. C. Hung et al. (Eds.): FC 2023, LNEE 1132, pp. 182–187, 2024.
https://doi.org/10.1007/978-981-99-9538-7_26

and the human maintenance method for the equipment was basically post maintenance, that is, failure analysis and maintenance were carried out after the equipment ran into problems. After the 20th century, due to the development of mass production, especially the appearance of the assembly line production mode, the technical level and complexity of the equipment itself have been greatly improved, and the impact of equipment failures on production has increased significantly [3]. Thus, regular maintenance has emerged, so that accidents can be handled before they occur. In the 1960s, the U.S. military realized a series of drawbacks of regular maintenance and began to change regular maintenance into predictive maintenance, that is, it began to monitor the equipment during normal operation to find potential fault factors and take measures as early as possible to prevent sudden accidents [4].

2 Formation of Fault Diagnosis Based on Neural Network

In 1989, Venkat Venkatasubramanian and King Chan of Purdue University in the United States used artificial neural networks in fault diagnosis and compared them with knowledge-based expert systems. The equipment they diagnosed was the fault of a fluidized catalytic cracking unit, identified 18 symptoms and 13 fault types, and there were 5–27 nodes in the hidden layer. The algorithm used is the back propagation algorithm, which has obtained the ideal effect; It can correctly determine 94%–98% of the fault causes [5]. The drawback is that the training time is too long, and The data input during training is not real-time; It is much more difficult for ANN to map continuous variables than Boolean variables. Despite these limitations, they are still the first time to successfully apply ANN to pattern matching and fault diagnosis. Compared with traditional expert systems, neural networks have the following unique advantages:

1) Artificial neural networks can run in real time, which is a challenge for expert systems.
2) Artificial neural networks can directly use time series data, while expert systems need to convert numerical values into symbolic information.

Professor Yu Heqi of Northeastern University and others have deeply studied the artificial neural network method of mechanical equipment fault diagnosis; Qu Liangsheng, a professor of Xi'an Jiaotong University, and others made a comprehensive study on various faults of large rotating machinery by using artificial neural networks; Zhong Binglin of Southeast University and others also studied the expression, association, memory ability and network structure of neural network for given knowledge, and pointed out that neural network showed great application potential in mechanical fault diagnosis. In addition, some people have studied various problems of neural network used in equipment fault diagnosis and obtained certain results.

3 Introduction to Mine Hoist System

3.1 Composition of Mine Hoisting Equipment

As a typical mechanical and electrical equipment, the mine hoisting equipment is responsible for the important task of lifting coal, known as the "mine throat". It is an important production equipment connecting the underground and the ground. It occupies an important position in the overall comprehensive mechanized production. The hoist is an important part of it, and it is also an important large fixed mechanical equipment of the mine. Generally speaking, the main shaft hoisting equipment (mostly skip hoisting) only lifts the useful minerals mined underground from the shaft bottom yard to the ground; The auxiliary shaft hoisting equipment shall complete such operations as lifting gangue, lowering materials, lifting personnel and equipment, etc. The effective lifting capacity of the lifting container of modern lifting equipment has reached 30 to 50 tons at a time, and its speed in the shaft can reach the running speed of the railway train - 20 to 25 m per second (shaft). The maximum capacity of the drive motor of a hoist has reached 10000 to 15000 kw. If the safety and reliability of its operation are ignored, the production of the whole mine will not only be brought to a standstill, but may even lead to unimaginable consequences. The main components of the mine hoist are shown in Fig. 1.

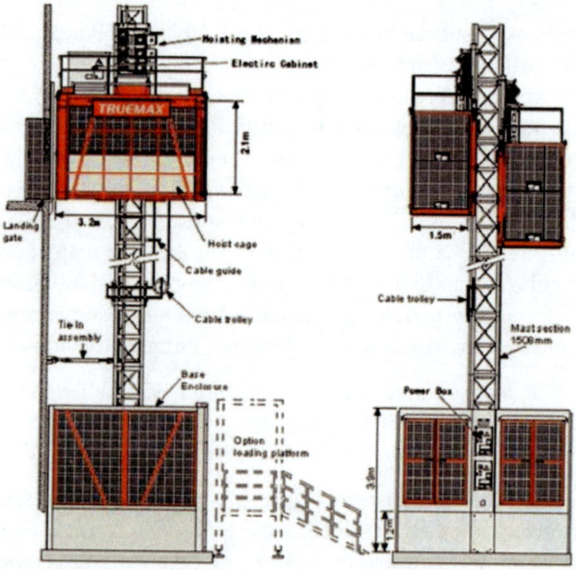

Fig. 1. Structural Diagram of Mine Hoist

At present, the main structural forms of China's production include: single rope winding type, single drum and double drum mine hoists; friction type, multi rope floor type and tower type multi rope friction type hoists. The driving mode is designed as required, and hydraulic drive mine hoist is also used in the mine. The commonly used mine hoists in China are single rope winding type and multi rope friction type. Compared with the countries with more developed mining industry in the world, China's mines have smaller mine types, shallow mine lifting heights, more coal mines, and less other mines (such as metal mines and non-metallic mines). In addition, inclined shaft lifting accounts for a large proportion. Therefore, in the 1960s, single rope wound mine hoists were used more, because single rope wound mine hoists are more suitable for such mines.

3.2 Composition of Mine Hoist Monitoring System

The abominability of the environment and the limitations of research, some protections have not reached the expected effect, resulting in many serious casualties and equipment damage accidents of the hoisting system in coal mine production.

(1) On line monitoring of electrical parameters of electrical equipment

Through the monitoring and curve analysis of the signal quantity of the given link and feedback link in the speed loop and current loop, it will be helpful to predict and judge the link and location of the fault; By testing the rectifier current and voltage of the main circuit and drawing the change curve; It can be used as the first-hand data for waveform identification, fault analysis, technical measurement and the basis for judging whether the system is overloaded.

(2) On line detection of pressure and current in hydraulic system

The monitoring and curve analysis of the system parameters such as the oil pressure of the main circuit of the hydraulic system, the control current of the electro-hydraulic proportional valve and the nitrogen pressure of the accumulator will help to predict and judge the fault link; The curves of oil pressure of main circuit and control current of electro-hydraulic proportional valve in the working process and the curves of secondary brake oil pressure of hydraulic main circuit in the emergency braking process can be used as the first-hand data for technical measurement. In addition, the status parameters of the hydraulic system.

4 Fault Diagnosis System of Mine Hoist Based on BP Algorithm

In comprehensive consideration, we will select the LMBP algorithm with fast convergence speed as the diagnosis algorithm of our fault diagnosis system based on neural network.

$$E(w, b) = \sum_{j=0}^{n-1} (d_j - y_j)^2 \quad (1)$$

As the main equipment of mine hoist system includes hydraulic station, reducer and brake, the operation of each equipment is relatively independent. Therefore, we use multiple data acquisition boards to collect the parameters and performance indicators of the main equipment, and transmit them to our fault diagnosis system through the Internet. Accordingly, in the user operation interface, we have also developed diagnostic modules corresponding to the main equipment, which are used to monitor the operation and fault diagnosis of each equipment in real time, as shown in Fig. 2.

In the user interaction interface, the current operating program receives the data from the lower computer through the Internet, and then delivers the real-time data to the background data processing system, diagnoses the current data through the fault diagnosis algorithm, and transmits the results to the foreground interface for use. The user can select the real-time diagnosis of a module by operating the "dialog box". On the other hand, the fault diagnosis system will display the received real-time data and diagnosis results together according to the user's requirements.

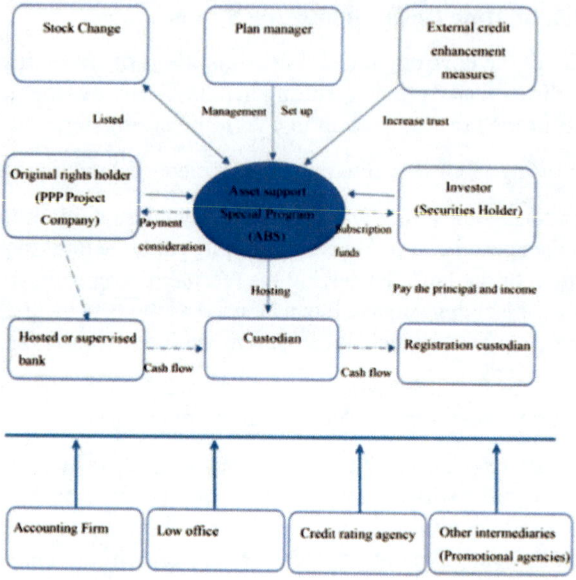

Fig. 2. Overall framework of fault diagnosis system

5 Conclusion

This article introduces the design and research of a fault diagnosis system for mine hoists based on BP algorithm, aiming to improve the reliability and safety of mine hoists.

In coal mine production, once the mine hoist malfunctions, it will not only cause production losses, but also bring safety accidents such as casualties. Therefore, real-time monitoring and diagnosis of mine hoists is particularly important. This article proposes a method for diagnosing mine hoist faults using BP algorithm. This article first collected operational data of the mine hoist and extracted fault data as the training set of the model. Then, a BP neural network model was established on the MATLAB software platform, and the model parameters were optimized and adjusted. Finally, use this model for real-time monitoring. The experimental results show that the mine hoist fault diagnosis system can effectively detect and diagnose faults of the mine hoist, and has high accuracy and robustness. The system can also realize data visualization, so that users can intuitively understand the operation status of the mine hoist, take corresponding maintenance measures in time, and further improve the operation efficiency and safety of the mine hoist.

In summary, the fault diagnosis system for mine hoist based on BP algorithm proposed in this article has good practical value and application prospects, and is expected to become one of the important technical research directions in coal mine production.

Acknowledgements. 2022 fundamental science foundation of Heilongjiang province universities. Research on detection system of disc brake of mine hoist.

References

1. Yang, C.: Research on rolling bearing fault diagnosis based on TVFEMD-SVD and random forest algorithm. IOP Publishing Ltd. (2022)
2. Hui, L., Yu, J., Zhao, A., Wang, F., Yang, X., Zhou, M.: Research on multi-heat-source scheduling strategy of heating system based on improved fireworks algorithm (2021)
3. Zhang, X., Jiang, J.: Research on frame design of sports image analysis system based on feature extraction algorithm (2021)
4. Huixin, G., Gang, Z., Yang, C., et al.: Research on edge detection and image segmentation of cabinet region based on edge computing joint image detection algorithm. Int. J. Reliab. Q. Saf. Eng. (2022)
5. Chen, D., Cheng, P.: Development of design system for product pattern design based on Kansei engineering and BP neural network. Int. J. Clothing Sci. Technol. **34**(3), 335–346 (2022)

Research and Application of Animation Visualization Based on HTML5 Algorithm

Guoji Liu(✉)

Software Engineering Institute of Guangzhou, Guangzhou 510990, Guangdong, China
lgj0328@tom.com

Abstract. The history of animation began in the late 19th century. At that time, it was a new way to present films. In those days, there was no screen or projector. The only way to watch movies is through magic lanterns and shadow plays. These are very simple ways to display images on paper and cloth, without sound effects. In this era, people's lives are changing rapidly due to the progress of technology. Animation is an art form that uses drawings, models, pictures, videos, or other methods to create motion without using camera motion or movie recording techniques, such as live action movie production. As technology develops and becomes more complex, so does technology. The most important thing everyone uses is the visual representation of things, because without it, they cannot easily understand. In this paper, we will introduce the research and application of animation visualization based on HTML5 algorithm.

Keywords: HTML5 technology · Visual animation · graphic display

1 Introduction

Animation can also make data vivid and intuitive. With the explosive growth of information data, data visualization, as an important means of data display and analysis, how to improve the efficiency of information transmission has become increasingly critical. There are many design methods to improve the effectiveness of static visualization charts in existing related research [1]. However, in some scenarios (such as describing the process, showing changes or unfolding narratives), it is difficult to express data laws efficiently by simply using static primitives and text. At this time, visual animation can be used to improve the perceived effectiveness of data and bring readers data experience that other visual means do not have. Scientific animation design can effectively attract and maintain readers' attention, and provide effective context information during the change process (for example, in the smooth transition process of changing elements, other elements maintain their original visual state, and will not re analyze and render due to the state switching of changing elements, resulting in flickering), so that readers can accurately identify and track the changes of elements and establish the relationship between element states, And then guide readers to understand the data information conveyed in visualization. For example, through the transition of the intermediate state, it can help readers to clarify the rules of data change, or show the accumulation process of data over time [2].

J. C. Hung et al. (Eds.): FC 2023, LNEE 1132, pp. 188–193, 2024.
https://doi.org/10.1007/978-981-99-9538-7_27

The lack of abstract description of the relationship between data, visual coding and time sequence in visual animation leads to the increase of user creation cost, and the consistency between output animation and data perception is not guaranteed. Users need to manually calculate the visual attribute values of each element at each time node according to the data. For example, in order to build an animation in which elements appear in sequence according to data values, users need to group elements according to data attribute values, and then define animation effects for each group and calculate animation time. This process will lead to an increase in user creation costs as the number of elements and the complexity of animation increase. Because the entities are not connected according to the data relationship, users can specify any animation parameters for different entities, which may lead to inconsistency between the output animation and data perception, resulting in ambiguity. During the creation process, it is difficult for users to imagine and understand the current animation content without previewing, and the interaction is complex and inefficient. For example, in the keyframe based interactive tool, it is necessary to repeatedly switch the time track corresponding to different visual attributes of each element to adjust the attribute values at each time point to achieve the animation effect [3]. At the same time, entities with the same animation parameters need to be repeatedly defined. Property changes at any time.

2 Related Work

2.1 HTML5 Technical Algorithm

HTML5 outline algorithm refers to an algorithm that extracts document outline by processing h1-h6 header elements in section, article, aside, nav and other block elements. As a SEOER who deeply studies semantic tags, it should be able to have a keen insight into the important role this algorithm will play in the SEO process. And apply it to the daily process of website optimization [4]. As shown in Fig. 1 below, we can understand all divs as section elements and generate a new outline structure.

Define block<section>label:

Section represents a block, which is used to identify sections in a document. It is often used to partition content on a page, such as chapters, headers, footers, or other parts of the document.

Define the article block<article>tag:

Article represents an article, which is used to identify a complete, independent and forwarding content in the page.

Define the sidebar<aside>label:

The side represents the side, which is used to identify the content other than the content. The content of the aside should be related to the content nearby. For example, the auxiliary information part of the current page or article can include references, side ads, navigation bars, and other similar parts different from the main content related to the current page or main content.

To define navigation<nav>tags:

Nav stands for navigation bar, which is used to identify the link group of page navigation. A page can have multiple nav elements as navigation for the whole or different parts of the page. Specific application fields include: main menu navigation, sidebar navigation, page turning navigation, etc.

```
<section>

   <h1>Bob Dylan
  Albums</h1>

   <p>Some text</p>
   <section>

     <h2>Blood on the Tracks</h2>

     <p>Some text</p>

   </section>
   <section>

     <h2>Highway 61 Revisited</h2>

     <p>Some text</p>

   </section>
   <p>Some text</p>

</section>
```

Fig. 1. Section element outline structure

After mastering HTML5+CSS3 technology, you can write a static page. If you want to achieve dynamic effects, you need to use JavaScript technology [5].

2.2 Animation Examples

The expressiveness of visual animation is determined by two orthogonal factors (visualization and animation), so this section selects animation examples that can show the diversity of design space. On the one hand, these examples include all kinds of visual charts created by D3, Vegas lite and Characterizer; On the other hand, they cover a series of animation types proposed by Amini et al., including creation, deconstruction, circulation, accumulation, transition, detail scaling, annotation, and multi view animation. Since each animation instance may integrate multiple animation types, this section will distinguish between instances by visual chart type and animation type.

DP1: Provide an intuitive and clear way to display visual animation. As a universal tool, the existing animation creation tools based on keyframes will show the animation on the pure timeline interface in an abstract way. For example, in Adobe After Effects and Adobe Premiere, keyframes are represented as a diamond node on the timeline.

This abstract way makes it almost impossible for users to understand the animation they are currently creating without previewing it. In contrast, HTML5 uses explicit visual components to represent animation primitives and corresponding animation effects (such as time, animation effects, etc.) to promote users' understanding of the currently specified animation process. In addition, HTML5 also contains an animation creation panel, which integrates storyboards and timelines to display the animation process in an intuitive and accurate way in accordance with the visual standard expression of animation. The process of using explicit visual components in HTML5 is shown in Fig. 2.

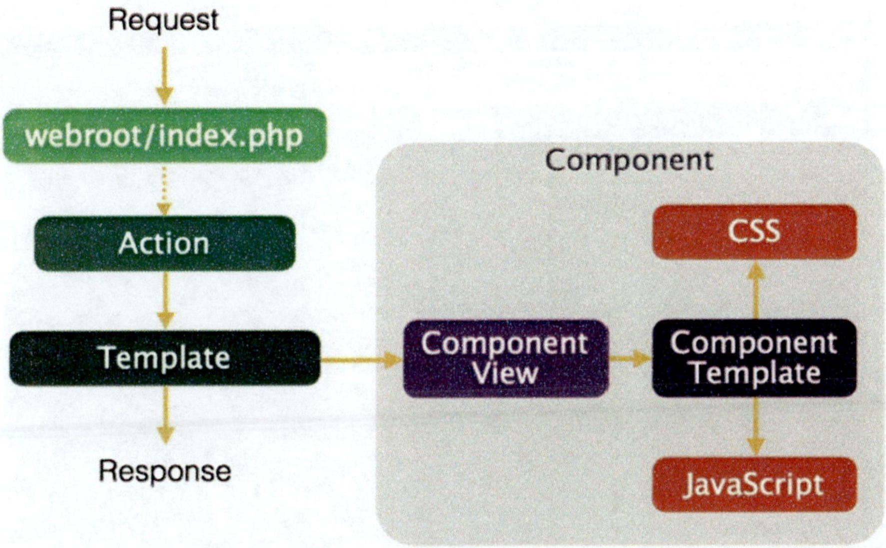

Fig. 2. HTML5 uses explicit visual component flow

3 Research and Application of Animation Visualization Based on HTML5 Algorithm

HTML5 is designed to help beginners who lack programming skills or are unfamiliar with animation to create rich graphic animation. In the process of using HTML5 to create animation, users first provide the dSVG file of the visualization chart as input, and complete the animation through key frame construction, key frame sequence generation, and animation parameter definition and adjustment. After completing the construction of key frames and key frame sequences through a series of selection operations with the help of the automatic completion function, users can adjust the animation process by directly interacting with the animation process described by the visual specification. The optimized Canis and its compiler provide good support for real-time preview and updating the description of visual specifications in the interaction process. At the same time, the visual specifications that are easy to understand and the intuitive interaction mode formed with the help of automatic completion algorithm also greatly reduce the learning and interaction costs of users on the basis of ensuring the expressiveness of animation.

HTML5 almost completely replaces the use of Flash. Due to the performance of mobile application developers, ease of use and the cutting of open standards, it has also been greatly adjusted. HTML5's predecessor, HTML4, has many improvements, one of which is the inclusion of canvas elements for instant rendering of graphics on web pages. As shown in Fig. 3 below, HTML5 technology animation visualization key frame extraction.

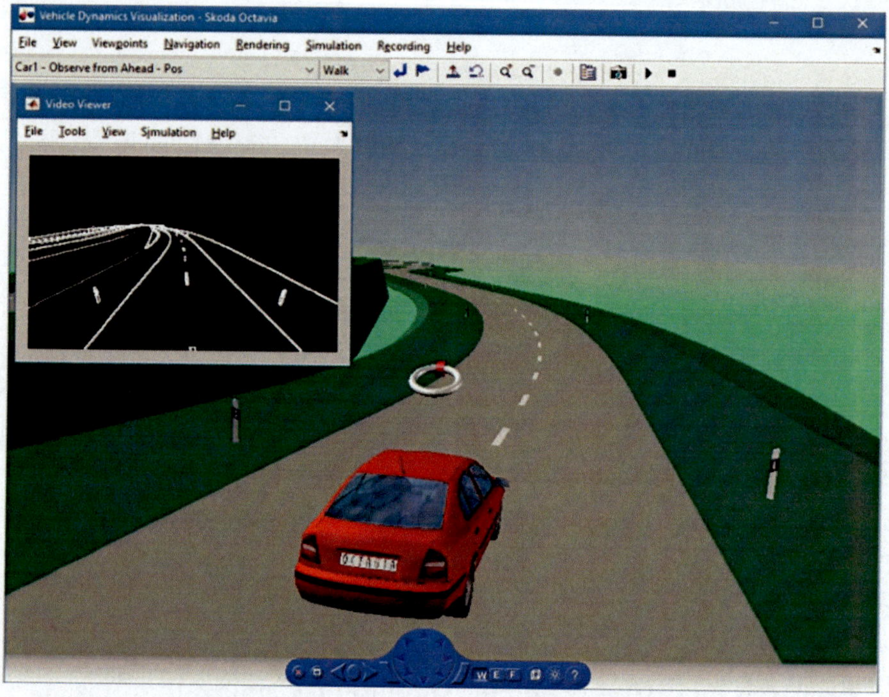

Fig. 3. HTML5 technology animation visualization key frame extraction

4 Conclusion

Html5 is the latest technology in web page development. Html5 has many advantages, such as faster load time, less bandwidth consumption and compatibility with mobile devices. In this article, we will discuss how to use animation visualization based on Html5 algorithm. After confirming the information structure, the animation arrangement part is relatively simple. InfoMotion provides a series of commonly used playing forms for users to choose. By default, the animation is played in the form of repeating units one by one.

References

1. Habibi, H.: Implementation of ICARE learning model using visualization animation on biotechnology course. AIP Conf. Proc. **1911**(1), 20027 (2017)
2. Wenbo, L.I., Pan, H., Han, Q., et al.: Visualization of medical image classification based on interactive model. Journal of Computer Applications (2016)
3. Lu, D., Zhu, D., Wang, Z., et al.: Efficient level of detail for texture-based flow visualization. Comput. Anim. Virtual Worlds **27**(2), 123–140 (2016)
4. Wang, H., Shen-Shen, L.I., Xiao-Yu, H.E., et al.: Design and Implementation of Interactive Mask Erasing Animation Based on Personalized Customization. J. Graph. **40**(3), 479–480 (2019)
5. Gehrer, D., Viola, I.: Visualization of molecular machinery using agent-based animation. In: Spring Conference on Computer Graphics, pp. 1–10. ACM (2017)

Order Allocation Optimization and Genetic Algorithm in Logistics Service Supply Chain

Langqin Liu(✉)

Changjaing polythecnic, Wuhan 430070, China
4545486@qq.com

Abstract. Logistics service integrators usually charge a certain proportion of transaction costs from subcontractors according to the assigned order value, set the transaction costs as a linear function of the transaction amount, and constructed a new mixed integer programming model for order allocation optimization of logistics service supply chain. The optimization goal is to minimize the transaction costs, procurement costs, the quantity of logistics capacity of short service and delayed supply. It is a research on the practical teaching design of applied university mental health curriculum in the virtual simulation scene. The main purpose of this study is to explore how to improve teaching methods and technologies through the application of new technologies such as virtual reality (VR) technology, which has been widely used in the field of education, such as learning games and training projects. In addition, it aims to study whether VR can be used as an effective tool to improve students' mental health during their study. Researchers believe that virtual reality will help us better understand human psychology and behavior, and better understand various psychological barriers.

Keywords: Supply chain · Logistics services · Order allocation optimization · Genetic algorithm

1 Introduction

At the end of 1990s, science and technology brought about rapid development, and the production capacity of traditional manufacturing industry also brought about huge growth. As a result, the world ushered in an era of excess physical products. The number of rich products is no longer the most important demand of people. People gradually turned their attention to personalized and attractive services, which made the service industry develop rapidly. Makes the boundary between it and the related traditional manufacturing industry become blurred [1]. Transportation, warehousing, and other links of goods, and is one of the important factors that affect the level of logistics service and operating costs. Faced with the continuous increase of e-commerce business and the increasing number of orders, allocation optimization problems have become increasingly complex and critical. In response to this issue, genetic algorithms are widely used in order allocation optimization. Genetic algorithm is a method of solving optimization problems by simulating the evolutionary process, in order to obtain the global optimal solution.

J. C. Hung et al. (Eds.): FC 2023, LNEE 1132, pp. 194–198, 2024.
https://doi.org/10.1007/978-981-99-9538-7_28

In order allocation optimization, genetic algorithm has the following advantages: First, genetic algorithm can construct fitness function according to optimization requirements, and optimize according to different objectives; Secondly, genetic algorithms can quickly find multiple optimization solutions that meet constraint conditions, effectively reducing computational time [2]; Thirdly, the optimized order allocation scheme through genetic algorithm can be targeted for practical optimization, improving the operational efficiency and service quality of logistics services.

Therefore, this article aims to study order allocation optimization methods based on genetic algorithms. First, the order allocation model is proposed, and the fitness function is established. Then, the genetic algorithm is used to optimize to obtain the optimal order allocation scheme. Finally, the effectiveness and feasibility of this method were verified through simulation experiments. The experimental results indicate that the proposed order allocation optimization method based on genetic algorithm can improve the operational efficiency and service quality of the logistics service supply chain, and has practical application value.

In summary, the goal of this article is to propose and study an order allocation optimization method based on genetic algorithms. This method can provide new ideas and methods for optimizing the logistics service supply chain, and provide more decision-making support for logistics service supply chain management.

2 Related Work

2.1 Service Supply Chain

At the end of the 1990s, science and technology brought about rapid development, and the production capacity of traditional manufacturing also brought about huge growth. As a result, the world ushered in an era of excess physical products [3]. The number of rich products is no longer the most important demand of people. People gradually turned their attention to personalized and attractive services, which made the service industry develop rapidly. The rapid development of the service industry makes the boundary between it and the related traditional manufacturing industry become blurred.

Service supply chain refers to a comprehensive system of activities that deliver service products from suppliers to end users to meet customer needs. Unlike traditional material supply chains, the products of service supply chains are services, suppliers are service providers, and users are service recipients. The characteristic of service supply chain lies in the unstorability and ubiquity of services, and require more coordination and integration.

The meet customer needs while reducing operational costs and improving service quality and performance levels. Effective management of the service supply chain can bring many benefits to enterprises, such as improving customer satisfaction, reducing inventory costs, shortening product lifecycle, improving price competitiveness, and so on [4]. The optimization of the service supply chain needs to consider various factors such as information sharing, resource coordination, and timeliness between different links, so appropriate management methods and technologies need to be adopted.

Currently, the use of information technology for integration and coordination in service supply chain management is receiving increasing attention. For example, customer relationship management technology (CRM) and enterprise resource planning software (ERP) are used to manage the supply chain, realize information sharing and real-time monitoring, and through intelligent algorithms and optimization technology. In the future, face more challenges and opportunities, requiring continuous exploration and innovation.

2.2 Logistics Service Supply Chain

The process of planning, procurement, production, delivery, after-sales and other aspects of logistics services through collaboration and coordination among various links. In the trend of globalization and digitization, the importance of logistics service supply chains has increased, while also facing more challenges and opportunities.

The core coordinate the relationships between various links, so that the entire logistics service supply chain can be optimized according to factors such as customer needs and costs. Optimizing order allocation can improve the operational efficiency and service quality of logistics services.

The management of logistics service supply chain needs to consider various factors, such as order processing and scheduling, transportation, warehousing, and so on. Therefore, in logistics service supply chain management, it is necessary to adopt various advanced technologies and methods, such as RFID.

The optimization and management are closely related to socio-economic development. In the context of highly interconnected global supply chains, the ability to optimize logistics service supply chains will bring greater competitive advantages and market opportunities. Therefore, the research and application an indispensable part of today's logistics industry and economic development.The problem description is shown in Fig. 1.

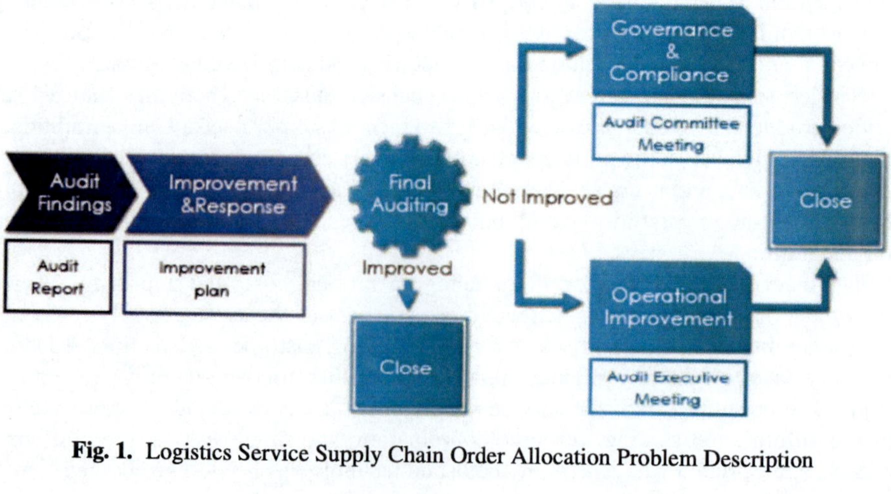

Fig. 1. Logistics Service Supply Chain Order Allocation Problem Description

3 Logistics Service Supply Chain Order Allocation

Then the satisfaction expression of the distribution result of the jth logistics service category of the ith logistics service provider is:

$$d_{ij} = \begin{cases} Q_{ij} \\ x_{ij} \\ \frac{x_{ij}}{Q_{ij}} \end{cases} \tag{1}$$

In the current competitive environment, logistics service integrators are more willing to choose logistics service providers with higher management level and service capabilities, and establish long-term contracts in the process of cooperation. Therefore, different logistics service providers have different degrees of importance, which is related to the long-term revenue potential, the possibility of establishing long-term partnerships, and the level of capabilities.

For example, a customer proposes to transport a batch of furniture from Shenyang to Beijing, which requires transportation services of 200 standard vehicles and storage services of 2000 square meters. That is, transportation: storage = 1:10. At this time, the matching proportion relationship is dimensioned, For transportation: warehousing = 1:10 (vehicle/square meter). It can distribute it completely according to the matching proportion, that is, it distributes the order for 100 standard cars to the provider, and requires the provider to provide 1000 square meters of warehousing service. However, if the provider does not have the corresponding warehousing service capacity, the provider cannot complete the service order. At this time, logistics service integrators need to give play to their integration characteristics. At this time, the allocated order quantity has an incomplete matching relationship, However, integrators also need to minimize the mismatch of different service orders allocated.

4 Optimization Research

In a certain planning period, the logistics service integrator accepts logistics tasks from multiple customers. The logistics tasks can be transportation services, warehousing services, or the subdivision of services such as truck transportation and air transportation. The logistics integrator will summarize the received tasks to obtain the total logistics capacity of various tasks. Then, considering the service price, transaction cost, service capacity of each subcontractor and the characteristics of the order task itself, the logistics integrator will formulate the order task allocation plan of each subcontractor, form a service order indicating each logistics task of each subcontractor in each time cycle, and issue it to each subcontractor.

The steps of applying genetic algorithm to solve this problem are shown in Fig. 2.

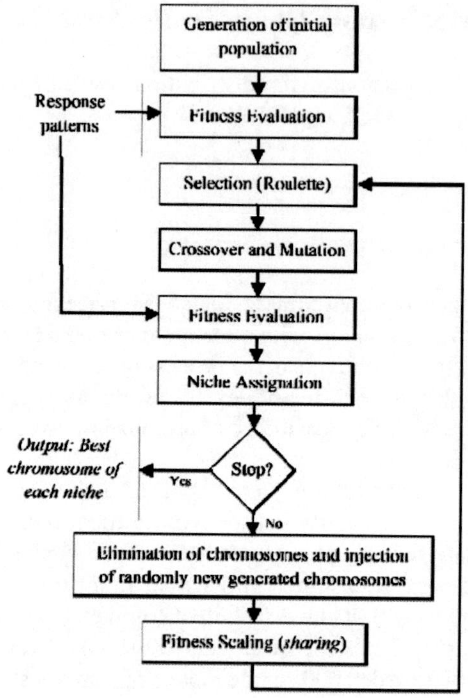

Fig. 2. Genetic algorithm solution steps

5 Conclusion

The algorithm designed can obtain satisfactory solutions with high quality in an acceptable time. For large-scale order allocation optimization problems, the solving time and results of genetic algorithm are better than those of LINGO software. Considering linear transaction costs is a dynamic and complex decision-making system, and the reputation of subcontractors is also a significant factor. How to objectively evaluate subcontractors according to their historical records and incorporate them into the order allocation system will be a focus of future research.

References

1. Balseiro, S.R., Lu, H, Mirrokni, V., et al.: From online optimization to pid controllers: mirror descent with momentum (2022)
2. Shimada, D.: Optimization Apparatus, Optimization Method, and Optimization Program:, US20210271729A1 (2021)
3. Varaprasad, J., Radha, R.K.: Optimal allocation of solar photovoltaic distributed generation in electrical distribution networks using Archimedes optimization algorithm. Clean Energy **2**, 2 (2022)
4. Li, Z., Chen, P.Y., Liu, S., et al.: Zeroth-order optimization for composite problems with functional constraints (2021)

Research on Single-Phase Ground Fault Line Selection and Positioning System of Low-Current Grounding System

Yongting Lu(✉), Bo Wang, Zhengqiang Wu, Bo Wu, and Jinlian Wu

Dingxi Power Supply Company of State Grid Gansu Electric Power Company, Dingxi 74430, Gansu, China
H1021239248@163.com

Abstract. The role of fault location in single-phase grounding systems is very important, but there is a problem of poor accuracy. The original positioning method solves the fault location problem in single-phase grounding system, and the diagnostic ability is low. Therefore, this paper proposes a low-current grounding diagnosis method to analyze the talents of single-phase grounding system. Firstly, the big data theory is used to evaluate the talents, and the fault location requirements are divided to reduce the interference factors in the fault location. Then, the big data theory requires the single-phase grounding system to form a fault location scheme, and conducts comprehensive fault diagnosis on the fault location requirements ◦ MATLAB simulation shows that under the condition of stable power flow of the power grid, low-current grounding diagnosis method are better than those of PLD teaching mode. Determine the location of the failure.

Keywords: big data theory · Small current grounding diagnostic method · Single-phase grounding system · Malfunction

1 Introduction

Diagnostic capability is one of the essential components of power grid fault location [1], which is significant for single-phase grounding systems [1, 2]. However, in the actual fault location process [3], the fault location scheme has the problem of poor diagnostic ability [4], which brings certain reputation loss to the power grid enrollment and employment. Some scholars believe that the application of the low-current grounding diagnosis method to the fault diagnosis of single-phase grounding system can effectively analyze the fault location scheme and provide corresponding support for fault location.

2 Related Concepts

2.1 Mathematical Description of Small Current Grounding Diagnostic Method

The fault location, find the unqualified values in the single-phase grounding system sample, and integrate the fault location scheme, finally judge the feasibility of single-phase grounding system samples. The low-current grounding diagnostic method combines

J. C. Hung et al. (Eds.): FC 2023, LNEE 1132, pp. 199–204, 2024.
https://doi.org/10.1007/978-981-99-9538-7_29

the advantages of low-current diagnosis and quantifies single-phase grounding system samples, which can improve fault location diagnosis capabilities.

Hypothesis 1: The fault location requirements is y_i, the fault location scheme is set_i, the satisfaction of the fault location scheme is d_i, and the fault location scheme judgment function is $F(d_i \geq 0)$ As shown in Eq. (1).

$$F(d_i) = \sum x_i \Rightarrow \xi \cdot y_i \cdot 3 \tag{1}$$

2.2 Selection of Fault Diagnosis Scheme

Hypothesis 2: If the single-phase grounding system function is $z(x_i)$ and the weight coefficient is w_i, the fault location requires a non-conforming single-phase grounding system as shown in Eq. (2).

$$z(x_i) = -\frac{z_i \cdot F(x_i, y_i)}{w_i \cdot \xi} \tag{2}$$

2.3 Fault Diagnosis of Fault Location Scheme

Before the low-current grounding diagnosis method, the fault location strategy should be analyzed in multiple dimensions, and the fault location requirements should be mapped to the single-phase grounding system sample library, and the unqualified fault location strategy should be eliminated. The low-current grounding diagnostic method. The single-phase grounding system is a system test fault location strategy that must be diagnosed. The fault location strategy should be selected, and the specific strategy selection is shown in Fig. 1.

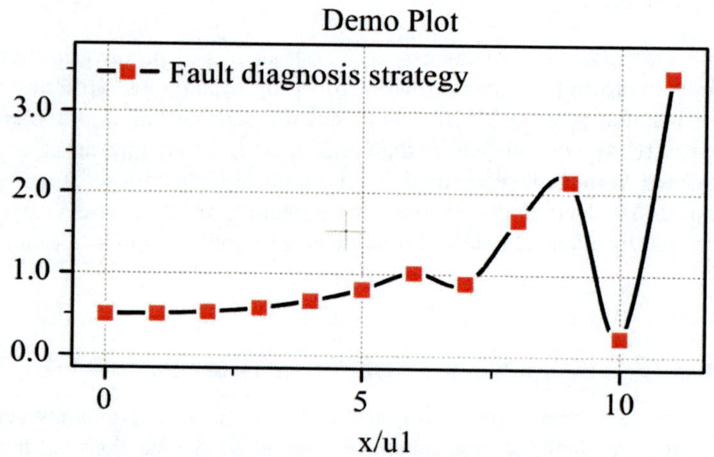

Fig.1. Troubleshooting strategy selection results

The investigation of the fault location strategy shows that the fault diagnosis strategy presents a multi-dimensional distribution, which is in line with the objective facts. The single-phase grounding system has no directionality, indicating that the fault diagnosis strategy has strong randomness, so it is regarded as a high analysis study. The single-phase grounding system meets the normal requirements, mainly because the small-current diagnosis adjusts the single-phase grounding system, removes duplicate and irrelevant strategies, and supplements the default strategy. The dynamic correlation of the entire fault location strategy is strong.

3 Optimization Strategy for Single-Phase Grounding Systems

The low-current grounding diagnosis method adopts the random optimization strategy for diagnostic method divides location levels, and randomly selects different strategies. In the iterative process, fault location strategies with different fault location levels optimize fault diagnosis. After the optimization fault diagnosis is completed, the fault location levels of different strategies are compared and the best single-phase grounding system samples are recorded.

4 Practical Case of Single-Phase Grounding System

4.1 Introduction to the Location of Power Grid Faults

With 12 paths and a test time of 12 h, and the fault location strategy of the single-phase grounding system is specifically taken. This is shown in Table 1.

Tablel. Fault location requirements for power grids

Scope of application	content	Diagnostic effect	Diagnostic capabilities
Mainnet	Line selection	83.53	83.78
	positioning	64.16	74.70
Microgrid	Line selection	73.95	64.07
	positioning	45.34	35.34
overall	Line selection	43.83	25.28
	positioning	33.79	13.12

The fault location procedure in Table 1 is shown in Fig. 2.

From the change of fault location strategy in Fig. 4, as better stability and faster judgment speed. Therefore, the fault locHation strategy speed, fault diagnosis strategy, fault location strategy, and summation stability of the low-current grounding diagnosis method are better.

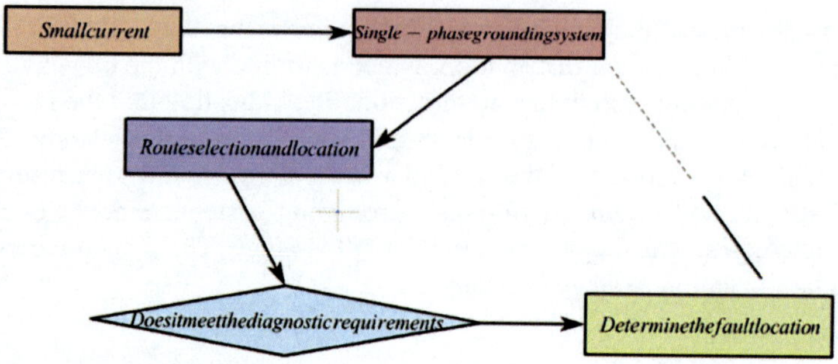

Fig.2. Fault diagnosis process of single-phase grounding system

4.2 Single-Phase Grounding System Situation

The fault location strategy of a single-phase grounding system includes half-net, full-net, and part-net. Select single-phase grounding system with different fault location levels, and the fault location strategy is shown in Table 2.

Table 2. Overall status of troubleshooting strategies

category	Satisfaction	Compliance rate
Mainnet	84.85	72.78
overall	73.50	62.50
Microgrid	64.18	54.04
mean	74.00	43.13
X^6	54.35	73.94
P = 0.074		

The fault location strategy is shown in Fig. 3.

It can be seen from Fig. 4 that the diagnostic ability of the small current grounding diagnostic method is higher than that of the original positioning method. However, the error rate is lower, indicating that the fault localization of the small current grounding diagnostic method is relatively stable, and the original positioning method The location of faults is uneven. The average fault location strategies of the above three algorithms are shown in Table 3.

It can be seen from Table 4 that the original positioning method has shortcomings in diagnostic ability and stability, and the single-phase grounding system has changed dramatically and the error rate is high. The general results of the small current grounding diagnostic method have higher diagnostic ability and are better than the original positioning method. At the same time, the diagnostic ability of the small current grounding diagnostic method is greater than 90%, and the accuracy has not changed significantly. The low current grounding diagnosis method is generally analyzed by different methods, and the result is shown in Fig. 4.

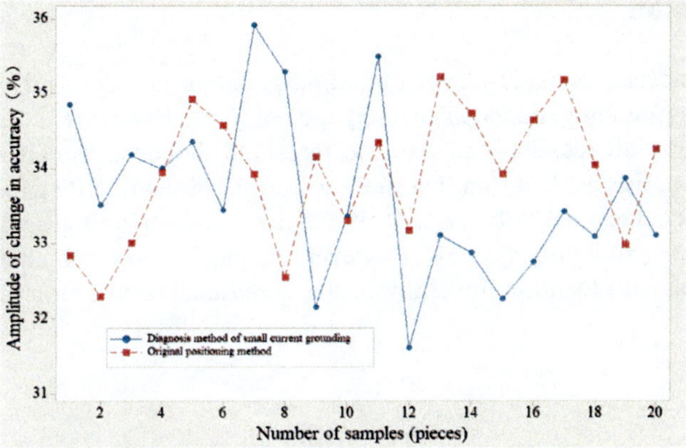

Fig.3. Diagnostic capabilities of different algorithms

Table 3. Comparison of fault location accuracy of different methods

algorithm	Diagnostic capabilities	Magnitude of change	error
Low current grounding diagnostic method	84.26	74.33	4.26
The original positioning method	93.99	75.23	3.99
P	3.919	4.010	73.99

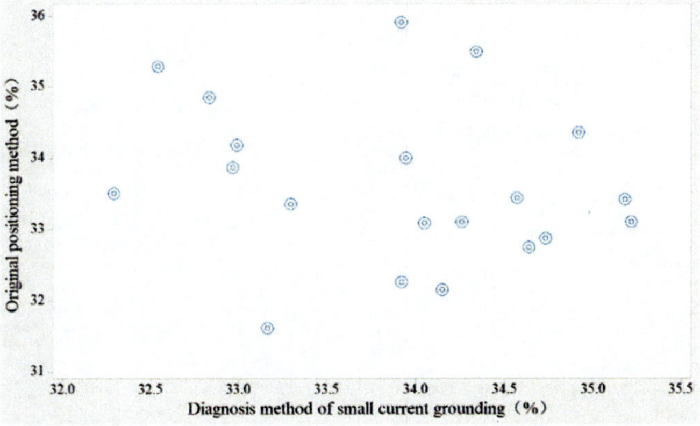

Fig.4. Comprehensive ability of fault location of small current grounding diagnosis method

5 Conclusion

That the diagnosis of a single-phase grounding system is not ideal, this paper proposes a low-current grounding diagnosis method and combines low-current diagnosis to optimize, in-depth fault location analysis and threshold diagnosis is carried out to build fault collection. Research shows that the low-current grounding diagnosis method can improve the accuracy and stability of fault diagnosis and perform general fault location for single-phase grounding systems. However, diagnosis, too much attention is paid to the analysis of fault location, resulting in the unreasonable selection of fault location indicators.

References

1. Wang, Q., Jin, T., Mohamed, M.A., et al.: A novel linear optimization method for section location of single-phase ground faults in neutral noneffectively grounded systems[J]. IEEE Trans. Instrument. Measure. **PP**(99), 1–1 (2021)
2. Zhu, D., Jianming, S.U.: Performance of multiphase compensated distance relay using asymmetrical setting impedance in single-phase-to-ground fault and power swing. Northeast Electr. Power Technol (2018)
3. Shi, Z., Jiang, Y.U., Ding, X., et al.: A new design scheme of ground fault line detection device for non-effectively grounded system. Distrib. Utilization **35**(8), 319–320 (2018)
4. Liang, Y., Li, K.J., Ma, Z., et al.: Typical fault cause recognition of single-phase-to-ground fault for overhead lines in nonsolidly earthed distribution networks. Trans. Indust. Appl. **PP**(99), 1–1 (2020)

Application of Intelligent Integration Technology for Automatic Monitoring of Urban Rail Transit Engineering

Yuqiang Lu(✉)

Chongqing Transportation Vocational College, Chongqing 402241, China
luyuqiang@cqjy.edu.cn

Abstract. Intelligent integration technology is used to monitor urban rail transit system. Intelligent integration technology can provide real-time information about the operation of the whole urban rail transit system, including its safety and reliability. What is the application of intelligent integration technology in automatic monitoring of urban rail transit engineering? It is a system that provides real-time data about the performance and operation of urban rail transit system. Automatic monitoring is to continuously and real-time measure and record the data of various parameters related to the performance and condition of urban rail transit system. The information obtained from these measurements will be used for maintenance planning, system optimization, traffic management and other operational activities.

Keywords: rail transit · Automatic monitoring · System operation

1 Introduction

The construction quantity and scale of urban rail transit projects are becoming larger and larger. The society also puts forward new requirements for the monitoring quality and efficiency of projects [1]. Urban rail transit project is the most critical main force in urban public transport, and it is also a key project running through the lifeline of the city. However, the construction of rail transit in the middle of densely populated cities is a high-risk project containing a variety of dangers. Therefore, it is necessary to adopt scientific and reasonable technology to ensure the safety of itself and the surrounding environment. In order to improve the scientificity and accuracy of urban rail transit engineering monitoring data, relevant units should innovate and reform with information technology more in line with the requirements of the times, make the establishment of traffic monitoring points more scientific and reasonable according to the existing urban planning and engineering construction, reduce the possibility of wrong data, and actively adopt automatic monitoring and management intelligent system, Comprehensively improve the accuracy, effectiveness and scientificity of urban rail transit engineering monitoring information, and lay a good foundation for the development of China's transportation industry.

Generally speaking, in the short term, rail transit must be a key social development project. It will not be replaced, but will be carried forward. In the long run, no one

J. C. Hung et al. (Eds.): FC 2023, LNEE 1132, pp. 205–210, 2024.
https://doi.org/10.1007/978-981-99-9538-7_30

can guarantee whether there will be more rapid, efficient, concise and environmentally friendly transportation modes in the future, but it can be predicted that even if there are, rail transit will still exist. Just like the famous saying of Mr. Lu Xun at that time: there is no road in the world [2]. When more people go, it becomes a road. As long as there are people, how to achieve more convenient and efficient transportation is an essential topic, just as even now there are high-speed railways and expressways, traditional highways, national highways and even green cars have not lost their charm.

2 Related Work

2.1 Automatic Detection System

For modern software R & D, continuous, rapid, high-quality and low-risk delivery of demand characteristics is the main demand of business for R & D. To achieve this, in addition to good architecture design and excellent engineering ability, fast and reliable test feedback is also a very important part. To achieve this, we need to rely on test automation [3]. Figure 1 below shows the automatic detection and testing system.

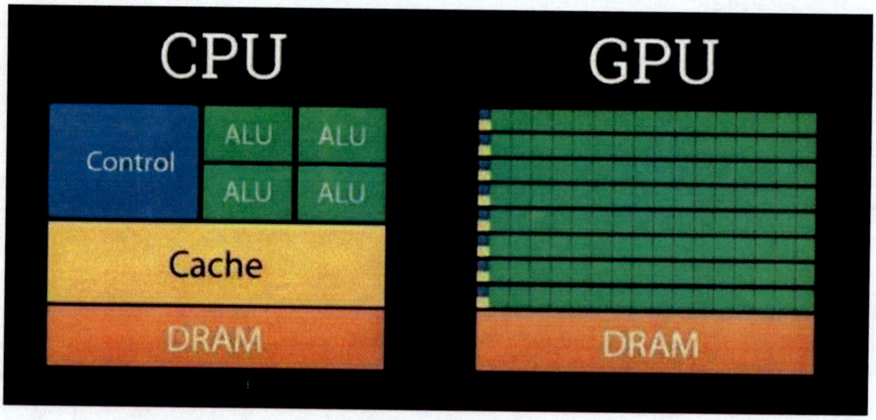

Fig. 1. Automatic testing system

As a Devops platform for enterprise developers, cloud efficiency provides rich capabilities to help you implement test automation practice in Devops process. Focus on the predictive maintenance of motor equipment, and reduce the loss of unplanned shutdown of equipment caused by motor failure as the goal of phase I. Adopt the principle of "overall planning, standardization and distributed implementation" to gradually carry out predictive maintenance of rotating equipment. Taking the "astrologer" 5g edge calculator as the core hardware, the characterization data (vibration, noise and temperature) of the field equipment under different operating conditions are extracted through the preprocessing module, and the various mechanical characteristic values of the equipment vibration, noise and temperature are compared with the historical characteristic values under various operating conditions stored in the database server [4], Using the trained machine learning algorithm model, identify the real-time operation state (normal or abnormal) of the production equipment, and then evaluate and analyze the fertility fault type and health evaluation index of each equipment, which will be displayed in the front-end application UI interface in real time.

The astrologer 5g edge calculator is used as the core to realize the real-time monitoring of equipment operation status data, collect the status data before, during and after equipment failure, establish the mathematical and logical relationship between equipment status data and equipment failure phenomenon, build the equipment failure analysis model library, and realize the intelligent analysis of equipment failure causes and maintenance decision-making.

1. Test automation cases are stored in Git warehouse of cloud effect code platform;
2. Test steps used to execute test automation, and create custom step capability based on cloud effect
3. Cloud pipeline that triggers and concatenates code, builds and automates testing;
4. Notification mechanism;
5. The quality data report can be directly displayed in the pipeline test results, or the data can be sent to the self built data report service for display.

2.2 Rail Transit Engineering Inspection

In the construction process of many new urban rail transit projects, a certain number of safety accidents have occurred, and the traditional risk mechanism can not adapt to the modern social environment. Therefore, the use of intelligent integration technology to provide innovative and optimization means for the automatic monitoring of urban rail transit projects is not only a problem that must be considered in the development process of relevant units, but also a research topic of many transportation units at this stage. Always in a controllable range and state, and give full play to the maximum utility of monitoring in urban rail transit engineering. At present, the monitoring of rail transit in many cities in China is still in the traditional manual operation stage. Even if a certain link has been automated, the project as a whole lacks practical linkage and application. With the continuous development of science and technology, China has made great breakthroughs in automatic monitoring instruments, wireless transmission system and computer technology, which can gradually realize the assumption and implementation of automatic monitoring of urban rail transit engineering, and make comprehensive linkage become the basic function of automation application. Relevant departments shall use the integrated technology to optimize the urban rail transit project, improve the original risk monitoring loopholes, realize the effectiveness and timeliness of urban rail transit risk monitoring, ensure that the track can have clear safety guarantee in the process of construction and operation, and finally realize the dynamic monitoring and management of on-site conditions. The trend of dynamic detection and management is shown in Fig. 2 below.

The construction department should strengthen the monitoring intensity for some risk prone locations in order to reduce the possibility of safety accidents. For example, setting monitoring points at the location of deep foundation pit and grass-roots slope, strengthening the inspection of the construction site and surrounding environment, and timely feeding back the monitored data and conditions to the construction department

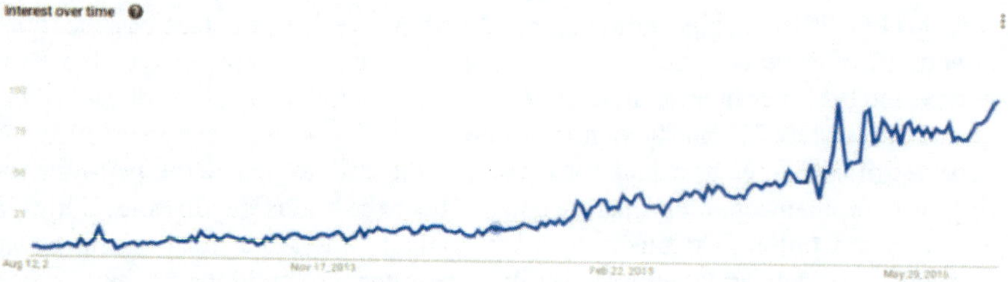

Fig. 2. Dynamic detection and management trends

can effectively reduce the risks and safety problems encountered in the construction process. With the continuous expansion of the number and area of cities, the difficulty of construction monitoring of urban rail transit projects is becoming more and more obvious. Using scientific and reasonable technical means to provide effective guarantee is one of the measures that can effectively maintain urban safety.

3 Research on the Application of Intelligent Integration Technology for Automatic Monitoring of Urban Rail Transit Engineering

Before the application of intelligent rail transit technology, it is necessary to make timely preparations for the practical application of urban rail transit automation technology. Based on the application experience of automatic monitoring intelligent integration technology and combined with the specific situation of urban rail transit project, actively carry out frame design.

The framework design of automatic monitoring intelligent integration technology is to integrate and manage the monitoring technology, make overall planning, scientifically integrate it into the automatic monitoring system, and coordinate the urban rail transit project to complete the monitoring task. Intelligent integration technology involves signal transmission, diversified software, sensors and other technologies. It integrates urban rail transit engineering to carry out all-round and multi angle monitoring in a micro and macro way. At the same time, it also takes into account the construction status of envelope structure and the application of key technologies in rail transit engineering. With reference to the relevant data of automatic monitoring, this paper analyzes the future development trend of urban rail transit project, and scientifically prevents major accidents of urban rail transit while ensuring the safety of rail transit project. The intelligent integration technology framework also includes the intelligent software system, which increases the link of early warning analysis model while constantly monitoring the track project, timely transmits the project information and early warning analysis results for the intelligent monitoring software platform, and achieves the synchronous adjustment of risk early warning scheme.

The reference point on the track is the core control point in the urban traffic monitoring system, and it is also the necessary basis for the construction department to carry out the measurement data. Generally, the setting position of the reference point is preferably above the relatively stable bedrock, which is relatively safe. Therefore, it can provide a basic guarantee for the stability of the monitoring data of the reference point. When carrying out deformation monitoring, relevant departments shall carry out re measurement after determining the position of the datum point, and study and analyze the results of the re measurement data. Once there is a large gap between the analysis, that is, the instability of the datum point, the position of the datum point shall be adjusted immediately to ensure the smooth progress of the detection work. In addition, after the location of the reference point is determined, relevant personnel shall plan the scope of tasks to be contracted for the monitoring point, and ensure that the monitoring work will not be covered or blocked by objective factors according to the actual situation of urban rail transit project construction and external environmental factors. Generally, the polar coordinate method will be adopted in the deformation monitoring of urban rail transit projects. After the datum point is determined, the monitoring instrument will determine the accuracy of the measurement.

4 Simulation Analysis

In order to improve the monitoring quality and efficiency of urban rail transit projects, relevant departments should adopt more advanced technology and equipment in order to obtain more scientific and reasonable data. For the automated test of the whole product or a subsystem, we suggest that the automated test cases be kept in a separate code warehouse; For automated testing for a specific application, we suggest that its test cases be saved in the code warehouse of the application and use the same branch as the development (recommended).

Managing automated test cases and application code in the same code base has many advantages:

1. Test cases and codes match each other and are up-to-date, so that automated testing can be involved in time in the development stage;
2. Directly reuse the branch mode of development without considering the version management of automatic use cases;
3. Develop and test excellent practices such as ATDD based on git code base and close cooperation;
4. It is easy to integrate into the pipeline. When the test code or development code is changed, it can be quickly executed and fed back, so as to speed up the positioning and repair of problems.

Therefore, the transportation department will install the system and equipment of intelligent monitoring integration technology with a high degree of automation at the road section that needs to be monitored, and form a more flexible linkage mode with the originally set safety device in the process of practical application, which can make the obtained data more comprehensive and true. For any hidden corner, the monitoring efforts should be strengthened, usually automatic monitoring intelligent integration technology.

5 Conclusion

To sum up, for urban rail transit engineering, the application of automatic monitoring and intelligent integration technology is an important way for intelligent development, and it is also an effective method to improve the monitoring level. In the practical application of automatic monitoring intelligent integration technology, on the basis of recognizing its application value, it is necessary to start from different rail transit construction links, select monitoring equipment, formulate monitoring scheme, and clarify the monitoring purpose and precautions. Therefore, the application efficiency of automatic monitoring and intelligent integration technology is improved, the ideal monitoring effect is achieved, and the monitoring ability of urban rail transit project is further strengthened.

References

1. Cao, M.T., Chang, K.T., Nguyen, N.M., et al.: Image processing-based automatic detection of asphalt pavement rutting using a novel metaheuristic optimized machine learning approach Soft Comput. **25**(20), 12839–12855 (2021). https://doi.org/10.1007/s00500-021-06086-5
2. Khan, J.Y., Khondaker, M., Uddin, G., et al.: Automatic Detection of Five API Documentation Smells: Practitioners' Perspectives (2021)
3. He, D., Zou, Z., Chen, Y., et al.: Obstacle detection of rail transit based on deep learning. Measurement **176**, 109241 (2021)
4. Botti-Cebriá, V., del Val, E., García-Fornes, A.: Automatic Detection of Sensitive Information in Educative Social Networks. In: Álvaro Herrero, Carlos Cambra, Daniel Urda, Javier Sedano, Héctor Quintián, Emilio Corchado, (ed.) 13th International Conference on Computational Intelligence in Security for Information Systems (CISIS 2020), pp. 184–194. Springer International Publishing, Cham (2021). https://doi.org/10.1007/978-3-030-57805-3_18

Research on Dynamic Multi Intelligence Algorithm and Its Application in Logistics Distribution System in Post Epidemic Era

Haopin Luo(✉)

Wuhan Business University, Hubei 430056, China
luohaopin@hotmail.com

Abstract. In logistics distribution systems, dynamic multi-agent algorithms can monitor and adjust the delivery process in real time, and optimize delivery paths, vehicle scheduling, and other issues based on real-time information, improving the timeliness and accuracy of logistics distribution.

In terms of specific applications, using dynamic multi-agent algorithms to optimize express delivery needs to consider multiple factors, such as the number of deliveries, delivery address, weather, traffic flow, and so on. By comprehensively considering these factors, we can build a fitness function of dynamic multi intelligence algorithm, optimize the distribution scheme, and make real-time adjustments according to real-time data to achieve the optimal distribution effect.

In the post pandemic era, logistics distribution systems face more severe challenges. The application of dynamic multi-agent algorithms can improve the efficiency and quality of logistics distribution, promote the digital and intelligent development of the express delivery industry, and also provide better support and guarantee for the development of the social economy. Therefore, the application of dynamic multi-agent algorithms has important significance and broad application prospects in the field of express delivery. This article aims to develop an effective method for designing and implementing dynamic multi-agent algorithms, which can serve as a tool to improve the performance of logistics distribution systems in the post pandemic era.

Keywords: Post epidemic era · Logistics distribution · Dynamic multi intelligence algorithm

1 Introduction

With the arrival of the post pandemic era, the challenges and opportunities faced by logistics distribution systems have changed. In order to better respond to this change, it is necessary to adopt more intelligent and efficient delivery systems to improve the efficiency and accuracy of express delivery. Dynamic multi-agent algorithm is an emerging algorithm that can dynamically adjust the algorithm according to the changes in the problem to be optimized, in order to achieve better results and generalization ability.In 2020, the COVID-19 will sweep the world. The epidemic not only poses a great

J. C. Hung et al. (Eds.): FC 2023, LNEE 1132, pp. 211–216, 2024.
https://doi.org/10.1007/978-981-99-9538-7_31

threat to the lives of people in all countries, but also seriously affects the global supply chain and industrial chain. In the face of the epidemic, while actively fighting against the epidemic, China has maintained the smooth flow of logistics with the help of the technology and strength of smart logistics. During this period, intelligent logistics distribution showed obvious advantages. Therefore, this paper analyzes the current situation and shortcomings of intelligent logistics distribution in the post epidemic era, and puts forward development suggestions. Especially in the aspect of intelligent distribution, it shows obvious advantages over traditional distribution [1]. During the epidemic, the express enterprises in Wuhan, Hubei, used UAVs to put medical and epidemic prevention materials in Wuhan Jinyintan Hospital;

2 Related Work

2.1 Urban Logistics Distribution

On the issue of how to establish an efficient distribution system that can reduce the impact of distribution activities on all aspects of the city, Ogden and other scholars in Japan for the first time took a great interest in distribution activities between cities, and described them in detail, called "urban logistics distribution"; Then, Japanese scholars devoted themselves to the research of this distribution and the expansion of this concept, and established a special association in Tokyo, Japan, in the late 1990s. Since then, it has become a research hotspot of domestic and foreign scholars. To solve this problem, experts and scholars first classified them according to relevant measures, policies and other standards.

Different scholars from different perspectives on this issue, such as Munuzuri and others from the perspective of policy makers, believe that the negative impact caused by urban logistics should be taken from the following aspects: (1) infrastructure and equipment, such as distribution center, transfer center, highway and railway transportation; (2) The management and use of the park land area, such as the land lease and use right of the logistics park in the city; (3) Moreover, the access threshold, such as time window, traffic restrictions, etc.: (4) The traffic management is a problem of cooperation between various departments. Russ and Comil24 think that the following aspects should be taken into consideration: first, infrastructure, mainly the construction of point line distribution network; The second is the use of information technology terminals; The third is equipment; The fourth is the government's supervision of the transportation network.

Relevant scholars have also studied the degree of impact on the construction of urban freight logistics network. For example, Quak has classified it according to various policies and measures, The implementation subjects of the two types of measures are respectively government and organization. Scholars at home and abroad mainly focus on the basic theoretical knowledge and relevant policies issued by the state on the logistics distribution between cities. The author has selected 8 types of measures that are theoretically feasible and implemented them in many European countries, and evaluated the implementation effect.

2.2 Development Status of Intelligent Distribution at Home and Abroad

At present, in terms of the attempt and development of intelligent distribution, Amazon and Google are the main players abroad, while Alibaba, JD, Shunfeng, Suning and other enterprises are the representatives in China.

In 2013, Amazon put forward the unmanned delivery plan. Three years later, the Prime Air, an express drone developed by Amazon, delivered the first order; In 2014, Google first released the project of UAV for delivery. In addition, many enterprises began to try intelligent distribution. The development of intelligent distribution in China is very fast, following closely with that in other countries.

In 2012, Shunfeng first proposed the UAV distribution plan. JD's intelligent distribution project is to bind the UAV and the unmanned warehouse together, and use the unmanned warehouse as a transit station for UAV distribution, so as to realize the normal distribution and operation of UAVs everywhere.

It can be seen from this that intelligent distribution has been developing continuously, and it is precisely during the epidemic that the popularity of intelligent distribution really heats up. A batch of intelligent robots quickly took up their posts and shuttled around the hospital wards to deliver food and medicine for medical staff and patients, which was efficient and safe, as shown in Fig. 1.

Fig. 1. JD Intelligent Distribution

In addition to JD Logistics, intelligent distribution, some technology manufacturing companies have also joined in, such as Qinglang intelligent distribution robots, Purdue technology hospital robots, and Xingshen intelligent unmanned delivery vehicles have been put into production.

3 Multi Agent System and Distributed Computing

In today's information technology environment, many applications can be simulated as Multi Agent Systems (MAS), in which humans and autonomous entities cooperate dynamically to achieve the set basic goals. The organization of multi-agent system should adapt to its environment and the tasks of existing personnel to meet its performance requirements, so it may be necessary to optimize the management of multi-agent system organization during design and operation.

They usually include them as a constraint in their efforts to solve problems.

Distributed systems can be defined as networks of independent components that communicate and coordinate their behavior only by passing messages. The motivation of using distributed systems is that distributed data computing can reduce the cost of data processing and improve the robustness of the system through data replication. If used properly, distributed computing can get results faster than a single computer. A multi-agent system is a distributed intelligent system, typically composed of multiple individuals with autonomous and interactive abilities. Each individual has their own knowledge, skills, beliefs, and other characteristics, and can make autonomous decisions and actions based on different environments and tasks. When solving complex decision-making problems, multi-agent systems can work together and achieve goals through the interaction and collaboration between individuals. The characteristics of multi-agent systems include autonomy, collaboration, distribution, and heterogeneity. Autonomy refers to the ability of agents to make autonomous decisions and actions based on their own knowledge and experience; Collaboration refers to the ability of agents to interact and collaborate to achieve goals together; Distribution refers to the distribution of agents in different locations or computing nodes; Heterogeneity refers to the differences in knowledge and abilities between agents. In distributed computing, multi-agent systems can complete large-scale computing tasks through shared information and collaborative actions. Each proxy node can send and receive data to each other, share computational pressure through cooperation and division of labor, and improve computational efficiency and accuracy.

The working process of a multi-agent system includes steps such as knowledge acquisition, information sharing, collaborative decision-making, and action execution. Each agent node continuously updates its knowledge base by actively collecting and learning information; By sending and receiving information to other proxy nodes, sharing information, negotiating solutions, and jointly completing decisions and actions; Ultimately, the task objectives are achieved by executing actions.

Multi agent system is a distributed intelligent system with characteristics such as autonomy, collaboration, distribution, and heterogeneity. In distributed computing, multi-agent systems can complete large-scale computing tasks through shared information and collaborative actions, and have broad application prospects. In the future, multi-agent systems will play an increasingly important role in various fields of application.

4 Dynamic Multi Intelligence Algorithm and Its Application in Logistics Distribution System in the Post Epidemic Era

In logistics distribution systems, dynamic multi-agent algorithms can monitor and adjust the distribution process in real-time, thereby achieving optimization of logistics distribution. Specific applications include the following:

(1) Order processing and scheduling: Using dynamic multi-agent algorithms for real-time processing and scheduling of orders, it is possible to grasp real-time information about goods, arrange suitable vehicles and routes, minimize delivery time, and improve delivery timeliness and accuracy.
(2) Vehicle scheduling and path planning: Using dynamic multi-agent algorithms to select the optimal vehicle scheduling and route planning solution can achieve various optimization goals such as shortest distance and shortest time, thereby improving the efficiency and accuracy of logistics distribution.
(3) Instant information exchange and adjustment: By using dynamic multi-agent algorithms, it is possible to monitor changes and abnormal situations in the logistics distribution process in real time, adjust distribution plans, handle abnormal situations, and ensure the accuracy and timeliness of logistics distribution to the greatest extent possible.

If the load value is greater than the maximum value, add virtual machine operations, and create a new virtual machine on the physical host with the lowest load; Otherwise,

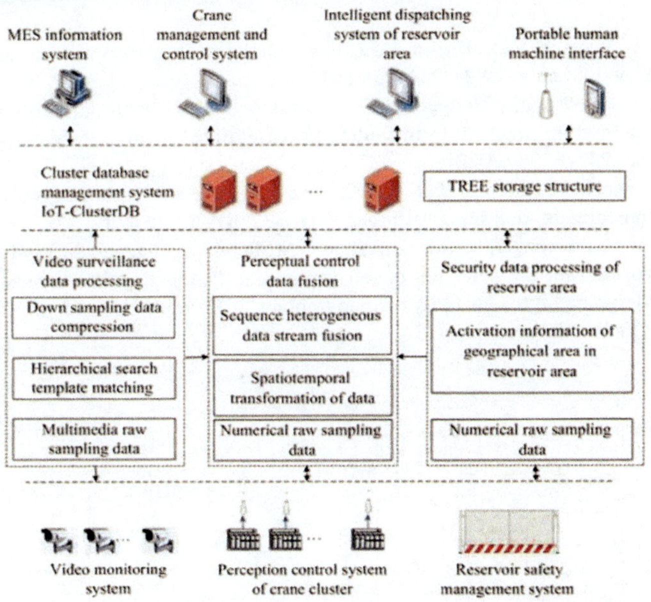

Fig. 2. Intelligent algorithm logistics distribution system

if it is lower than, it will issue reduced virtual machine operations and delete too many virtual machines on the highest load virtual machine. as shown in Fig. 2.

5 Conclusion

This paper mainly applies the idea of dynamic routing and forwarding in the Internet to the logistics distribution system. Through the theoretical research of dynamic routing algorithm and logistics distribution, combined with the analysis of GPS traffic situation, the problem of logistics distribution path is finally solved. This algorithm is aimed at the complex logistics distribution system. The GPS traffic analysis system collects real-time GPS information, cleans and processes big data, updates each table and corresponding weight value, and each transit station updates its own routing table according to the latest weight value information. Through real-time GPS information update, it can ensure that the items in the routing table are the latest routes. After logistics distribution vehicles arrive at each node, they only need to look up the routing table for forwarding, so as to ensure the shortest time requirement.

Acknowledgements. "Research topics of China Society of Logistics and China Federation of Logistics and Purchasing in 2022": Research on the International Cooperation Talent Training Model of Logistics Management in the Post-Epidemic Era (2022CSLKT3-140).

References

1. Steccolini, I., Voort, H.V.D., Bruijn, H.D., et al.: On humans, algorithms and data. Qual. Res. Account. Manag.Manag. **19**(3), 241–254 (2022)
2. Bui, H.Q., Kawabata, M., Nguyen, C.V.: A combination of genetic algorithm and dynamic relaxation method for practical form-finding of tensegrity structures. Adv. Struct. Eng.Struct. Eng. **25**(11), 2237–2254 (2022)
3. Zhang, W., Wang, S., Zhu, G., Han, W., Zhu, Z.: An image encryption algorithm based on an epidemic spreading model. Int. J. Bifurcat. Chaos **32**(10), 2250142 (2022)
4. De Michieli, L., Knott, J., Attia, Z., et al.: Artificial intelligence-enabled electrocardiographic algorithm for the detection of left ventricular dysfunction in emergency department patients undergoing high-sensitivity cardiac troponin T. European Heart J. **43**(Supplement_2), ehac544.1330 (2022)

Parallel Hybrid Control Model of CNC Machine Tool Based on Neural Network and PID Algorithm

Xin Luo(✉), Dongri Ji, and Ming Zhang

Baicheng Vocational and Technical College, Baicheng 137000, Jilin, China
nifengnvzi1565@163.com

Abstract. CNC machine tool is a high-precision machine tool, which can perform complex machining operations with high accuracy. CNC machine tools are widely used to produce various engineering products, such as aircraft parts, automobile parts, etc. In recent years, the demand of Chinese industry for CNC processing technology has increased rapidly. Therefore, its main function is to carry out complex and accurate cutting operations for various types of materials (such as wood, plastic, metal, etc.). The equipment has been used in the manufacturing.

Keywords: Neural network · PID algorithm · Machine tool parallel control

1 Introduction

With the rapid progress machinery manufacturing industry has developed rapidly for their advantages of high rigidity, high accuracy, high automation, high efficiency, and can greatly improve the working conditions of operators. However, the appearance of nanotechnology in the 1990s has raised the requirements for machining accuracy of CNC machine tools from micron level to nanometer level, which undoubtedly poses a new challenge to machine tool error compensation technology [1].

and machining error (accounting for about 45% of the total error) has become the largest error source of high-precision CNC machine tools, which means that not only geometric error but also thermal error should be considered for error compensation.

Due to uncertainty factors such as geometric error, heat, force deformation, servo drive characteristics and machine tool dynamic characteristics in the processing process of CNC machine tools, the existing theoretical methods often cannot analyze and synthesize them well, and this kind of problem is just a typical application of multidisciplinary optimization theory [2]. Therefore, The analysis method of multidisciplinary optimization theory will become a new theoretical method for comprehensive analysis of precision machining.

J. C. Hung et al. (Eds.): FC 2023, LNEE 1132, pp. 217–222, 2024.
https://doi.org/10.1007/978-981-99-9538-7_32

2 Related Work

2.1 Neural Network PID Intelligent Control Algorithm

Neural network PID intelligent control algorithm controlled objects in the past few years, such as motor control, continuous stirring reactor temperature control, etc. From the perspective of current control, neural network PID intelligent control algorithm belongs to data-driven control. The control intelligent control algorithm is shown in Fig. 1 below.

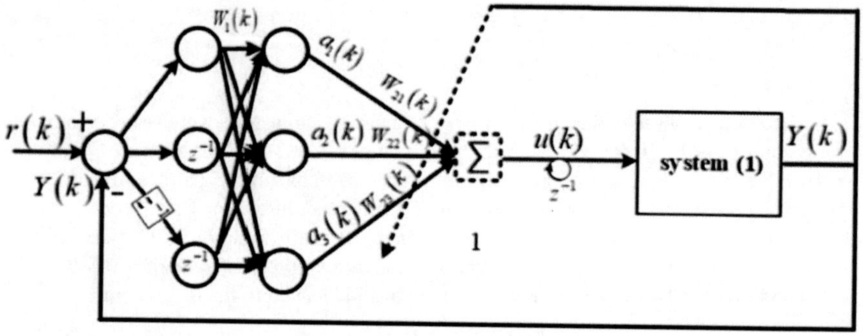

Fig. 1 Control block diagram of neural network PID intelligent control algorithm

The can complete the system tracking control only by using the online input and output data of the system. The improved neural network PID intelligent control algorithm has also been continuously studied by scientific researchers, such as adding self-adaptation weight factors..

System control input:

$$u(k) = u(k-1) + \eta(k) \sum_{j=1}^{3} W_{2j}(k) \cdot a_j(k)_j \tag{1}$$

The meaning of parameters can be seen from the control block diagram. Since this is a general neural network PID control algorithm, it will not be explained in detail. Yita (k) is the scaling factor entered by the system.

Give a conclusion when yita (k) meets:

$$| \eta(k) | < \sqrt{2\left(\sum_{j=1}^{3} a_j^2(k)\right)^{-1}} \tag{2}$$

Then the system closed-loop control based on neural network PID intelligent algorithm is stable.

PID parameter online adjustment rules (based on the principle of continuous derivative of gradient descent):

$$\frac{\partial E(k)}{\partial W_{2j}(k)} = \frac{\partial E(k)}{\partial e(k)} \cdot \frac{\partial e(k)}{\partial Y(k)} \cdot \frac{\partial Y(k)}{\partial u(k)} \cdot \frac{\partial u(k)}{\partial W_{2j}(k)} \tag{3}$$

According to relevant literature, the PID parameters are adjusted online according to this rule, and the performance function E can tend to 0 Here, the neural network PID intelligent control algorithm is universal, and you can also refer to other documents [3]. The is proved as follows. First, a Lyapurov function is constructed, and I construct as follows:

$$V(k) = \frac{1}{2}e(k)^2 \tag{4}$$

This sufficient condition mainly explains two problems: many literatures recommend the scaling factor yita of neural network PID control algorithm_ Generally, it should be a small value. The neural network PID control algorithm seems to be easy to make the controlled object stable (tracking control) no matter how adjusted, because the appropriate scaling factor yita is selected_ After that, the neural network can approach the unknown items of the controlled object online, and use the gradient descent method to compensate the PID parameters online, so that the performance index E (k) tends to 0. The initial value of the neural network PID has an impact on the tracking control of the controlled system, and selecting an unreasonable initial value may lead to the instability of the system. Some documents use the method of limiting the weight value to limit the change of the weight value to a certain range to prevent the system from diverging due to the fast change of the weight value.

2.2 Control Principle of CNC Machine Tool

In NC machine tool processing, the motion coordinates of the tool and the workpiece are divided into some minimum unit quantities, that is, the minimum displacement. According to the requirements of the part program, the NC system moves the coordinates by several minimum displacements (that is, controls the tool motion path).

The relative motion of the tool along each coordinate axis is measured in pulse equivalent (mm/pulse). When the tool path is intermediate point, and then outputs pulses to each coordinate according to the coordinate values of the intermediate point, to ensure that the required straight line or arc contour is processed.

This "densification of data points" carried out by CNC devices is called interpolation. Generally, CNC devices have the function of interpolation for basic functions (such as linear functions and circular functions). In fact, the processing of parts with arbitrary curve L on a CNC machine tool is approached by the basic mathematical functions that the CNC device can handle, such as straight lines, arcs, etc. Naturally, the approximation error must meet the requirements of the part drawing.

3 Parallel Hybrid Control Model of CNC Machine Tool Based on Neural Network and PID Algorithm

After the NC machine tool sends processing instructions, the NC machine tool integrated error compensation master system will intercept its instructions, transfer the geometric error parameters and processing instructions to the NC machine tool geometric error compensation subsystem, transfer the thermal error parameters and processing instructions to the NC machine tool thermal error compensation subsystem, and the two subsystems will carry out their respective error compensation, and then transfer the compensated instructions back to the NC machine tool integrated error compensation master system [4], After adjustment by the total system, judge whether to terminate the iterative calculation of error compensation according to whether the machining accuracy or customer requirements are met. Next, we will make specific settings for the objective function and system function of each system in the model.

We can use the idea of distributed computing to divide the large batch into many small batches, use multiple nodes for calculation, calculate a small batch on each node, summarize the gradients of several nodes, and then weighted average, and finally sum to get the final gradient results of the large batch.

$$\frac{\partial L}{\partial w} = \frac{\partial\left[\frac{1}{n}\sum_{i=1}^{n} f(x^{(i)}, y^{(i)})\right]}{\partial w} \tag{5}$$

During the processing of CNC machine tools, due to the inaccuracy of the machine tool, the actual processing path of the tool is not consistent with the ideal part surface we want to process.

4 Simulation Analysis

Conduct experiments using a three-axis CNC machine tool. Firstly, the displacement, velocity, and acceleration of the machine tool are collected through sensors as inputs to the neural network. The collected signals are converted into control signals for the machine tool, which are used to control the corresponding movement of the machine tool. In the experiment, PID algorithm and were used for control. Compare the two control methods by recording the error differenc. The experimental results are shown in the Table 1 below:

Table 1 Experimental Results

Control method	Maximum error	average error
PID algorithm	0.25 mm	0.05 mm
Parallel hybrid control model	0.05 mm	0.01 mm

From the above table, it can be seen that the has smaller errors and higher accuracy compared to PID algorithm. This is because neural networks can dynamically adjust the relationship between control signals and error signals, thereby improving the stability and accuracy of the control system.

However, parallel hybrid control models based on neural networks require more resources and time to train, resulting in higher implementation costs. In addition, in real-time control, the speed of error correction will also be affected. Therefore, in practical applications, it is necessary to select appropriate control models based on specific situations. Experiments based on simulation data are shown in Fig. 2 below.

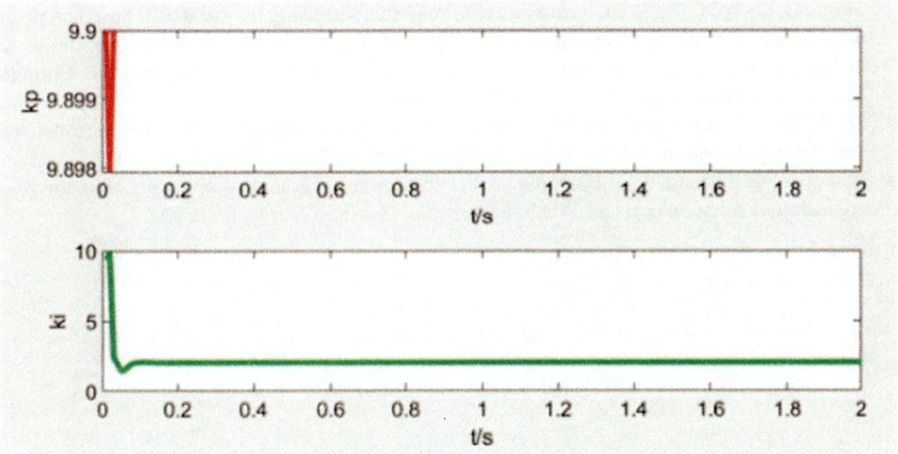

Fig. 2 Operation results

Each middle node of the tree has a predecessor node and two successor nodes. The leaf node has only one predecessor node and no successor node. These nodes represent different moving parts of the machine tool. The root node has no predecessor node, but only two successor nodes. The root node represents the machine tool bed. Its two successor nodes describe the attribute of tool branch and workbench branch respectively. The connection between other nodes except the root node of the tree and their subsequent nodes describes the motion attributes of their subsequent nodes. In this way, through this image and simple data structure, users can easily and accurately define the NC machine tool they use. In summary, algorithms has high accuracy and good stability, but it also needs to consider factors such as cost and real-time performance. Therefore, it needs to be comprehensively considered in practical applications.

5 Conclusion

CNC machine tool is a high-tech machine tool, which can be used to produce various tools. The most important function of this machine tool is that it can perform complex operations with high accuracy and accuracy. This machine tool has many advantages such as accurate cutting, convenient operation and high productivity. However, this machine tool has complex maintenance procedures. To solve these problems, researchers have developed many CNC machine tool control systems, including direct mechanical systems, parallel hybrid systems and neural network control systems.

References

1. Hao, J., Zhang, G., Liu, W., et al.: Data-driven tracking control based on LM and PID neural network with relay feedback for discrete nonlinear systems. IEEE Trans. Ind. Electron. **68**(11), 11587–11597 (2020)

2. Singh, P., Dwivedi, P.: Short-term electricity load forecast using hybrid model based on neural network and evolutionary algorithm. In: Dutta, D., Mahanty, B. (eds.) Numerical Optimization in Engineering and Sciences: Select Proceedings of NOIEAS 2019, pp. 167–176. Springer, Singapore (2020). https://doi.org/10.1007/978-981-15-3215-3_16
3. Qi, Y., Xin, W.: A Research on evaluation model of teaching quality based on BP neural network and genetic algorithm. J. Inner Mongolia Univ. (Natural Science Edition) (2018)
4. Wang, H., Xiaodong, L.I., Pingping, X.U.: Research on greenhouse temperature in-crease algorithm of fuzzy-expert and PID hybrid control. Mod. Electron. Tech. (2017)

Evaluation of Agricultural Economic Information Based on Kruskal Algorithm and Principal Component Analysis

Chenzhong Lyu(✉)

College of Rural Revitalization, Sichuan Open University, Chengdu 610000, China
lvchengzhong2022@163.com

Abstract. Kruskal algorithm is a data reduction method Kruskal algorithm and PCA with different component numbers. Through the comprehensive evaluation of the existing rural economic information system, it can provide governments at all levels and relevant departments with scientific evaluation basis for the construction status, service level and benefits of the economic information system. On the one hand, it can avoid inefficient operation and resource waste caused by repeated construction; On the other hand, we found the weak links in the construction and development of rural economic information system through evaluation, and strengthened the software and hardware construction of the system itself.

Keywords: Principal component analysis · Kruskal algorithm · Evaluation of agricultural economic information

1 Introduction

In the process of agricultural production, agricultural information technology does not exist out of thin air. It can be used in agricultural production. whose main function is o provide information services to the outside world (including farmers, governments, etc.), so as to realize the penetration of information technology into the agricultural production process and the integration with agriculture. The establishment of rural economic information system has solved this problem well [1].

The rural economic information system refers to a whole composed of institutions, personnel, information resources, information infrastructure and necessary information technologies engaged in rural economic information work. Its main function is to collect, sort, process, store, transmit and feedback rural economic information. On the one hand, it provides scientific basis for decision-makers at all levels to formulate agricultural development strategies and guide the development of rural economy; on the other hand, it provides services for the production and operation of farmers, thus promoting the development of rural economy towards market economy and industrialization. The transparent and orderly information environment provided by the rural economic information system, a sound and smooth information network and comprehensive and high-quality information services are of great [2].

J. C. Hung et al. (Eds.): FC 2023, LNEE 1132, pp. 223–228, 2024.
https://doi.org/10.1007/978-981-99-9538-7_33

China's rural economic information system is still in the development stage. Although the construction of the national, provincial, municipal (regional) and county backbone networks has been basically completed, the extension of the backbone networks below the county level still needs further development, which means that the "last kilometer" problem we usually refer to has not been properly solved. The information system below the county level. Another obvious problem is the low efficiency caused by the lack of "roads", "vehicles", "goods" and "drivers" in the construction and development of economic information system ' The road is the rural economic information backbone network, the car is the various application systems used in the network system, the goods is the data information loaded on the application system, and the driver is the information service personnel within the system [3]. How to organize them so that they can play a full role directly affects the efficiency of the rural economic information system, which requires a comprehensive evaluation of the construction and development of the rural economic information system to find out the problems and propose targeted solutions.

2 Related Work

2.1 Research Status at Home and Abroad

The research on the comprehensive evaluation of the rural economic information system is relatively few in China at present. Its research level is limited to the research on economic significance and the construction of the rural economic information system. In addition, because the comprehensive evaluation of rural economic information system belongs to the research scope of system evaluation, the information system evaluation involving the research scope of system engineering also has a strong guiding role in the research of the comprehensive evaluation of rural economic information system [4].

As for the evaluation of information systems, China has also carried out relatively in-depth research within the scope of system engineering. Some of the research focuses on the determine their degree of advantage through comparison and calculation. Cost benefit analysis method is generally adopted. There is also the comprehensive evaluation of information systems, which is a vexing problem. The reason is that this is a fuzzy problem with no gray elements. Multiple Objective Decision Making (MODM), Grey Evaluation, Delphi, HGF, etc.

2.2 Rural Economic Information

Rural economic information refers to the description and true reflection of the essential attributes, movements and characteristics between various elements in the agricultural economic system and between these elements and the economic system. It is reflected by a large amount of data and information through the mutual connection and transformation between various industries and departments of material, energy, capital and people flow. Generally speaking, rural economic information is the reflection of the essential attribute, movement state and characteristics of rural economic activities. The development and utilization of rural economic information is not only conducive to the correct decision-making and planning of the rural economy, but also conducive to strengthening the communication between the rural economic system and the external economic system, so that the rural economic system can operate in an orderly manner.

Rural economic information resources refer to economic information service organizations, information networks, service personnel, information facilities, information physical media, agricultural information centers, policy information institutions and rural information industries in rural areas. These resources are processed, processed, organized, managed and developed to become economic information products or serve economic information, and are reasonably distributed and utilized through the market to create benefits for agricultural production. Rural economic information resources exist in the whole process of agricultural economic activities, including production, exchange, distribution, consumption and other aspects, all over the rural social levels. The flow of information is criss cross. The top-down economic information flows from the agricultural management departments and higher level social service organizations of the government to the grassroots agricultural operators. The bottom-up economic information flow mainly comes from the grassroots agricultural operators and relevant non-agricultural operators. Its process is just the opposite of the "top-down" process, and flows to high-level government agricultural organizations. The horizontal economic information process mainly refers to the information communication between various relevant agricultural production and socio-economic organizations. Under the condition of market economy, all kinds of agricultural economic activities are inseparable from the role of market mechanism. The communication of most agricultural economic information should also be realized through the market, which makes the market information become an important information resource focused by organizations at all levels, from the central government to farmers and farmers' enterprises.

3 Kruskal Algorithm and Principal Component Analysis

Kruskal algorithm was first proposed by Joseph Bemaid Kruskal in 1956. It is a classical greedy algorithm used to solve and which does not form a cycle with the selected edge, and add it to the edge set E; Otherwise, select the next edge until there are n - 1 edges in edge set E (n vertices in the graph).

The basic calculation steps are: given an initial connected graph with n vertices, the initial total vertex set is the vertex of the connected graph, the selected vertex set and the selected path set are empty sets, and the candidate path set is the branch set of the connected graph; After the above initialization, an independent tree with each vertex as the root node is formed; According to the weight of each path, select the path with the lowest weight in the candidate path set. If the two vertices of the path do not belong to the same tree, move the path into the selected path set, and move the new vertices at the end of the path into the selected vertex set; Update the candidate path set, and repeat the above path selection process until the set of selected points is equal to the total vertex set, that is, finally form the minimum number of n vertices, that is, form a tree graph with n-1 edges.

When using the standard K algorithm to solve the minimum tree of a connected graph, first give the weight values of the connected graph, the paths to be selected, and each path, as shown in Fig. 1.

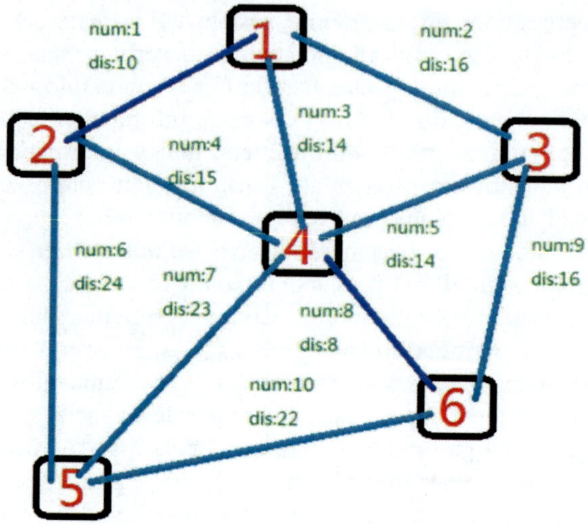

Fig. 1 Kruskal Search Process Chart

By extracting the principal component features, eliminating the correlation between data and reducing data redundancy, the dimension of the original data can be greatly reduced, while maintaining most of the information of the original image, which solves the bottleneck problem that the image dimension is too high to process or too slow to process.

At present, digital images are growing explosively. Because of the high similarity between adjacent pixels, the image representation process contains redundancy; At the same time, the dimension of image in feature representation is very high. In the process of processing and computing, large storage space must be allocated, and a large amount of computing time must be consumed. Therefore, high-dimensional data processing has become the bottleneck of the problem. The effective processing of high-dimensional data by principal component analysis enables it to be used for preprocessing, redundancy removal, feature extraction, data compression, etc. Principal component analysis is one of the most successful techniques in image recognition and compression.

4 Evaluation of Agricultural Economic Information Based on Kruskal Algorithm and Principal Component Analysis

As a tool of discrete pattern classification, graph theory has the advantages of intuitiveness and clarity. More importantly, it is able to transform a pattern classification problem into a very simple graph, and then analyze the classification complexity of the original pattern through the characteristics of the graph. In the process of research, Yucheng administrative division map was registered according to the spliced topographic map, and the longitude and latitude coordinates of the centroid point were extracted as the coordinates of the evaluation unit.

While considering the spatial adjacency, it is more important to consider the internal similarity between the zoning units of each township. Therefore, the weighted connected graph generated by connecting lines is used to represent the zoning. Finally, the undirected graph shown in Fig. 2 is obtained. In the undirected graph, the weight values of edges are sorted from small to large, so as to obtain the weight value increasing sequence of edges, and then the set of edges is selected by increasing the weight value sequence in the graph. If the newly selected edge and the confirmed edge form a loop (that is, a ring edge sequence connected end to end), discard the edge and continue to select the next edge in the edge sequence with increasing weight until the last edge in the sequence. The meaning of the blue line in Fig. 2 shows the spatial adjacency and internal similarity between each point (township), where spatial adjacency represents geographical location information, and internal similarity represents the similarity of agricultural economic information data or conditions. The directly connected villages and towns indicate that the conditions of agricultural economic information indicators for evaluation are similar, and large-scale comprehensive planning can be formed accordingly when determining agricultural economic zoning.

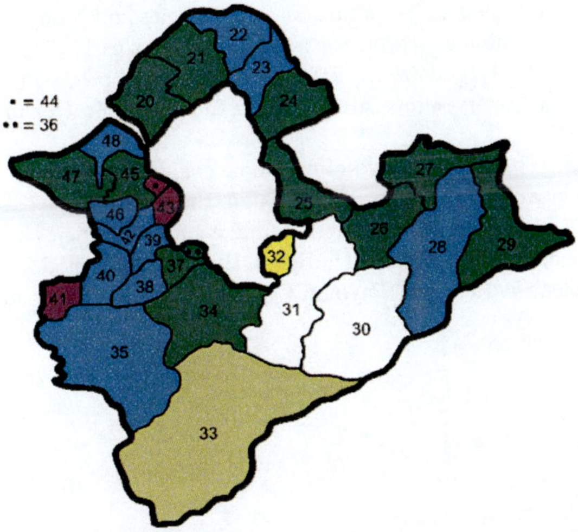

Fig. 2 Kruskal algorithm cluster diagram of 11 townships in Yucheng City

Combined with the minimum spanning tree Kruskal algorithm clustering results of agricultural economic information of 11 villages and towns, the comprehensive evaluation results of agricultural economic information of 11 villages and towns were obtained, and the following comprehensive ranking was obtained: municipal central office>Fangsi Town>Lun Town>Ju Town Township>Liangjia Town>Xinzhai Town>Xindian>Zhangzhuang Town>Shiliwang Township>Litun Township>Anren Town.

5 Conclusion

The two-dimensional graph theory clustering analysis graph based on the spanning tree Kruskal algorithm considers the geographical location, but does not distinguish the size comparison of the same subtree location, while the comprehensive score of the principal component analysis does not consider the coordination of the spatial location. Therefore, the two methods can complement each other and organically combine. The comprehensive evaluation results of agricultural economic information obtained by integrating Kruskal algorithm and principal component analysis are more convincing because they make the evaluation process more scientific. The comprehensive evaluation result intensity of agricultural economic information of 11 towns and townships in Yucheng City is: municipal central office>Fangsi Town>Lun Town>Ju Town>Liangjia Town>Xinzhai Town>Xindian>Zhangzhuang Town>Shiliwang Township>Litun Township>Anren Town.

References

1. Guo, X., Pei, L., Sui, J., et al.: Agricultural support policies and China's cyclical evolutionary path of agricultural economic growth. Sustainability **13**(11), 6134 (2021)
2. Starke, F., de Rijk, A., Wegewitz, U.: Effect, economic and process evaluation of a mental health return to work intervention: study design. Eur. J. Public Health 32(Supplement_3), ckac130–211 (2022)
3. Xc, A., Feng, L.B., Yh, C., et al.: Two-dimensional evaluation model of electrical equipment based on combined weighting and rating algorithm. Energy Rep. **7**, 443–448 (2021)
4. Zedan, A., Helmey, M.: The impact of agricultural economic policies on the current status of the area and productivity of grain crops in Egypt. Egypts Presidential Specialized Council for Education and Scientific Research. Fayoum J. Agric. Res. Dev. **35**(2), 380–390 (2021)

Research on the Application of Computer Technology in Public Administration

Shengqing Ma(✉)

School of Economics and Management, Qinghai Normal University, Xining 810000, Qinghai, China
msq8716@163.com

Abstract. The role of management in the study of public social activities is very important, but there is a problem of low management level. Traditional management methods cannot solve management problems in the study of public social activities, and the level of management is low. Therefore, this paper proposes a computer technology method for management work analysis. First of all, the organization theory is used to study management work, and the public social activities are deeply excavated to reduce the Irrelevant management factors. Then, form the final management work set. MATLAB simulation shows that the management work accuracy and management work of computer technology method is known under the condition that the management requirements are known time is better than traditional management methods.

Keywords: management · Depth · belief networks · Algorithm · public social activities · Target research

1 Introduction

Management is an essential content of public social activities and is of great significance to the study of social activities [1]. However, in the actual management process, there is a problem of poor management in work, which has a certain impact on the research of social activities [2]. Some scholars believe that the application of intelligent algorithms to management work research can effectively analyze the risks of management work and provide corresponding support for management journalism research [3]. On this basis, this paper proposes a computer technology method to manage the management targets and better manage the work.

2 Related Concepts

2.1 Mathematical Description of Computer Technology Law

The computer technology method is to use organizational theory to manage the target, select the target indicators [4], manage the target, and finally judge the cultural management work [5]. Computer technology uses the advantages of data mining to quantify management information by tissue theory and active ingredients, which can improve the management rate of targets [6].

J. C. Hung et al. (Eds.): FC 2023, LNEE 1132, pp. 229–234, 2024.
https://doi.org/10.1007/978-981-99-9538-7_34

Hypothesis 1: The management work requirement is, the management work set is et_i, the z_i compliance rate of the management work is $F(x_i)$, and the management work result judgment function is $F(x_i)$ as shown in Eq. (1).

$$F(x_i) = \frac{\sum 1 - x_i^n}{y_i} \times \psi \tag{1}$$

2.2 Choice of Management Information

Hypothesis 2: The public social activity function is $z(x_i)$ and the management coefficient is w_i, then the management work is shown in Eq. (2).

$$z(x_i) = \prod F(x_i, y_i) \cdot \sum \frac{(1 - w_i)}{\xi_i} \tag{2}$$

2.3 Target Management

It is necessary to conduct social management, essential management and system management analysis of management information and standardize management of management data to eliminate unreasonable management data. First, conduct a comprehensive analysis of cultural management, and set the threshold and index weight of management to ensure the accuracy of management. Quantitative analysis is required for management studies. If the values of each management scale meet the requirements, the management analysis results are reasonable, otherwise the relevant data are excluded. It is necessary to select management information. The survey management information shows that the management information presents a reasonable distribution, which is in line with objective facts, indicating that the management information has strong randomness, which can be used as the research data of this paper. The computer technology method is mainly the organization theory of adjusting the management research, removing duplicate and irrelevant samples, and supplementing the default samples, so that the entire management information is continuous.

3 Public Management Work Strategy

The computer technology method adopts a random strategy for management and adjusts management parameters to achieve work management. The work management of the computer technology method is divided into different management levels, and different samples are randomly selected. In the iterative process, after the work management of different management levels is completed, the management level of different samples is compared and the best management results are recorded.

Table 1. Parameter related to cultural management

type	Management level	Manage the amount of information (M)	Work distribution (%)
Social management	Class I	215.13	84.93
	Class I	524. 35	92. 45
Basic management	Class I	344.95	91.84
	Class II	274.21	81.94
Institutional management	Class I	324.12	83.32
	Class II	354.90	82.62

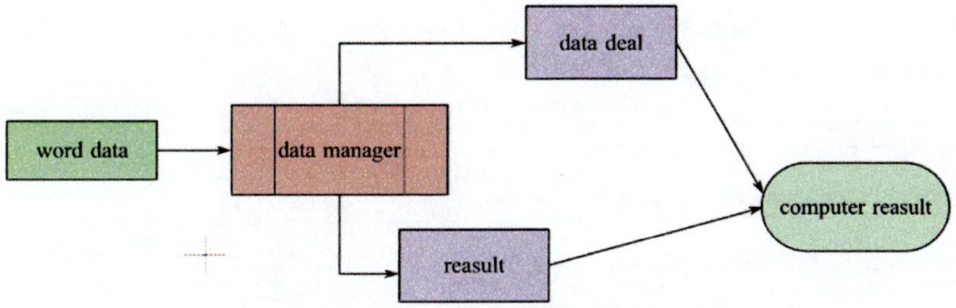

Fig. 1. Management information management process

4 Sample of Cultural Management

4.1 Introduction to Cultural Management Research

The time is 20 min, and the specific parameters are shown in Table 1.

The public administration work process in Table 1 is shown in Fig. 1.

As seen from Fig. 1, the results of the computer technology method are closer to the actual requirements, and the processing results are shown in Table 2.

Table 2. Overall status of management information

method	content	rationality	Error adjustment range
Computer Technology Act	Work adjustments	92.71	4.42
	Policy selection	93. 47	3.32
Traditional management methods	Work adjustments	63.23	6.34
	Policy selection	62.39	5.98
	X2	15.203	11.102

Compared with the traditional management method, the computer technology method rationality and error margin of the management work. Through the change of management information in Fig. 1, public social activities is mainly social management, supplemented by institutional management, and the Computer Technology Act is regulated.

4.2 The Results of the Management of Public Social Activities by Computer Technology

The management information includes social management, system management, and basic management. After the pre-selection of the computer technology method, preliminary management information is obtained, and the feasibility of management information is analyzed. In order to verify the management effect more accurately, different management levels were selected for the study, and the results are shown in Table 3.

Table 3. Overall status of management information

content	Manage compliance rates	Manage satisfaction rates
Social management	93.31	92.46
Institutional management	89.27	95.42
Basic management	83.21	81. 04
mean	87. 39	84.18
X2	5.633	11.802

4.3 Accuracy and Stability of Management Work

In order to compliance rate of the computer technology method, the accuracy and stability compared with the traditional management method are shown in Fig. 2.

It can be seen from Fig. 2 that the accuracy of the computer technology method management method, but the adjustment rate is lower, indicating that the management of the computer technology method is relatively stable. The management results of traditional management methods are not satisfactory. The average management information of the above three algorithms is shown in Table 4.

Table 3 shows that the traditional management method has deficiencies in management and stability in the management research of the direct role of public social activities, and the management research has undergone substantial changes, the error rate is high.The management of the comprehensive results of the computer technology method is higher than that of the traditional management method. At the same time, the management of the computer technology law is greater than 96%, and the adjustment rate has not changed significantly. The computer technology method is comprehensively analyzed by different methods, and result 4 is shown.

The reason is that the computer technology method increases the management coefficient and sets the threshold of management to exclude administrative work data that does not meet the requirements.

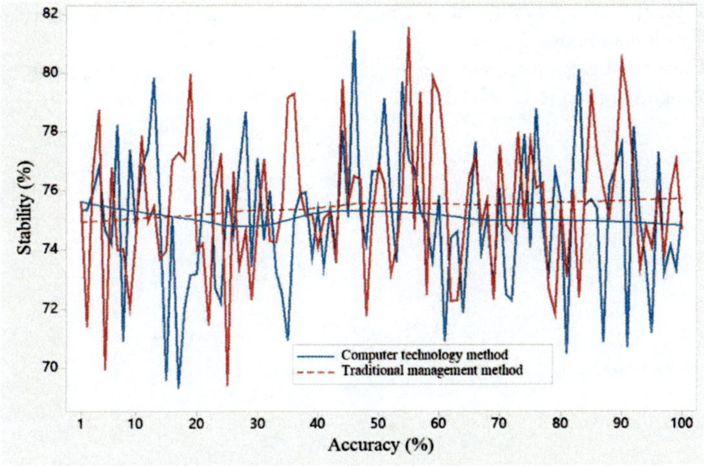

Fig. 2. Management of different algorithms

Table 4. Comparison of management accuracy of different methods

algorithm	accuracy	stability	Adjustment rate
Computer Technology Act	96.93	96.28	2.98
Traditional management methods	85.32	83.69	4.38
P	11.417	32.061	12.172

5 Conclusion

In view of the inaccuracy of management work, this paper proposes a computer technology method and combines organizational theory to manage public social activities. At the same time, the content and effect of the direct management of public social activities are analyzed in depth, and a management collection is constructed. Research shows that computer can directly affect public social activities management is management protensively. However, in the process of computer technology law, too much attention is paid to management analysis, resulting in the unreasonable management data processing.

References

1. Moin, P., Davis, S.H.: Biomimetic Survival Hydrodynamics and Flow Sensing (2016)
2. Danielson, M., Hansson, K., Cars, G., et al.: Deliberation, Representation, Equity: Research Approaches, Tools and Algorithms for Participatory Processes (2017)
3. Shendo, M.V., Sviridova, E.V.: Technologies of digital economy as tools for improving quality of life, business efficiency and public administration. Vestnik Astrakhan State Tech. Univ. Ser. Econ. **2019**(4), 29–36 (2019)

4. Ni, D., Cheng, S.: On the ethics of information technology application in public management. Administration and Law (2019)
5. Fu, J.: The innovative application of new technology in e-government and its enlightenment to China. Administration Reform (2019)

Intelligent Service Platform for Epidemic Prevention and Control Based on Big Data Technology

Yizhe Meng[1](✉), Zhikai Yang[1], Wenna Bao[2], and Xueming Wu[1]

[1] Zhangjiajie Campus of Jishou University, Zhangjiajie 427000, China
771941329@qq.com
[2] Shaziao Campus of Jishou University, Jishou 416000, China

Abstract. The intelligent service platform for big data technology is a new approach to infectious disease management. It aims using big data technology. It integrates epidemiological, clinical and laboratory information as well as social network analysis tools, so as to detect and prevent the epidemic as early as possible by predicting its occurrence or spread to public health authorities. The project aims to develop an intelligent service platform to support the rapid collection of big data from various sources such as health departments and other relevant institutions across China, so as to promote the management of epidemics. It will also provide real-time information sharing between local governments and other relevant parties. In addition, it will provide customized services for local governments, such as epidemiological forecasting tools and national systems for monitoring disease outbreaks throughout the country.

Keywords: Intelligent service platform · Big data technology · Epidemic services

1 Introduction

At present, with the spread of COVID-19 variant strains, the epidemic situation across the country is grim. Although the spread of COVID-19 has been basically curbed, the prevention and control situation is gradually improving, and all parts of the country are accelerating the resumption of production. However, the epidemic prevention work cannot be slackened. Markets, hospitals, office buildings, construction sites, communities and other places are densely populated and highly mobile.In the face of large-scale infectious diseases epidemic, the society is still in the process of exploration, and there are also difficulties in epidemic prevention in today's society: artificial body temperature detection is easy to be infected: during body temperature detection and introduction, artificial contact body temperature detection is insufficient in protection, and front-line supervisors have a high risk of exposure to viruses, and can not grasp the health status of residents in real time [1]. High cost of health search: when registering for community access, residents fill in the registration manually, which is easy for people to gather,

J. C. Hung et al. (Eds.): FC 2023, LNEE 1132, pp. 235–240, 2024.
https://doi.org/10.1007/978-981-99-9538-7_35

manual registration errors occur frequently, and information distortion is difficult to trace. When viewing the health code, the lawless personnel muddle through through the screenshots of other people's QR codes, which can not be found in time, and is likely to lay hidden dangers for the outbreak of the epidemic.

The regional access module of the intelligent epidemic prevention solution is composed of face recognition and temperature measurement equipment. The non sensing access can realize non-contact screening in airports, stations, subways, schools, office buildings, communities and other scenarios, realize high temperature, non mask abnormal event alarm, health code exception and personnel recognition, quickly identify personnel information and register temperature measurement, reducing the possibility of epidemic transmission, At the same time, it provides background records, which can be intelligently analyzed for easy management once exceptions are found. At the same time, it can store and display the real-time information of the correspondents, so that the managers can know the regional situation in time, making the city more convenient and safe.

2 Related Work

2.1 Scientific and Technological Warfare Methods

Community epidemic prevention has played a huge role in this war without gunpowder smoke, and measures such as closed management, access control, temperature measurement and disinfection have effectively curbed the spread of the epidemic. However, behind this, there are problems such as the lack of community manpower and materials, the difficulty in implementing management measures, and the lack of personnel information collection. In order to solve these problems, we have developed a community epidemic prevention and control platform by relying on cloud platform, 3D visualization, mobile applications, intelligent temperature/face/ID card recognition equipment and other technical means, providing a comprehensive and effective information management tool for the community, and helping the community achieve the community epidemic prevention and control information management [2].

At present, community management and control mainly focus on crowd monitoring, body temperature monitoring and community environment disinfection and sterilization. It mainly relies on the way of requiring the quarantined to sign a commitment, consciously isolate at home, and social health workers to monitor at home. With the increase of people returning to Shenzhen, the spread of the epidemic is on the rise, and the original management and control mode and means urgently need the intervention of new means [3]. The world's leading intelligent location service provider has made use of its own technical advantages to urgently customize and develop the "intelligent epidemic control platform". Through high-precision indoor positioning technology and the building capacity of the prevention and control system, it has realized the real-time location information and health status reporting of isolated people at home, and technology has assisted in the meticulous, efficient and low-risk management of epidemic prevention. It is currently the largest indoor geographic location information platform in China, and has the large-scale indoor positioning capability, and the positioning can be accurate to the indoor meter level. It truly realizes the efficient control of isolated personnel without contact or less contact in the process of epidemic control through technology [4]. Compared with the traditional man to man management, the intelligent epidemic control bracelet has effectively improved the management efficiency of community grid administrators by more than 10 times, saved at least 10 times more social resources, and greatly reduced the probability of community grid administrators being infected and the probability of epidemic spreading. Figure 1 below shows the epidemic situation database information.

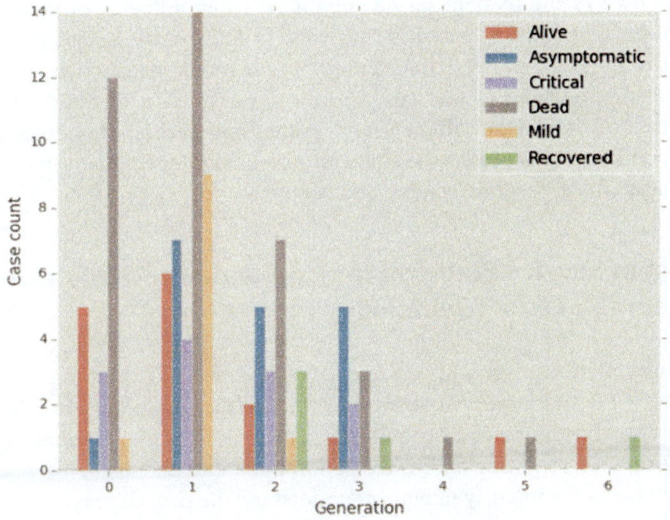

Fig. 1. Epidemic database information

The intelligent epidemic control solution (epidemic management system) can realize the functions of real-time monitoring of the position of the quarantined personnel, alarm of the isolation area, health sign monitoring, instant messaging, and equipment monitoring, and is applicable to the home isolation management of suspected patients in the community. It can effectively solve the lack of auxiliary decision-making means for governments at all levels in the process of epidemic control, the difficulty in confirming epidemic related information, and uneven regional control forces Poor access to self isolated information. At present, the "Epidemic Management System" has been fully implemented. Up to now, thousands of isolated people have passed the product management in Shenzhen, where they can view the location and health status of the isolated people online at any time and anywhere, communicate the isolation status online, and achieve efficient synchronization of epidemic information and non-contact intelligent management.

2.2 Service Platform Indicators

These applications are obviously built on electronic information systems. Without the construction of information systems, these three types of electronic publishing platforms will not have specific entities to rely on. Content is included in the form. Without form, there is no content carrier. Therefore, many scholars often regard system quality or system service quality as the primary indicator of the indicator system in the design of service quality evaluation indicators for websites or mobile application platforms. The service quality of the mobile application platform system will have a significant impact on user satisfaction, and then affect the user's evaluation of the service quality of the application platform. In the social platform, the connotation and concept of system service quality are basically consistent with the interpretation of other scholars in the service quality evaluation index system. System service quality is often reflected in system operation, interface design, use comfort and other aspects. In the specific social platform service quality research, other scholars' interpretation and index selection of system service quality are also applicable to this study. Both the aesthetics of the platform and the response speed of the system affect users' evaluation of their service quality. In this paper, the secondary indicators of system service quality specifically include stability, aesthetics, ease of use, responsiveness, and security.

3 Intelligent Service Platform for Epidemic Prevention and Control Based on Big Data Technology

The community platform is mainly composed of import and export monitoring system, mobile phone APP and large visualization screen. The community import and export monitoring relies on the original monitoring equipment of the community. The APP provides information reporting access for community staff and residents, and the visualization screen helps community managers understand the overall prevention and control situation. The combination of the three will help the community to quickly complete information collection, carry out patrols, rehearsals, isolation and other prevention and control work, start from the source link of personnel input to achieve intelligent and visual community epidemic prevention.In order to minimize the contact of personnel, we provide remote implementation services. We can quickly complete the deployment of a cell in 1–3 days through the "contactless" method and provide detailed operation instructions to ensure that users can become proficient after simple learning.

The rapid deployment of the platform takes into account the large differences in the informatization degree of each community, the different management levels and epidemic prevention strategies, but also the needs of the epidemic prevention platform. This set of community epidemic prevention and control platform provides two forms of standardized deployment and customized deployment. Standardize the deployment community, install the APP to achieve data entry, and realize the background data management and front-end visual situation viewing through the cloud platform. The system is developed using mainstream technology, cross platform and cross device, and can be quickly implemented in communities without information foundation. The system architecture is flexible. For communities that have deployed management platforms, it can be rapidly integrated and expanded, and has customized development capabilities such as APP page embedding and existing data on the big screen. The import and export monitoring system uses cameras, temperature measuring instruments, infrared thermal imaging cameras and other hardware equipment in combination with manual registration and other methods to record the face recognition, body temperature and other information of the incoming and outgoing personnel. To ensure that the flow information and health information of the incoming and outgoing personnel are fully mastered. The large visual screen displays the real-time situation of community entrance and exit, mainly including personnel thermal imaging, community entrance and exit personnel information, real-time monitoring video and statistics of the number of people entering and leaving the community. It not only clearly shows the flow of people entering and leaving the community, but also helps managers quickly grasp the flow situation of community personnel through a large screen. Even in the event of an emergency, you can quickly call the real-time video of the camera to view the actual situation, to ensure that the event is resolved in a timely manner.

Provide information reporting access for ordinary residents, encourage residents to report their personal information on a regular basis, and build a strict information reporting system during epidemic prevention and control.

- Provide entry for reporting epidemic related events. Users can provide feedback and evidence when finding suspicious situations, which can be verified by community personnel
- Provide an entry for reporting the shortage of materials, timely collect the information of residents who lack materials, and serve as a reference for the situation of community materials, so as to facilitate unified deployment
- For the community that controls travel, residents can report daily access demand and each access situation, which is convenient for the community to issue passes and count personnel access data
- Provide other help information reporting entry and emergency contact number, users can ask for help with one click

This epidemic situation mainly involves various questions and answers on virus, prevention, development and other major scenarios. For the epidemic related natural language questions input by users, the key elements (concepts, entities, attributes, various operators, etc.) are identified, and the subgraph matching method is combined to transform them into computer understandable query intentions to obtain answers. The difficulty is that different users have different words, sentences and expressions. It is necessary to combine AI technologies such as knowledge mapping, deep learning and transfer learning to achieve accurate identification of users' intentions.

4 Conclusion

The deployment of the community epidemic prevention platform can not only help managers to grasp the epidemic prevention situation in the community at any time and anywhere, but also achieve efficient scheduling and command through docking with the APP. Provide internal information bulletin, education and publicity, external information release, data reporting, work plan and implementation, and help the community effectively and orderly carry out prevention and control work.

References

1. Xie, Z., Zhang, X., Shu, H., et al.: Application research of guangzhou one-stop travel service platform based on big data technology. Mod. Inf. Technol. (2019)
2. Liang, B., Yang, Z.: Exploration and application of college students' management model during the epidemic based on big data technology. Math. Probl. Eng. **2022** (2022)
3. Li, X., Wang, S.: Research of big data analysis platform based on positioning information. In: Journal of Physics: Conference Series (2020)
4. Wu, J., Wang, J., Nicholas, S., et al.: Application of big data technology for COVID-19 prevention and control in China: lessons and recommendations. J. Med. Internet Res. **22**(10), e21980 (2020)

Dynamic Path Planning of Robot Based on Depth Learning

Chenhua Ouyang, Shudi Wei(✉), and Zhong Chen

Hengyang Normal University, Hengyang 421002, Hunan, China
dnncro@163.com

Abstract. Multi joint serial industrial robot is widely used in industrial production because of its convenient operation, accurate positioning, flexible execution and other advantages. The working principle of industrial robot is mainly to remember the running track through manual teaching, and use control to make it reach the specified position and pose. Generally, industrial robots only follow one or more fixed routes in actual pipeline work. Dynamic path planning is a method types of robots, such as wheeled mobile robots or underwater robots (UUV). The key idea behind our method is to use the depth map generated by a stereo camera, rather than simply measuring the line of sight distance between two points as traditional methods. The depth map provides information about the relative position between two objects, which is crucial for finding a free path.

Keywords: robot · Deep learning · Dynamic path planning

1 Introduction

At present, the more successful and feasible traditional robot path planning methods mainly include genetic algorithm, artificial potential field method, fuzzy algorithm, A * algorithm, etc. But these methods all need to fully understand the environment information of the robot system, and then use some rules or algorithms to plan its path. Environments. Therefore, the research on how to improve the self-learning ability and adaptability of robot system in path planning is particularly critical [1].

Artificial intelligence is a branch of computer science. It tries to understand the essence of intelligence, so that computers can think like people!. As a part of artificial intelligence, machine learning is not a way to manually code software routines with specific instructions to complete specific tasks, but a way to "train" algorithms so that it can learn how to complete tasks. The reinforcement learning (RL) method is between supervised learning and unsupervised learning, and it does not need to give any marked information in any state. The then evaluates the actions selected for implementation through the feedback reinforcement signal. Finally, a satisfactory solution is obtained by using constant trial and error and selection [2].

With the increase of application fields and the complexity of tasks, if some tasks are completed only by a single robot, the efficiency is very low, and even may not be completed. Therefore, people are also increasingly considering the use of multiple

J. C. Hung et al. (Eds.): FC 2023, LNEE 1132, pp. 241–246, 2024.
https://doi.org/10.1007/978-981-99-9538-7_36

robots to coordinate tasks that are difficult for a single robot to complete. For multi robot systems (MRS).

2 Related Work

2.1 Path Planning Method for Single Robot

The path search space based on the known environment information can generally be simplified into a two-dimensional model. The current representation methods can be roughly divided into three categories: grid representation, geometric information representation and topological graph representation [3]. The grid representation is to divide the whole working environment into several small squares, namely grids, according to the same size, and explain whether there are obstacles for each grid; It is characterized by easy creation and maintenance. If there is a directly connected path between nodes, it is equivalent to the arc connecting nodes in the graph. It can achieve fast path planning. However, when there are two very similar places, this method is difficult to determine whether this is the same node. Finally, on the constructed environment map, use the path finding algorithm (such as A * algorithm, etc.) to search for an optimal path without collision.

2.2 Multi Robot Path Planning Method

For the must be considered as a whole. Nowadays, there are four main types of path planning for multi robot systems: fully centralized path planning, incompletely centralized path planning, incompletely decentralized path planning, and completely decentralized path planning.

The centralized method is one of the simplest path planning methods for multiple robots, which is suitable for situations with a small number of robots. This method typically involves assigning tasks and planning paths through a central control unit. The central control unit will optimize and plan based on the robot's information and task requirements, and then assign tasks and routes to each robot. The advantage of this method is that it is simple and feasible, and is not affected by the interaction between robots.

The distributed method is suitable for situations with a large number of robots. In this method, each robot has its own decision-making and path planning abilities, and they collaborate to complete tasks and allocate paths. In distributed methods, robots achieve task allocation and route planning through communication and collaboration. Due to each robot having its own decision-making power, distributed methods are more flexible and fault-tolerant.

Collaborative path planning is a popular multi robot path planning method. In this method, robots collaborate to complete tasks and path planning, and adjust their routes in real-time based on task allocation and environmental changes. This method can achieve task planning and path allocation through the interaction and collaboration between robots, thereby improving the efficiency and security of the entire system.

Artificial intelligence technology is playing an increasingly important role in path planning for multiple robots. For example, optimizing robot path planning through techniques such as machine learning, genetic algorithms, and deep learning. This method can automatically discover path planning schemes from a large amount of data and adjust the path scheme according to real-time environmental changes.

In short, multi robot path planning is a complex problem that requires assigning tasks and planning routes among multiple robots. Different multi robot path planning methods have their own advantages and disadvantages, and in practical applications, it is necessary to choose the appropriate method based on the specific situation.

3 Deep Learning Theory

Neural network is a mathematical model of propagation between bionic neurons. It mainly has two propagation formats, one is feedforward and the other is feedback. The feedforward type mainly reflects the mapping relationship between a group of data and fits the objective function. At present, common perceptrons, BP, RBF, etc. are feedforward neural networks [4].

BP neural network belongs to the category of basic deep learning. Its full name is Back Propagation neural network, which is translated as back propagation neural network. This is also the biggest feature of BP neural network. It can not only forward transmit data, but also back transmit the error of result evaluation. All BP neural networks have a three-layer structure, including the input layer, hidden layer and output layer, as shown in Fig. 1.

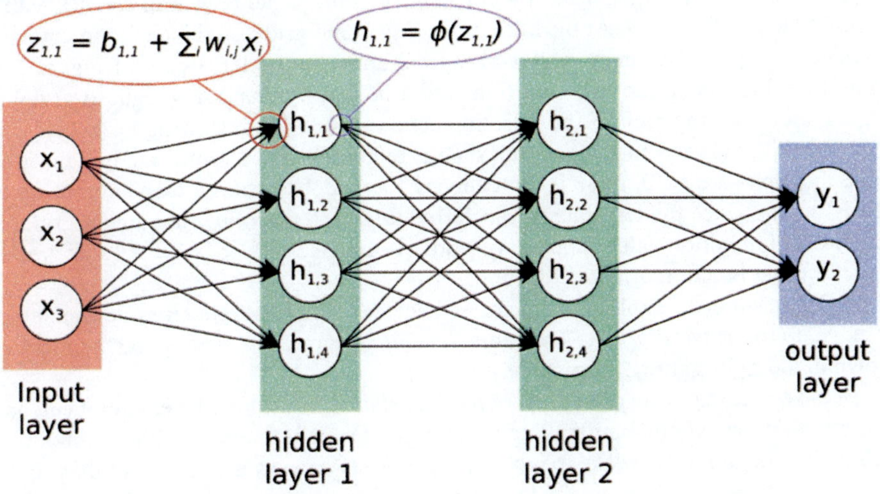

Fig. 1. Three layer structure of BP neural network

The output layer receives information variables from the hidden layer for further processing. At the same time, the most basic function of the output layer is to output the network results.

In BP algorithm, S-type function is generally selected as the transfer function. S-type function can realize nonlinear mapping between input data and output data of training data, and concentrate the mapping of output results in the range of (−1, 1), reducing the different effects caused by large data span.

$$E(w, b) = \sum_{j=0}^{n-1} (d_j - y_j)^2 \tag{1}$$

The propagation process of BP algorithm is divided into forward and backward propagation, the core of which is back propagation, and its learning algorithm is gradient descent method. The essence is to find the minimum error value. The difference is that the gradient descent method is applied to the BP neural network, which has its own network structure and connection mode, so it has formed its own algorithm, namely BP algorithm.

4 Robot Dynamic Path Planning Method Based on Deep Learning

The greedy strategy balances exploration and utilization. The disadvantage of this method mentioned above is that it may converge to a suboptimal solution or even a common solution. This defect will also affect the results in the single robot path planning.

For example, in the environment of Fig. 2, the yellow dot grid 4 in the upper right corner is set as the target state, and the red block grid 1 in the lower right corner is set as the starting state. Assuming that the initialization value function is both 0, when the robot reaches the target state, it will receive an immediate reward of r = 10, with a discount factor of y = 0.9. At the same time, grid 2 and grid 3 in the environment have obtained the optimal strategy, as shown by the arrow in the left figure of Figure 3-1. Then the value function of the selection action in grid 3 transferring right to grid 4 is 10, and the value function of the selection action in grid 2 transferring up to grid 3 is $0.9 \times 10 = 9$, then the current starting state is red block grid 1. Since the initialization value functions are all 0, there is no tendency for grid 1 to choose the action of up or left. However, since the left state is transferred to grid 2, the value function of grid 2 is 90, and the left action value function of grid 1 is updated to $9 \times 0.9 = 8.1$, far greater than the value function of upward action.

In Q-Learning, the exploration factor will decrease Then the robot may never attempt to move up to grid 4 at grid 1. Therefore, it will converge to a suboptimal solution, as shown in the right figure of Fig. 2.

At the beginning, because the robot has just started to explore the environment, and the number of successful path finding is not high, it should maintain a high exploration factor at the beginning ε, Make the robot continuously try new actions; With the robot's understanding of the environmental information, the number of successful path finding will exceed a threshold, which proves that the state action value function of the robot tends to be stable after that. But at this time, the optimal solution may not have been found, and it may have fallen into the suboptimal solution. At this time, the number of different paths found by the robot is added. If the number is very small, it means that the robot has been repeating the old path, and the exploration factor should be increased

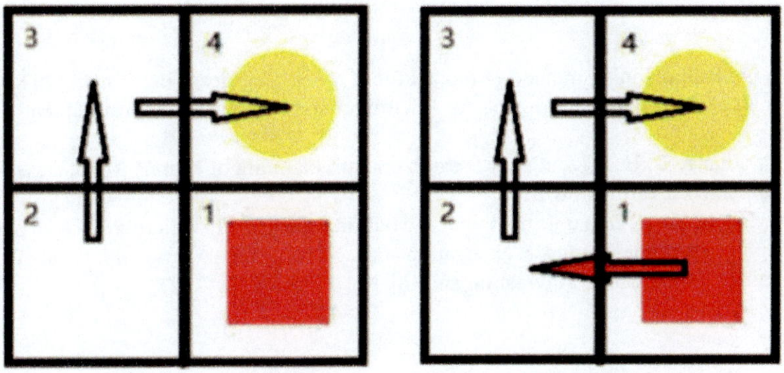

Fig. 2. Example Diagram of Robot Dynamic Path Planning

ε, Force the robot to try new actions; If the number is greater than a threshold, it is proved that the robot has basically found all the paths to reach the target state, and the exploration factor should be slowly reduced ε.

In this way, the above two bases are used to constantly update the exploration factors according to the learning effect ε The value of, so that ε- The greedy strategy has stronger adaptability and conforms to the operation rules.

5 Conclusion

Dynamic path planning for robots is currently an important issue in robot control. Its goal is to plan a suitable path for the robot to achieve its predetermined goals in the future. Unlike static path planning, dynamic path planning requires real-time response to environmental changes during movement and corresponding actions to ensure the safety and effectiveness of the robot. In recent years, deep learning technology has been widely used in robot dynamic path planning. This article will summarize the dynamic path planning of robots based on deep learning. Deep learning is a machine learning method based on multi-layer neural networks. In the dynamic path planning of robots, deep learning technology is mainly achieved through neural network models. Sensors are usually implanted in robots, and by collecting and processing a large amount of sensor data, the robot's environment is modeled. Deep learning technology is used to extract features for classification and recognition, thereby achieving dynamic path planning. In practical applications, deep learning based dynamic path planning for robots has been widely applied in fields such as robot navigation, environmental perception, and decision-making. By modeling and analyzing the environment, deep learning models can predict future environmental changes and adjust robot path design to ensure the efficiency and safety of robots in most situations.

Acknowledgements. Education Department of Hunan Province of China (No. 20C0290).

References

1. Jiang, Q.: Path planning method of mobile robot based on Q-learning. JPhCS (2022)
2. Song, K.: A Path Planning Method for Environmental Robot Based on Intelligent Algorithm (2021)
3. Shi, Z., Zhang, T., Han, L., et al.: Research on Path Planning of Mobile Robot Based on Deep Reinforcement Learning (2021)
4. Qi, Y., Liu, J., Yu, J.: Dynamic modeling and hybrid fireworks algorithm-based path planning of an amphibious robot. Guidance, Navigation and Control: An International Journal of Technical Committee on Guidance, Navigation and Control, CSAA **2**(1) (2022)

Research on Optimizing Face Recognition Algorithm Based on Adaboost Algorithm

Ning Pan(✉)

Hubei University Campus Construction and Information Office, Wuhan 430062, China
panda2005_163@163.com

Abstract. Face detection and recognition technology has always been a hot research direction, face detection and recognition technology has received extensive attention from society. The detection and recognition rate and running speed of the face detection and recognition system directly affect the user's experience, so it is very important to realize a fast and accurate face detection and recognition system. The adaboost method uses a set of classifiers (also called decision trees), which are trained for different subsets of data points (called splits).

Keywords: Face recognition · Adaboost algorithm · optimization

1 Introduction

Since the beginning of the 21st century, the way to ensure social security and information security has also been upgraded. In public places, video monitoring systems can be seen everywhere. By calling up video monitoring videos near traffic lights, the police can quickly confirm the identity of the target by playing back the video files, and find out the criminals, Social security is undergoing profound changes. It takes a long time to find valuable information from video surveillance videos, so how to quickly search for effective targets is the most critical [1]. If face detection and recognition technology is used to search for effective information in videos, it can help public security institutions greatly improve the efficiency of solving cases. The above examples reflect the application of face detection and recognition technology in social security. Digital cameras, access control systems, identity recognition, online payment and other fields. At present, many research institutes and Internet companies pay more attention to the research of face detection and recognition technology, and there will be more application scenarios of face detection and recognition technology in the future.

The resulting video conference, videophone, high-speed network technology and other image processing technologies have great application prospects [2]. For example, people can no longer be satisfied with just sitting in front of the computer for communication, and need to use human-computer interaction, remote operation, and even use ideas to control the machine. Followed by is related to the pattern recognition technology, graphics and image technology, virtual 3D technology and innovation. Face detection technology is an important part of them, and people's knowledge of face detection is also being intensively studied [3].

J. C. Hung et al. (Eds.): FC 2023, LNEE 1132, pp. 247–252, 2024.
https://doi.org/10.1007/978-981-99-9538-7_37

Face detection technology consists of image acquisition technology, image display technology and image processing technology. Among them, the sensor for image acquisition is composed of charge coupled device (CCD) with linear array and planar array, charge injection device (CID) with image sensing element capable of storing a small number of carriers, and MOS image sensor with self scanning photodiode array. From the perspective of technology development trend, whether CCD or MOS, the higher the resolution, the better the clarity, the smaller the volume, and the higher the cost performance of its image products. The development of the image display part has gone through three stages: CRT, liquid crystal and plasma. Because each of the three display modes has its own characteristics, it is difficult to determine which has an absolute advantage. There are many software used, each with its own characteristics [4]. The classic ones are MATLAB and C++, but the power consumption, volume, portability and other performance on the PC can not meet people's requirements in many occasions. The research on the optimization of face recognition algorithm based on Adaboost algorithm has great value.

2 Related Work

2.1 Research Status of Face Recognition

People began to study face detection and recognition technology very early, and has always been a hot research direction. Up to now, face detection and recognition technology has a research history of more than 40 years. With the in-depth research of more and more researchers, a large number of theoretical achievements emerged in the 1990s,At present, almost all the technicians engaged in relevant fields in domestic and foreign research institutions have studied face detection and recognition related technologies. In the 1970s, when we were in the early stage of face detection and recognition research, Goldston et al. proposed to use facial geometric features to achieve face detection and recognition, and designed a face detection and recognition system. The results show that the recognition using geometric features is greatly affected by factors such as facial occlusion and expression changes. Many statistical learning methods have been born, such as template matching, linear subspace, neural network and other methods. Since the 1990s, some related applications have appeared in the market. Nowadays, the research direction in the field of face detection and recognition is based on statistical learning methods and artificial neural network methods.

which generally include face contour features, gray level features, skin color features, texture features and organ features. Most of the faces are elliptical, and the positions of the five features are relatively fixed. Therefore, the facial contour and the five features form a figure with geometric relationship. You can determine whether the face is a face by comparing the geometric relationship. This method is based on geometric features; Although different people have different skin colors, the color of some parts of the human face is fixed, for example, the eyeballs and eyebrows of people are black, which is in obvious contrast with the lighter cheek parts. By distinguishing the colors of different areas of the face, gray level features is generated. In addition, there are many similar detection methods based on this kind of prior knowledge. The method based on feature knowledge often realizes face detection through a variety of prior knowledge, which not only improves the detection accuracy, but also accelerates the speed of face detection.

2.2 Comparison of Face Detection Methods

Face detection and recognition has always been a key research project of researchers. The existing face detection methods include: methods based on prior knowledge, feature invariant methods, template matching methods, representation based methods, etc., or innovative detection methods or improvement of existing detection methods. The facial recognition framework is shown in Fig. 1.

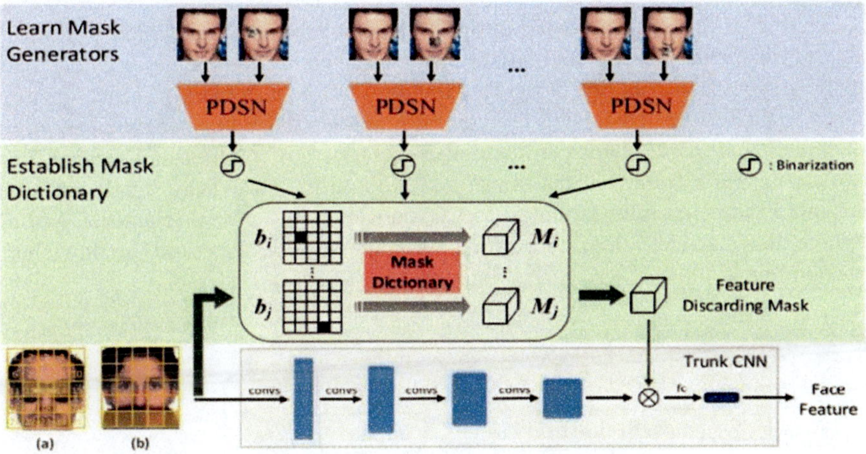

Fig. 1. Facial recognition framework

The method of face detection based on prior knowledge mainly uses existing knowledge and long-term experience to define face rules, then detects whether these face rules exist in pictures, backgrounds and video streams, and determines the candidate regions of faces according to the analysis of the similarity between the face to be detected and the defined face rules The method based on feature invariants is to extract the inherent structural features of the face by changing the lighting, expression, pose, angle and other methods, and comprehensively judge whether the area to be detected is a face. The boundary of the face feature is difficult to define due to the influence of lighting, occlusion, shadow, complex background and other factors, making edge detection difficult to achieve The method of template matching is mainly to calculate the matching degree between the input image and the standard template, so as to achieve the purpose of face detection. The edge, peak and valley values of the face image are unstable, and the process from theory to practical application is difficult to achieve. The representation based method needs to learn from the face sample training set according to a certain algorithm, so as to obtain the learning template, establish a classifier, and then scan the image to be detected. After multiple scans, it can determine whether the sampling window contains face features.

3 Adaboost Algorithm

Adaboost detection algorithm is a very classic face detection algorithm. Since Viola et al. proposed it in 2001, it has great significance for face detection, enabling this technology to achieve a turning point change. Many scholars are studying it and have made a lot of improvements.

Adaptive Boost is a learning algorithm, namely Adaptive Boosting. Given a series of positive and negative samples, it can be trained into a weak classifier. If a certain sample is correctly classified, when constructing the next training set, reduce its weight, otherwise increase its weight. Through this algorithm, each round will strengthen those difficult samples that are wrongly classified, and each round will select an optimal weak classifier.

Classifier is to classify specific things. Adaboost is a classifier, which is trained by a simple weak classifier. In Adaboost algorithm, face detection mainly uses the feature that the gray level distribution of the face image is different. It mainly detects gray level images. Haar features calculated by integral graph are used. Haar features are simple to calculate and fast to detect. In the detection process of Adaboost algorithm, firstly calculate the Haar feature of the image and propose the Haar feature with the lowest error rate in the algorithm training process, convert it into a weak classifier, and then combine the weak classifier into a strong classifier for face detection.The flow chart is shown in Fig. 2:

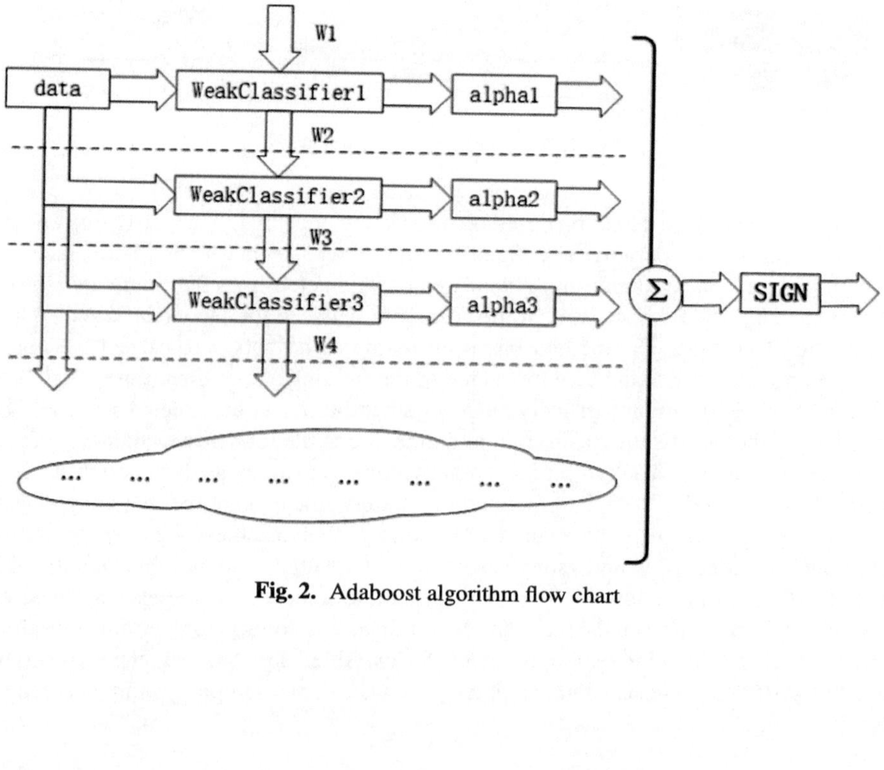

Fig. 2. Adaboost algorithm flow chart

4 Optimize Face Recognition Algorithm Based on Adaboost Algorithm

The strong classifier trained by Adaboost algorithm can be used for face detection in images. However, the size of the image to be detected and the classifier trained by Harr features are often different, which requires changing the size of the image in the detection process. There are two methods to use Adaboost for face detection:

One is to reduce the face image to be detected in proportion while keeping the feature detection window unchanged, and scan the image to be detected with the Haar feature window to detect the human face sub window. This kind of detection will change the image size, and the calculation of Haar eigenvalue is based on the integral graph, so the calculation will be large and the detection speed will be slow. As for 36 × 18 for 36 images × 18 of the detection window for detection, and the image needs to be sampled with a separation rate of 18 × 18.

The other is to enlarge the Haar feature window proportionally, while the image to be detected remains unchanged. From the above discussion, it is known that the change of Haar feature will not increase the calculation amount. Therefore, the detection speed of this method will be very high. This method is used in this paper. When the detection window is scaled up, the scale factor cannot. In the first several layers, most candidate windows are excluded, and the area that all pass through each level of classifier is the face, as shown in Fig. 3.

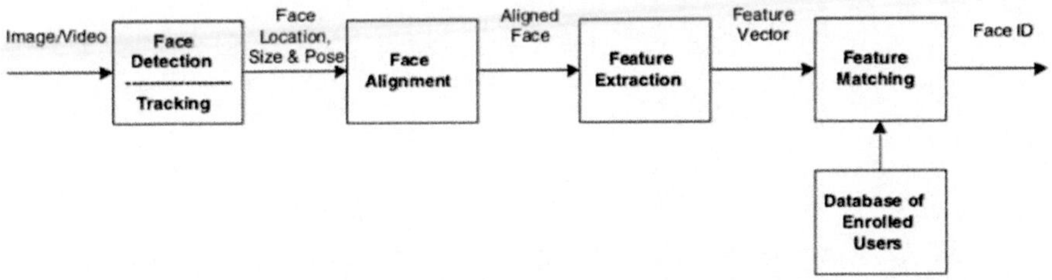

Fig. 3. Processing flow of face recognition system

The experimental results show that the facial recognition optimization method based on Adaboost algorithm proposed in this article has greatly improved accuracy and speed compared to traditional Adaboost algorithms. At the same time, this method also has higher robustness and anti-interference ability, which can better handle problems such as multi angle, lighting changes, and facial expression changes.

5 Conclusion

Carry out systematic research on face recognition methods. The Haar feature extraction method is selected as the main method of face detection, and the Adaboost algorithm is selected as the main algorithm of face recognition. It has been successfully tested on Android mobile phones. The approximation of different faces of the same person is about

90%. Tests show that adding classifiers cascades multiple weak classifiers into a strong classifier optimization algorithm. This article first analyzes the shortcomings of existing face recognition methods, such as slow computation speed and low accuracy. In response to these issues, this article proposes a face recognition optimization method based on the Adaboost algorithm. It improves the accuracy and robustness of the classifier by improving the weak classifier weight training process in the Adaboost algorithm, thereby effectively improving the face recognition effect.

References

1. Chen, R., Diao, X., Sun, Z., et al.: Research on emotion recognition based on GA-BP-adaboost algorithm (2022)
2. Wei, S., Li, X., Zhang, X.: Research on face emotional recognition algorithm (2021)
3. Yuan, B., Du, C., Wang, Z., et al.: Research on intelligent algorithm of identity authentication based on facial features. Wirel. Commun. Mob. Comput. (2021)
4. Thilagavathi, B., Suthendran, K., Srujanraju, K.: Evaluating the adaboost algorithm for biometric-based face recognition (2021)

Computer AR Technology to Help the Application of Cultural and Creative Product Development Research

Li Peng(✉) and Yangyang Yi

Hunan Automobile Engineering Vocational College, Changsha 412008, China
pengl1@126.com

Abstract. Using AR technology for tourism products research and development and application, especially should pay attention to the feeling of tourists experience, with cultural creativity as the starting point and the foothold, using AR technology, stimulate tourists a variety of feelings, stimulate their excitement, make they produce close feeling, constantly produce consumption desire, meet the actual needs of tourists.

Keywords: computer AR technology · cultural and creative product development · application

1 Introduction

Internet, the creative industry of tourism culture has attracted more and more attention. Tourism cultural creative products are a cultural symbols of a country, but also an important part of promoting the development of China's tourism. Introduced to the tourism market and achieved success to a certain extent. Take Dalian as an example. Dalian is a coastal city in eastern China, adjacent to the Yellow River and Bohai Sea, and has the reputation of "Northern Pearl", with profound architectural culture, square culture, seaside vacation culture and so on [1]. In this environment, how to better combine with the modern economic and social life, has become an urgent need for research and discussion topic.

2 Overview of the AR Techniques

AR technology is a new data technology that appeared in Japan in the 1990s. It can integrate reality and reality data together. It can integrate the corresponding pictures together through the perspective and position of the camera, and finally perfectly integrate the reality and reality together, so as to realize the interaction between the two.

The significance of computer AR technology to help cultural and creative products AR technology is a brand new electronic technology that can effectively integrate the real world and the virtual world, so as to achieve the interaction between people and things.

J. C. Hung et al. (Eds.): FC 2023, LNEE 1132, pp. 253–259, 2024.
https://doi.org/10.1007/978-981-99-9538-7_38

Let the user have a sense of immersive feeling. In addition to touch, the application of AR technology can also use visual synthesis, hearing and other sensory experiences, so as to achieve the "immersive" experience. Use AR technology, make the tourism cultural and creative products to get rid of the function of ornaments, improve the content of science and technology, more cultural connotation, but also more three-dimensional, vivid. Through a small memorial, people can feel a brand new travel cultural experience.

AR technology is due to its high virtualization [3] Convenience characteristics make the creativity of cultural and creative products more creative. Some designers start from the basic principles of AR technology and take advantage of its virtual role. Through a large number of ideas, AR technology is permeated into the cultural and creative design [2]. The AR technology diagram is shown in Fig. 1.

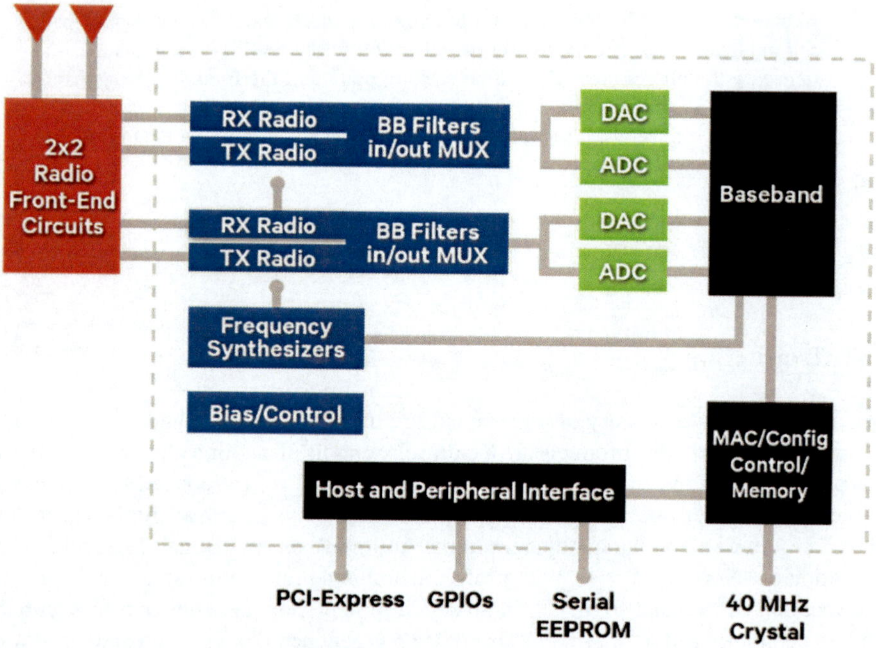

Fig. 1. AR technical diagram

For example, when you go to the scenic spot to select tourist souvenirs, scan postcards with your mobile phone, and scan the QR code with your mobile phone, you will find a 3D composite picture on the mobile phone, which is very realistic, along with music accompaniment, audio commentary and so on, so that tourists can be full of memories here. This is a novel tourism and cultural creative product, providing tourists with a brand new experience. In addition, they can also share their experience with others, so that they can get more happiness and satisfaction. Therefore, the application of AR technology in tourism cultural creative products has a great role in the development of tourism industry and tourism.

3 The Application Way of Computer AR Technology to Help the Development of Cultural and Creative Products

At present, under the background of the integration of culture and tourism, all places are trying to seek a synchronous development path along with the development, and develop cultural and creative commodities. Cultural commodities are a kind of flexible, convenient, diversified and personalized cultural commodities, which are of great significance in improving the regional cultural visibility and expanding the channel of the regional. Therefore, it is necessary to strengthen the development and optimization of cultural and creative products, and to build a real, instant and interactive cultural and creative viewing area based on AR technology. Through the construction of the "Metropolis" IP of "Romance k", inherit the Marine history of the region, through the combination of culture and tourism, promote the development of regional tourism, and promote the integration of regional culture and tourism [4] (Table 1).

Table 1. Parameter

1	To study and locate the regional characteristic culture	Cultural and creative design knowledge, simple AR programming and design, QR code design and product elements
2	Investigate the carrier of cultural and creative products	Through the creation, creation, express its spiritual connotation and emotion of the material products
3	Conduct interaction design	Using creative ideas such as manual drawing of the Palace Museum, the QR code of smart maps is added to regional museums and three-dimensional landmarks, allowing users to pass through mobile terminals
4	Test identification and release	Through the computer camera observation, test the development effect of the engine, to see if the computer has a picture mode, whether the compilation results are reflected in the development design
4	Focus on the development of intelligent cultural and tourism products	Local relevant departments cooperate with universities, publishers, social organizations and other organizations to jointly develop the "Coastal Odyssey" series, a virtual cultural and creative theme

(1) Research and positioning of regional characteristic culture

First of all, the most fundamental to accurate determination of the developed product is its humanistic value. In the specific design, according to different marketing conditions, so as to improve the attraction of tourism cultural creativity. Therefore, it is necessary to have a deep understanding of the regional characteristics, do a good job of research and analysis, especially in the tourism regional cultural origin, development process, social and historical environment, cultural characteristics and their role, to seek the application of cultural creativity. Secondly, the traditional handicraft and modern design elements are integrated together to make the cultural creativity more attractive and improve the use experience of users. AR technology provides strong support for regional cultural and creative design, and focuses on vigorously promoting the integration of smart tourism from both macro and micro aspects, creating a good social environment, and carrying out ideological innovation from the concept. And according to the different characteristics of different regions, according to the needs of different customers, to formulate the market strategy suitable for the local market, in order to facilitate the development of the region. From the big aspect, the local government to cooperate with the relevant institutions, colleges and universities, education, the AR technology into the product technology concept, such as the cultural institutions, study the intelligent brigade fusion may bring value, so can through watch AR technology production and product video, and micro film to attract many social groups. With the help of the media platform to promote smart tourism products, to expand its brand influence and brand image. At the macro level, universities, cultural units and relevant departments should also pay attention to the development of cultural and creative knowledge, and learn more about cultural and creative design knowledge, simple AR programming and design, QR code design and processing and analysis of product elements, so as to improve people's attention to the smart tourism industry. The process of smart tourism is shown in Fig. 2.

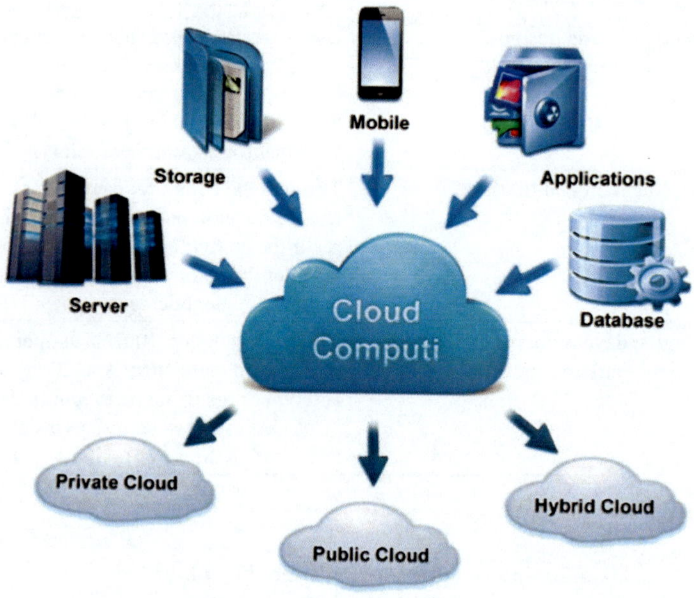

Fig. 2. Smart Tourism Process

(2) Investigate the carriers of cultural and creative products

At present, some people still have problems such as insufficient understanding and insufficient attention to the integration of AR technology into the cultural and tourism industry. The main reason is that in the long-term integration and development, the culture and tourism industry has formed a set of inherent integration concepts and models, and the public also has a high recognition of this traditional integration mode of culture and tourism. If AR is forcibly introduced as a new technical element, it will bring a thinking impact to the public. Therefore, in the new media era, people must constantly improve people's understanding of the cultural and tourism integration methods in the new media era, both in theory and in practice, so as to enhance the competitiveness of the urban tourism market. For some groups with low cultural level, the understanding of the concept of AR technology is not enough. The lack of in-depth research on AR technology is not conducive to the development of virtual reality interaction function, and exclthe layout form of simulation scenes and other unfamiliar content, which is unfavorable to the integration of AR technology and the tourism industry. Therefore, it is necessary to correctly understand and grasp the connotation and extension of the integration of the tourism industry in the new era, and reasonably choose the suitable cultural and creative carriers according to their own needs, so as to achieve the purpose of cultural and tourism integration. Cultural creation is a kind of material product that aims at specific groups in a specific period and expresses its spiritual connotation and emotion through creation. Using technology to transform abstract cultural consciousness into tangible audio and visual products, this is AR technology. Therefore, in the design of tourism products, it must be combined with the local unique regional resources, in order to achieve better publicity and promotion effect. Therefore, it is necessary to analyze and study the regional culture, integrate its basic characteristics into the carrier of the identification map, and form a regional culture and creative identification symbols with strong regional cultural characteristics.

(3) Interaction design

The implementation of AR technology requires interaction through some software. Among these technologies, the most critical thing is to have a good human-computer interaction, and this interaction system includes a 3D model and drawing. Current AR special effects on the market are mostly developed and designed based on Unity3D. Enter the prepared identification graphics, patterns and other information data into a virtual environment, and complete the interaction between users and goods through coding. Through each other, tourists can have a deeper understanding of the history and history of commodities.

Using AR technology to scan cultural and creative products and generate dynamic images can expand the extension of its culture. With the help of mobile phone connection and the application of big data, it can intuitively display and analyze the relevant information of products and trademarks, so as to provide more convenience for customers to go shopping. Moreover, the content of AR can be updated instantly in the background, thus preventing the uniformity of video programs, and thus ensuring the constant change of users. For example, the relevant units of local cultural creativity can use AR technology to build a four-dimensional virtual world, so that the public can hear it. In addition, there are some auxiliary space, mechanical interactive devices, voice remote control and other auxiliary means, as well as through voice remote control to appreciate cultural and creative works, can also be marked, to improve the perception of the public. In addition, the information collected by mobile phones can interact with other regions, countries and metropolises, so as to achieve global sharing, and bring more communication and communication channels to the people. At the same time, the use of the palace manual drawing, such as creative, the intelligent map of qr code to join the regional museum and 3 d landmarks, let users through the mobile terminal, can know the main location of the museum of geographical information and collection, so you can let the traditional ideas rooted in the people's heart, so as to promote the spread of our history and culture. Driven by AR technology, cultural and creative design departments can also use AR technology for simulation, and integrate artificial layout into intelligent design, so that the public can carry out cultural and creative design through the form of voice and hand-painting. The final virtual works are made into an electronic version and sent to the public to commemorate and enrich the creative expression of the creative works [5].

(4) Test identification and release

Implement two interactive digital information and graphics of editing control, and complete the engine development and design. Place an recognition map under the computer camera, check the development effect of the engine, see if the computer has a picture mode, and whether the compilation results are reflected in the development design. When the recognition figure appears on the screen, click the mouse, select the required information in the pop-up box, and type the corresponding code, and you can call the corresponding plug-in to perform this feature. If the display and identification map on the computer has been designed successfully, then the APK installation software can be sent to the mobile terminal to download and install the apk, thus scanning with the product.

4 Conclusion

The development of cultural and tourism integration industry must pay attention to the creativity of tourism products and conduct tourism development with the help of AR technology. It can be applied to sightseeing attractions to improve the experience and participation of tourists, and then promote the dissemination and publicity of culture. For example, relevant local departments cooperate with universities, publishing houses, and social organizations to jointly develop the "Coastal Wandering" series, virtual cultural and creative themes, such as "Square roaming", these cultural and creative works integrating social ideas, distinct personalities and intelligence. At the same time, we should cooperate with media companies, Internet enterprises and other aspects, and promote culture and tourism through a variety of media channels. At the same time, the major enterprises can also jointly form a wisdom creative team, specifically for the area of the existing cultural creative optimization and optimization, for example, through AR technology to create "eloquent shell", "the big fish bite fish, with dinosaurs as subject" ideas, make works more vivid, realistic, ornamental and collection.

Acknowledgements. Zhuzhou social science research project (Research on mutual promotion strategy of Intangible Cultural Heritage Inheritance and tourism cultural and creative product development in our cityzzsk20087).

References

1. Xu, Y., Chen, J.: Research on cultural and creative product development and design under the background of cultural and tourism integration. Think-tank Era (2), 9–10 (2020)
2. Li, Y.: Research on the design of characteristic tourism cultural and creative products based on regional culture. Art Mirror (11), 171–172 (2018)
3. Sun, S.: Research on the design and development strategy of tourism cultural and creative products under the perspective of cultural and tourism integration. Fujian Tea (9), 85 (2019)
4. Ma, H.: On the role of cultural and creative product design and development on the development of traditional culture. Fashion Tomorrow (5), 35 (2019)
5. Liu, X., Shi, Q.: —— takes the region as an example. Popular Lit. Art (17), 64–65 (2019)

Research on Distribution Network Monitoring and Fault Location Based on Edge Computing

Jia Qin(✉), Daquan Yu, and Hexin Wang

Heilongjiang University of Technology, Jixi 158100, Heilongjiang, China
qinjia2012@163.com

Abstract. The research of distribution network monitoring and fault location based on edge computing is a research, focusing on the design of low power consumption, high reliability and high fault tolerance distributed systems. The main purpose of this research is to develop a distributed system architecture for monitoring and locating faults in large-scale networks. The architecture consists of three components: fault location component (FLC), data acquisition component (DCC) and data analysis component (DAC). These three components are connected through an intermediate layer called the Service Interface Controller (SIC). SIC provides necessary interfaces between DCC, DAC and FLC.

Keywords: Distribution network monitoring · Edge calculation · fault location

1 Introduction

Whole power network, and directly contacts with electricity customers. Therefore, its reliable power supply capacity and power supply quality are the direct embodiment of the economic benefits of power enterprises, and correspond to the immeasurable social benefits [1].

The value is small, so it is not necessary to cut off the line immediately, and it is allowed to operate with fault for a period of time. In addition, abnormal over-voltage may occur when the fault occurs. When ferromagnetic resonance over-voltage occurs, its value can reach 4 times of phase voltage, and sometimes it can reach 3.5 times of phase voltage when intermittent arc grounding occurs. These high voltages affect the whole network. If they last too long, they will damage the equipment in the network, damage.

Most domestic people still use the method of manual line patrol. Due to the complex branches of the distribution network, in case of short circuit fault, only the circuit breaker at the outlet of the substation will trip, while in case of ground fault, it will not trip. In case of short circuit and ground fault, even if the trunk line is segmented by switches, only a limited number of sections can be isolated. It often takes a lot of manpower, material resources and time to find the specific fault location, It is difficult to adapt to the new requirements of distribution automation.

J. C. Hung et al. (Eds.): FC 2023, LNEE 1132, pp. 260–265, 2024.
https://doi.org/10.1007/978-981-99-9538-7_39

2 Related Work

2.1 Research Status

Material resources and financial resources to study distribution network fault location methods, and have made many research achievements. In a broad sense, fault location includes fault line selection and accurate location. In terms of specific implementation, the fault location method can be divided into wide area fault section location method using multiple line terminals (FTU) or fault indicator (FPI), and fault location method that directly uses the electrical quantity information measured at the line outlet to calculate the fault distance [2]. The former is used for fast fault isolation of urban distribution network with convenient transportation and high automation level; The latter is used for multi town distribution network with long power supply distance and difficult patrol inspection to find fault points.

Short circuit fault in power system refers to the short circuit between phases or between phases and ground that causes sharp increase of current, sharp decrease of voltage and further damage to electrical equipment.

2.2 Research Status of Edge Computing at Home and Abroad

Therefore, as edge devices generate more data, network bandwidth will become the bottleneck of cloud computing. To solve these problems, edge computing technology came into being. Edge computing proposed by Cisco refers to the use of localized idle computing and storage resources to implement delay sensitive tasks. Specifically, edge computing can first pre process tasks by utilizing limited resources (such as computing and storage cells), and then transfer them to the cloud for further processing. At the same time, it can also independently deal with applications that consume less energy and have lower latency than cloud computing.

3 Distribution Network Structure and Network Model

3.1 Distribution Network Structure

The distribution network is mainly composed of distribution lines and switchgear. The network composition is divided into overhead lines and cable lines according to the types of conductors. The proportion of cable lines used in developed countries such as Europe and the United States is relatively high, while the proportion of cable lines used in China was very low. With the needs of urban network transformation and urban development, the number of cable lines used has increased year by year [3].

Network connection can be divided into radial network. In the planning principle of urban distribution network, it is pointed out that the overhead lines should adopt the wiring mode of multi section and multi connection, and the contact points should operate in an open loop during normal operation. Sectional switch and interconnection switch are used here, both of which belong to load switch. Figure 1 shows several typical connection forms of overhead lines. The cable network structure is similar to the overhead line, but the sectioning and liaison functions are realized through the ring network cabinet.

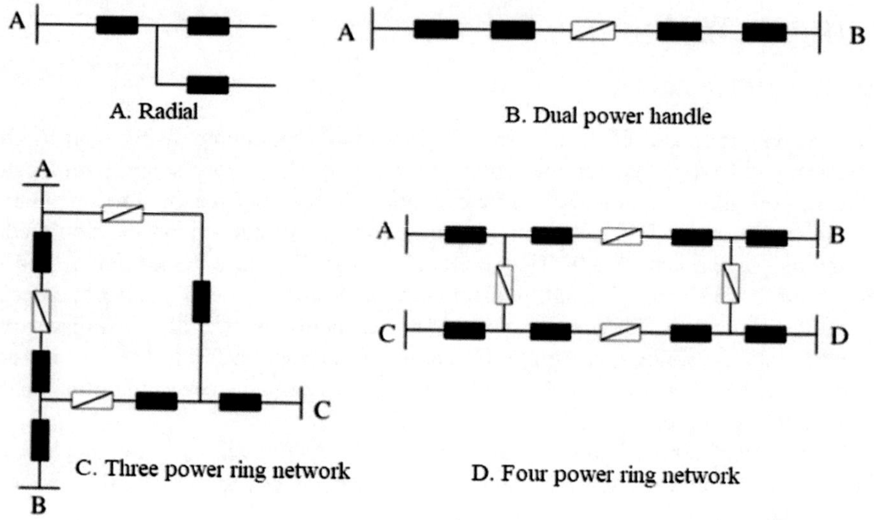

Fig. 1. Connection form of overhead line

In the process of design and optimization of distribution network, flexible connection mode shall be adopted according to the nature of load. Regardless of the connection mode, the following criteria shall be met: network connection shall be simple and flexible, with the purpose of improving system reliability and realizing automation;

The fault handling of distribution automation is closely related to the type of switchgear. Before discussing fault handling, briefly introduce some main switchgear in current distribution automation. Switches on distribution lines mainly include recloser, circuit breaker and sectionalizer. The recloser is a kind of equipment with high degree of automation.When it is confirmed that it is a fault current, it will automatically break the fault current according to the inverse time limit protection for a certain time, and automatically reclose for many times as required to restore power to the line [4]. If the fault is instantaneous, the line will resume normal power supply after reclosing of recloser; If the fault is a permanent fault, the recloser will complete the preset reclosing times (usually three times), confirm that the line fault is a permanent fault, and then automatically lock. The fault line will not be powered on until the fault is eliminated manually, and the recloser closing lock will be released again to restore to normal state.

3.2 Cable Network Fault Handling Mode

The possibility of transient fault of cable line is very small. Once the fault occurs, it is basically permanent.

However, in order to restore power supply in time, the circuit breaker at the outlet of the substation should still be configured with a reclosing function. The most common power supply mode of cable line can be roughly divided into ring network cabinet and switching post mode (this power supply mode can also be used for overhead lines, and multiple reclosing may be required at this time.

At present, most cable networks adopt the ring distribution mode of ring network cabinet. The ring network cabinet can be indoor or outdoor. The ring network cabinet generally has two incoming lines and multiple outgoing lines. The two incoming lines are supplied by two feeders led from two buses of one substation, or by two substations. Open loop during normal operation. In case of fault, fault location is carried out by relying on the information provided by the fault indicator, and then fault isolation is carried out by relying on the load switch of the ring network cabinet and power supply is restored to the non fault section.

4 Research on Distribution Network Monitoring and Fault Location Based on Edge Computing

The edge computing node simply processes some raw data generated by network edge devices and provides complex data to cloud service providers, thus reducing the relay of edge data, as shown in Fig. 2.

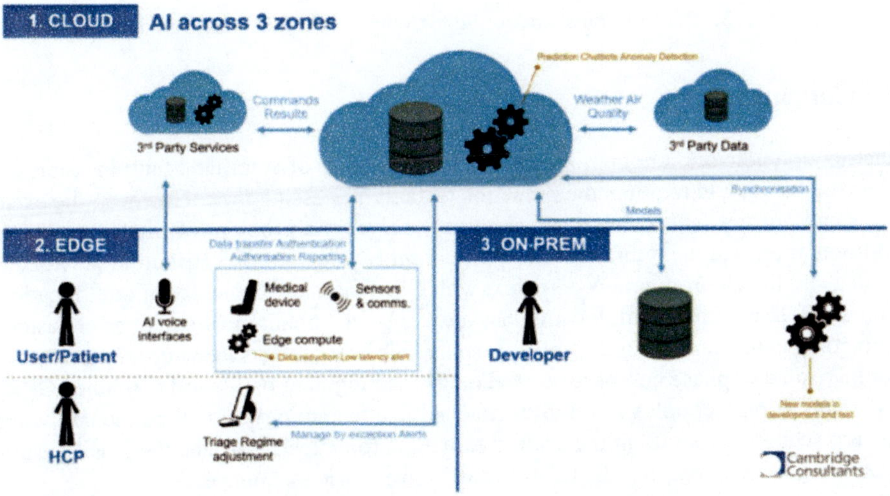

Fig. 2. Location of edge computing in OICT fusion

The fault transient and steady-state data obtained from each node will be very large. If all of them are uploaded to the master station for processing, not only will communication congestion and huge computing burden be caused to the master station. Compared with the traditional master station system and the cloud edge collaboration system designed, the effectiveness and reliability of the system are verified through simulation. The general framework of the edge computing collaborative system applied in the industrial field is shown in Fig. 3.

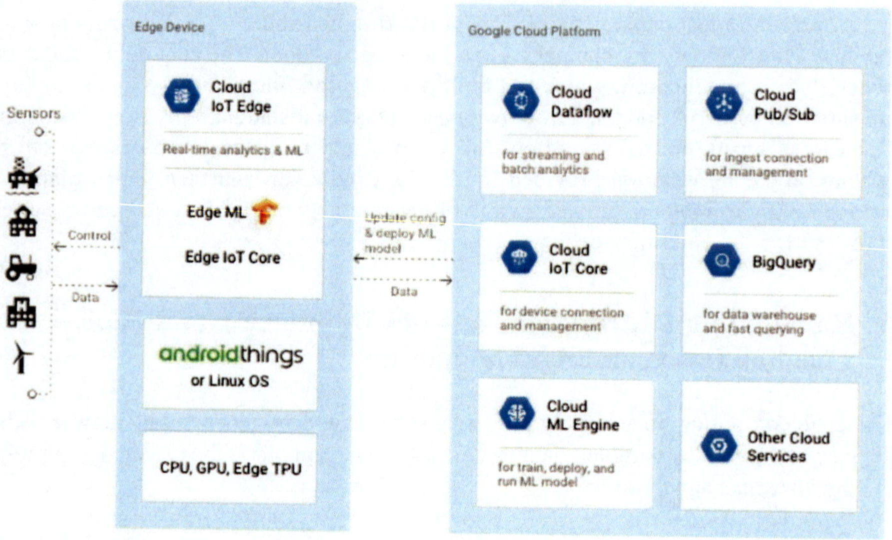

Fig. 3. General Framework of Edge Computing Collaborative System

5 Conclusion

Distribution network automation requires the capability of automatic fault location and pre-processing, and requires the reduction of fault processing time. However, the existing small current fault detection system only solves the problem of line selection. In addition, there are many branches of power supply lines in the distribution network in China, and the environment is complex, which increases the difficulty of fault location. Therefore, line current fault location has always been a blank in China's power system. According to the fault grounding phenomenon, this paper selects the wide area amplitude comparison and phase comparison method as the judgment basis, and combines GPRS communication technology and GPS time service technology to realize the collection of zero sequence vectors in the wide area, and automatically locates the small current grounding fault by judging the relationship between zero sequence vectors.

Acknowledgements. 2022 Project of Fundamental Scientific Research Professional Expenses for Undergraduate Universities in Heilongjiang Province, The project name《Research on distribution network Monitoring and fault location based on edge computing》.

References

1. Yildiz, H., Talluri, S., Xie, X., et al.: Multivariate process control methods for evaluating and monitoring distribution network efficiency. Int. J. Prod. Res. (2021, forthcoming)
2. Ju, Z., Xing, W., Duan, W., et al.: Research on status monitoring and fault alarm device of drop-out fuse. In: Journal of Physics: Conference Series, vol. 1952, no. 3, p. 032084, 9 pp. (2021)

3. Dods, C.: Network monitoring based on distribution of false account credentials, US11032318B2 (2021)
4. Xiangzhenhe: Research on fault location method of distribution network. In: IOP Conference Series: Earth and Environmental Science, vol. 647, no. 1, p. 012049, 6 pp. (2021)

Microbial Fermentation Simulation Based on Swarm Intelligence Algorithm

Qin Qin(✉)

Yunnan University of Business Management, Kunming 650106, China
mingtiandehong99@163.com

Abstract. The role of fermentation analysis in microbial research is very important, but there is a problem of low analysis accuracy. The real-world statistical method cannot solve the problems of yeast evolution and harmful bacteria identification in fermentation research, and the analysis accuracy is low. Therefore, this paper proposes a crowd intelligence algorithm to construct a fermentation simulation model. Firstly, the group theory is used to divide the fermentation process, and the method is selected according to the reading requirements to realize the preliminary observation of the fermentation data. Then, a collection of fermentation studies is intelligently formed and data mining analysis is performed on the yeast. MATLAB simulation shows that under certain requirements, the swarm intelligence algorithm's optimization degree and simulation stability are better than the realistic statistical method.

Keywords: group theory · stability · fermentation studies · simulation algorithms

1 Introduction

Fermentation research is an important way in the field of yeast and plays a very important role in improving microbial research [1]. However, in the microbial fermentation simulation, there is a problem of low analytical accuracy, and it is impossible to study yeast fermentation process accurately [2]. Some scholars believe that the application of swarm intelligence algorithm to microbial research can effectively carry out observation, research and analysis [3], and provide corresponding support for microbial research. On this basis, this paper proposes a swarm intelligence algorithm to simulate microbial research and verify model implementation results.

2 Related Concepts

2.1 Mathematical Description of Swarm Intelligence Algorithms

The crowd intelligence algorithm uses the fermentation process[4], fermentation method and proportion of harmful bacteria to simulate microbial fermentation, and finds harmful bacteria in fermentation research according to the fermentation indicators in microbial research, and forms a fermentation process table—microbial fermentation mini-is

J. C. Hung et al. (Eds.): FC 2023, LNEE 1132, pp. 266–271, 2024.
https://doi.org/10.1007/978-981-99-9538-7_40

performed by integrating microbial fermentation. The crowd intelligence algorithm to simulate the fermentation process results can improve fermentation research.

Hypothesis 1: The fermentation data is $\sum x_i$, the fermentation set is y_i, the fermentation process is x_i, the degree of intelligence is o_i, and the judgment function $f(x_i)$ of fermentation is as shown in Eq. (1).

$$f(x_i) = \lim_{i \to \infty} \sum x_i \cdot y_i \cup \xi^2 \tag{1}$$

2.2 Harmful Bacteria Presentation Process

Hypothesis 2: The simulation function is $f(x_i)$ and the proportion of harmful bacteria is z_i, then the fermentation study selection is shown in Eq. (2).

$$f(x_i) = \frac{\sqrt[3]{x^2 - 4x\xi}}{1 - x^4} \cup z_i \cdot \xi \tag{2}$$

2.3 Microbial Study of the Fermentation Simulation Process

The stability and rationality of fermentation research were analyzed standardly, and microbial fermentation was mapped into the simulation set to determine the effect of abnormal fermentation process. First, microbial fermentation is comprehensively analyzed, and the constraints and weights of the fermentation process simulation are set to support the accurate analysis of the swarm intelligence algorithm. The fermentation data needs to be preprocessed to show that the fermentation process is effective, otherwise the population theory is re-initiated. Degree of microbial fermentation simulation, the results are shown in Fig. 1.

The analysis of microbial fermentation simulation is uniform, which is in line with the objective facts. The selection method is not directional, indicating that the analysis of crowd intelligence algorithm has a high degree of simulation. Microbial fermentation simulation mainly uses population theory to adjust the process, remove harmful bacteria, and revise the fermentation process to make the whole fermentation process more reasonable.

3 Simulation Between Different Fermentation Processes

The swarm intelligence algorithm adopts the simulation degree judgment of stability, and adjusts the corresponding microbial research plan relationship to realize fermentation simulation. The crowd intelligence algorithm simulates and divides the fermentation process and randomly selects different fermentation processes for verification. In the data mining process, the fermentation process requires correlation processing with fermentation research. After the simulation is completed, different methods are compared for fermentation simulation, and the fermentation simulation results with the highest rationality are recorded.

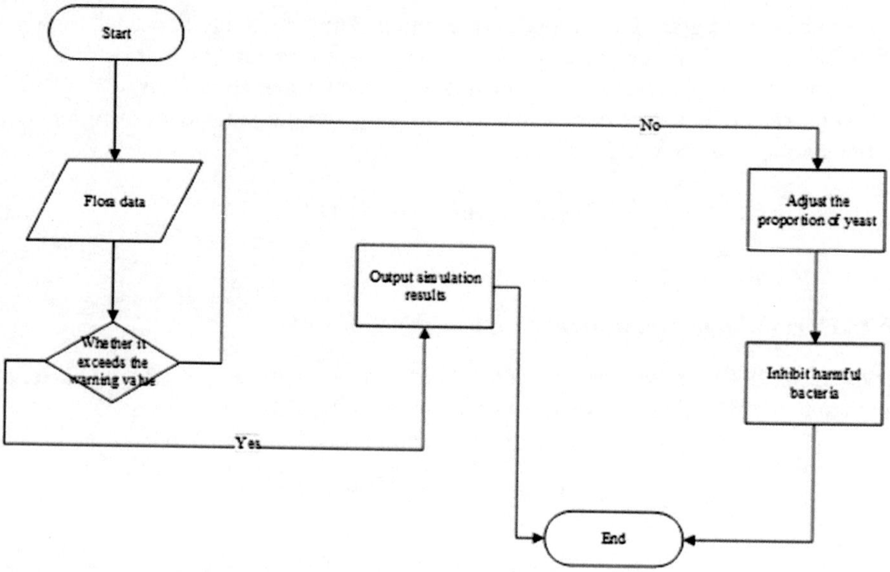

Fig. 1. Swarm intelligence algorithm analysis process

4 Actual Examples of Microbial Fermentation Simulation

4.1 Microbiological Research

In order to facilitate yeast analysis, the fermentation studies of different semesters are the research objects of this paper, as shown in Table 1.

Table 1. Data characteristics of different fermentation processes

Strain	authentication	Amount of data	Analysis accuracy	Constraints
Bifidobacteria	United States	74.91	74.97	74.73
	Europe	74.66	76.55	72.65
Lactobacillus	United States	76.08	72.98	75.31
	Europe	75.52	76.31	76.77
	United States	74.61	77.06	74.52
	Europe	76.71	74.51	74.82

The different fermentation processes simulated in Table 1 are shown in Table 2. single fermentation simulation, the fermentation simulation results of the swarm intelligence algorithm are closer to the actual stability. In terms of fermentation simulation, selection rate and simulation degree of fermentation process, crowd intelligent algorithm fermentation simulation. From the changes in the fermentation process in

Table 2. Simulates the processing of indicators

Strain	Degree of fermentation	Simulation results	Analog adjustment amount
Bifidobacteria	75.96	75.46	76.12
Lactobacillus	76.75	75.69	74.84
Lactobacillus	75.45	74.29	75.12
error	5.90	5.60	4.08

Fig. 4, simulation degree and faster judgment speed. Therefore, the fermentation process of the swarm intelligence algorithm has better processing speed, stability and simulation degree.

4.2 Digital Fermentation Research Ratio

The fermentation process includes the number of probiotics, the number of harmful bacteria, and the proportion of strains. After the standard constraint screening of the crowd intelligence algorithm, the preliminary fermentation simulation results were obtained, and the fermentation simulation results were analyzed. In order to verify the effect more accurately, different fermentation schemes were selected and the overall stability of the fermentation simulation was calculated, as shown in Table 3.

Table 3. Overall picture of fermentation simulation

Analog scale	The degree of simulation	Satisfaction rate
Bifidobacteria	77.43	75.60
Lactobacillus	75.89	76.13
Lactobacillus	75.99	75.12
mean	74.27	76.42
X^2	2.570	7.834
P = 0.002		

4.3 Stability and Rationality of Fermentation Simulation

The stability and rationality of the fermentation simulation are compared with the fermentation simulation, and the results are shown in Fig. 2.

The fermentation simulation stability of the swarm intelligence algorithm is shorter than that of the real statistical method, but the error rate is lower, indicating that the choice of the swarm intelligence algorithm is relatively stable, while the digital fermentation research of the realistic statistical method is uneven. The reasonableness of the above algorithm is shown in Table 4.

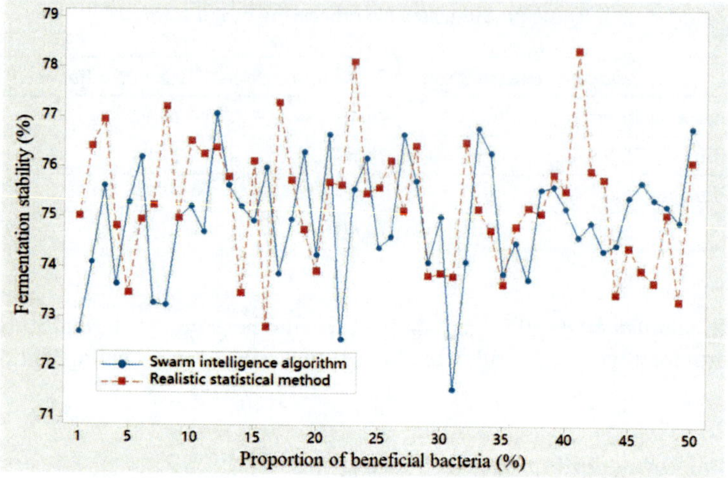

Fig. 2. Fermentation simulation stability of different algorithms

Swarm intelligence algorithm.
Realistic statistical method.

Table 4. Comparison of fermentation simulation results of different methods

algorithm	stability	rationality	error
Swarm intelligence algorithms	74.27	76.24	75.50
Realistic statistical method	75.36	73.95	75.45
P	5.827	7.353	7.859

The realistic statistical method have shortcomings in the stability and rationality of the simulation, and the simulation results change greatly and the error is high. The swarm intelligence algorithm has high stability and is better than the reality statistical method. At the same time, the simulation stability of the crowd intelligence algorithm is greater than 90%, and the error has not changed significantly.The results are shown in Fig. 3.

Intelligence algorithm are significantly better than the realistic statistical method, and the reason is that the swarm intelligence algorithm increases the stability adjustment coefficient and sets the simulation conditions to eliminate the results.

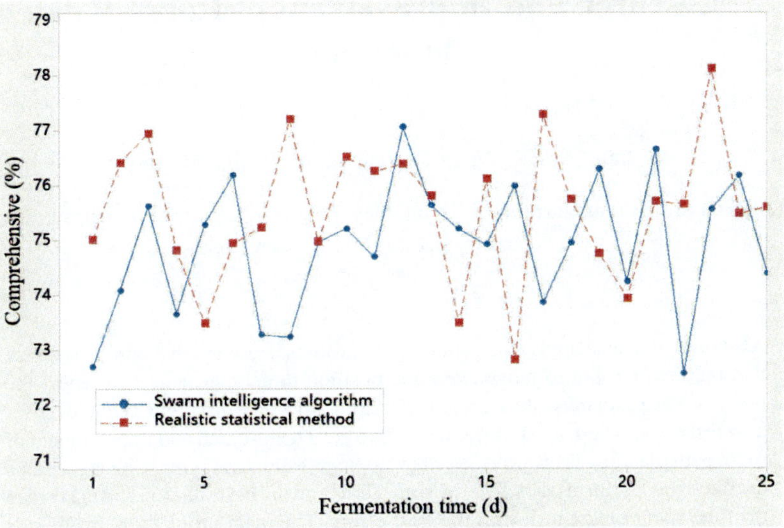

Fig. 3. Comprehensive analysis results of swarm intelligence algorithm

5 Conclusion

In view of the increasing requirements of class scheduling, this paper proposes a crowd intelligence algorithm and combines the swarm theory to analyze the yeast in colleges comprehensively. At the same time, the simulation of fermentation process is analyzed in depth, and a collection of fermentation studies is constructed. Research shows that the swarm intelligence algorithm can improve the accuracy and rationality of simulation.

References

1. Liang, J.C., Li-Juan, L.I.: Multi-objective swarm intelligence algorithm and dynamic optimization of double-layer spherical shell. Spatial Structures (2016)
2. Zong, C.G., Pei-Pei, Y.U., Fei, L.P., et al.: Research on AGV Scheduling Fusion Algorithm Based on Multi-Level Coding Particle Swarm-Genetic Algorithm. Mechanical & Electrical Engineering Technology (2016)
3. Pei, Z.L., Xing, J.S.: Improved Convolutional Neural Network Algorithm Based on Particle Swarm Optimization. Journal of Shanxi Normal University (Natural Science Edition), (2017)
4. Zhou, L., Peng, W., Zou, F., et al.: Particle Swarm Optimization based on self-adaptive mutation. Computer Engineering & Applications (2016)

Research and Improvement of Robot Path Planning

Yan Qin(✉), Wang Ouyang, and Minliang Gong

ShaZhou Professional Institute of Technology, Zhangjiagang 215600, Jiangsu, China
qinyan2021@126.com

Abstract. We also apply this method to our robots. Robot path planning is a task that requires a robot to move from one position to another, and then repeat the process. The robot must always know where it is, in which direction it should move, how fast it should go, and how far it needs to go. The research and improvement of robot path planning is the process of developing, testing and improving algorithms to effectively perform tasks. The main goal is to find the best solution and maximize the time spent on the task with the least errors. The main problem in this field is to find a strong enough solution to deal with unexpected situations or obstacles that may occur during path planning, but it is also fast enough to avoid wasting too much time on repetitive tasks, such as re planning when problems occur.

Keywords: Ant algorithm · Path planning · robot

1 Introduction

The so-called path planning, as the name implies, is to arrange the walking route of the mobile robot reasonably according to certain standards. When there are multiple obstacles in the robot's working environment, planned for the robot, so as to provide technical support for the robot to achieve autonomous movement [1]. The application of intelligent algorithms in the field of path planning research has achieved remarkable results. However, any intelligent algorithm often has various defects when solving path planning problems, such as high complexity of the algorithm, credibility of the results [2]. Especially for robots with complex working environment, it is very easy for algorithms to be difficult to solve. The above situation is the main challenge of intelligent algorithm in path planning research. Many researchers have done a lot of research on these problems.

This paper proposes introduces the pheromone updating strategy, engineer ant strategy and path back optimization strategy based on the double feedback mechanism, The problem of poor smoothness and low security of the planned path is solved, which ensures that the robot can work efficiently.

J. C. Hung et al. (Eds.): FC 2023, LNEE 1132, pp. 272–276, 2024.
https://doi.org/10.1007/978-981-99-9538-7_41

2 Related Work

2.1 Ant Colony Algorithm

In real life, the strategy of ants' foraging is: in the cave where ants live in groups, the "ant boss" first sends some ants to look for food aimlessly, and once an ant finds food, it immediately returns to the cave to report the situation to the "ant boss", and the ant boss will send other ants to look for food along the way the ant has walked.

As more and more ants find food along the path that the ant has walked, there are more and more "pheromones" on this path. The subsequent ants no longer "aimlessly" look for food, but directly follow the "road that the previous generation has walked" to find food through the "pheromones" that perceive the taste of food getting stronger and stronger.

Suppose another ant finds food closer to the cave when he is on the ground, and returns to the cave to report to the "Ant Boss". The "Ant Boss" sends ants to look for food and says to the ants: No matter how far or near the food is, just find the food [3]. Therefore, the subsequent ants face a choice. Although some ants still choose food from a distance, more and more ants will choose food from a distance.

As fewer and fewer ants choose distant food, there are fewer and fewer pheromones on that path. More and more ants will choose this near path because there are more and more "pheromones" on this path, and subsequent ants will not hesitate to follow this "shortcut" to find food.

ACO algorithm first produces a number of artificial ants, then generates and volatilizes "pheromones" in the search process through "artificial ants", and finally obtains the final solution of the problem through continuous iteration.

Taking the problem in the problem description as an example, suppose that there are 10 ants in the distribution center, and each ant knows the distance between the distribution center and customers. In addition, compared with others; Assume that the "pheromone" on the line segment between each two points is 1 initially, that is, the elements on the "pheromone" matrix of 11 rows and 11 columns are all 1. As shown in Fig. 1 below, the ant colony path minimum path planning.

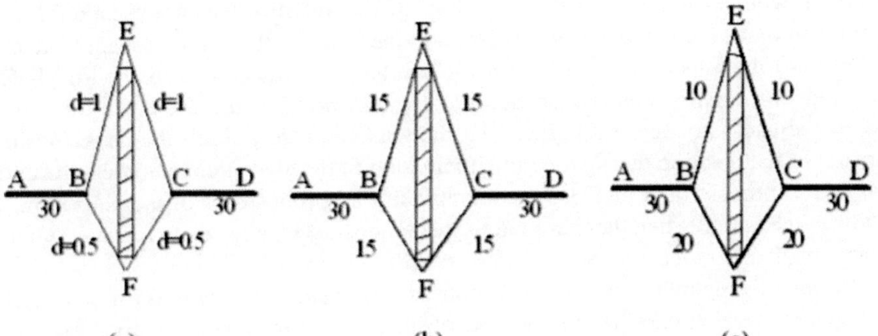

Fig. 1. Ant Colony Path Minimum Path Planning

2.2 Modeling of Path Planning Environment

The goal of path planning is to enable the robot to start from the starting node g egin and safely avoid obstacles to reach the known target node gender along the shortest path.

Set the robot workspace as a two-dimensional structure space and mark it as AS, and the position and size of obstacles are known, and the position and size of obstacles do not change during the robot movement.

Use grids with the same size to divide the AS. The grid size is limited to the free movement of the robot within it. Let the range of free movement of the robot be [o, R.].

3 Research and Improvement of Robot Path Planning Based on Ant Colony Algorithm

When the first ant k arrives at the target node ge ma, because k arrives first and takes the least time, the path it gets is optimal in the current round of ant colony optimization. Repeat until the specified number of algebra times is completed or other set conditions are met.

$$Q = (E(t) - M_2C), \left(\Gamma_{p1}C + M_1C + M_2C\right)\dot{x}_d(t) = Z, \Gamma_{p1}C = F \tag{1}$$

$$\begin{cases} E(t)\dot{x}_d(t) = f(t, x_d(t)) + B(t)u_d(t) + d_d(t) \\ y_d(t) = C(t)x_d(t) \end{cases} \tag{2}$$

when the environment is more complex. In some specific complex environments, for example, when there are some complex obstacles such as U-shaped obstacles in the environment, some ants searching the path may fall into these U-shaped obstacles, thus there is no successor node to choose from, which makes the entire algorithm stagnant [5]. As shown in Fig. 2 below, ant colony algorithm robot path planning.

Another common method to deal with the stagnation problem is to use the ant's fallback strategy: It is determined that the ant is in the trap. At this time, the ant's tabu list is modified to make the ant return to the previous grid of the current grid, and determine whether the ant is still in the trap. If the ant still does not escape the trap, continue to make the ant return to a grid and judge whether the ant has escaped the trap. Repeat until the ants escape the trap[6]. When an ant retreats, it adds the grid before retreating to the ant's tabu list, so that the ant will not fall into the same trap again. The ant fallback strategy can ensure that the ants can safely reach the target position from the initial position during the implementation of the algorithm, complete the entire path search process, and increase the adaptability and robustness of the algorithm. Its disadvantage is that when there are many special obstacles, it needs to repeatedly make the ants back and judge whether the ants are still in the trap.

The specific method of the complete convex processing strategy is as follows: during the grid environment initialization, special obstacles are treated specifically. First, judge whether the special obstacles follow the boundary and the U-shaped opening faces the boundary, or whether the special obstacles are connected with other obstacles and the U-shaped opening is completely blocked by the connected obstacles. If yes, complete

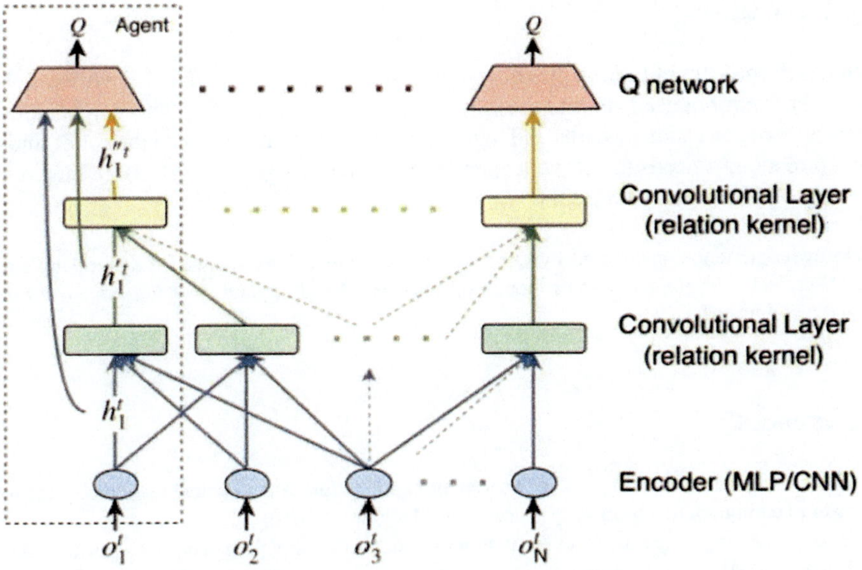

Fig. 2. Ant colony algorithm for robot path planning

convex processing is performed on the special obstacles that meet the conditions, otherwise, general convex processing is performed on the obstacles [8]. In this way, the problem of traps generated between obstacles or between obstacles and environmental boundaries can be eliminated.

The complete convex processing strategy can eliminate the stagnation phenomenon of the algorithm, so that the entire ant colony can continue to search the path.

4 Simulation Analysis

The node set that each ant passes through in an iteration process represents a feasible path (feasible solution), which means that the large. When the number of ants is too large, the pheromone strength of some paths cannot be effectively distinguished, which significantly weakens the information positive feedback effect and slows down the global convergence speed;

Load 30 during experiment $\times$ 30 The starting point coordinates of the environment model path are (0, 0) and the destination coordinates are (29, 29). The value of m is 10 to 28, sampling once every 2, and other parameter settings are run 10 times according to the formula (1) and formula (2) program.

In terms of planning time, the planning time increases linearly with the number of ants. It is easy to understand that for each additional ant, the algorithm will calculate the path for one more ant in each iteration, and the path generation of ants is almost the same.

5 Conclusion

What is the process of finding the robot path from one point to another. It involves several steps: 1) The robot needs to find the way from one place to another. This can be done by using sensors or vision systems and algorithms that calculate. 2) Once the robot finds its own direction, it needs to follow a specific path to avoid getting lost. 3) To ensure that the robot follows this path correctly.

Acknowledgements. sponsored by Qing Lan Project.The achievement of the company's horizontal project "Development and research of automatic handling and spraying system for robot visual recognition".

References

1. Ning, W., Shuai, G., Sheng, B.: Research and application of the critical technology for robot path planning applied to mobile processing. Machinery (2016)
2. Zhou, R., Zhang, Z., Zhai, X., et al.: Practical research on robot path planning algorithm. Mach. Electron. (2016)
3. Yue, X.U., Lin, Z.X., Yao, J.M., et al.: Target search path fuzzy control of robot navigation. Comput. Simul. (2016)
4. San-Zhu, W.U., University, X.S.: Research and implementation of path planning for small robots based on two-point method. Comput. Telecommun. (2019)
5. Wang, Z. , Shen, J.: Based on the regional coverage of cleaning robot path planning. Electron. Sci. Technol. (2017)

Application Practice of 5G Meta-universe in Cultural Tourism Industry Based on Improved AHP Algorithm

Yunjia Qu[1(✉)], Fangyuan Chen[2], and Zeshi Jiang[1]

[1] School of Economics and Management, Harbin University of Science and Technology, Harbin 150080, Heilongjiang, China
quyunjia666@163.com

[2] Hainan Tropical Ocean University, Sanya 572099, Hainan, China

Abstract. The cultural tourism industry is a pillar industry, and its development should keep pace with The Times. Therefore, the industry is actively building the 5G metauniverse, but the construction process is difficult and lacks a clear direction. In view of the current situation, this paper will conduct research on the basis of the improved AHP algorithm. First, it introduces the basic concept of the improved AHP algorithm, and puts forward practical strategies based on the improved AHP algorithm for the current situation of the construction of the cultural tourism industry 5G meta-universe. Through research, the improved AHP algorithm helps the cultural tourism industry find the direction of 5G meta-universe construction, so that the 5G meta-universe can play a full role.

Keywords: Improved AHP algorithm · Cultural tourism industry · 5G meta-universe

1 Introduction

Cultural tourism industry is a pillar industry and one of the traditional industries. In the early development, the industry showed strong economic driving ability, people found that the economic driving ability of the industry gradually decreased, this phenomenon has attracted people's attention, and related fields have conducted research on it. It is found that the main reason for this phenomenon is that the services of the cultural tourism industry are gradually inconsistent with the demands of modern people. That is to say, modern people have different aesthetics and tourism needs and attach great importance to their own tourism experience. However, the services of the cultural tourism industry still maintain the traditional situation and cannot bring convenient and good tourism experience to people. At the same time, the content is often disjointed from modern aesthetics and demands. In this case, the proposal of 5G metauniverse has attracted the attention of the cultural tourism industry. In theory, the construction of 5G metauniverse can provide individuals with personalized technical services by relying on technical means, so as to optimize the tourism experience, satisfy the aesthetics and needs, and restore the economic driving capacity of the cultural tourism industry as usual. However, how to do this in the actual environment is still a problem to be solved.

J. C. Hung et al. (Eds.): FC 2023, LNEE 1132, pp. 277–283, 2024.
https://doi.org/10.1007/978-981-99-9538-7_42

2 Improve the Basic Concept of AHP Algorithm

There is an orderly environment between the levels, representing that the problem is sorted out clearly. Then, a judgment structure is established according to people's subjective cognition of objective facts. In this way, the expert opinions can be combined with the judgment results of analysts to judge the importance of relevant factors in a certain level. During this period, the importance will be quantitatively described. After the completion, the relative importance order weight of factors at each level can be calculated by mathematical method. The higher the ranking of the factors, the more important and urgent, enabling people to make reasonable arrangements for the next work, such as the implementation of one plan before the implementation of other plans [1–3].

The AHP algorithm is widely used in the commercial field at first, which is commonly seen in the quality determination of the enterprise's own decision scheme. That is, the enterprise often uses the AHP algorithm to divide the relevant factors of the decision scheme into several levels, which are usually named by the criteria, the scheme, etc., and then carries out quantitative and qualitative analysis on each level and the factors within the level, and combines with each other to make the final judgment. The judgment result can not only show the priority and weight ranking of related work in the scheme, but also tell the enterprise what results a scheme can achieve finally through the final result, and the enterprise can make a judgment according to its own expectations [4].

In fact, AHP algorithm tends to be perfect in many cases. The reason is that it can establish a recursive hierarchical structure of problems, as shown in Fig. 1. Under this structure, the analysis and judgment process of problems will become very simple, and the comparison process between factors will be more convenient. At the same time, this structure can also transform many factors that cannot be quantified into operational indicators, so that the judgment results are more perfect. Most importantly, the AHP algorithm. so the judgment results under the algorithm are more objective, reasonable and accurate. However, AHP algorithm also has its own shortcomings, such as the judgment matrix of the algorithm depends on people, if the subjective and objective of people are not unified, there is a certain deviation between the subjective and the correct answer, the result will be inaccurate, which indicates that the algorithm is susceptible to the influence of human factors. In addition, AHP algorithm has other advantages and disadvantages, as shown in Table 1.

From this point of view, the defects of AHP algorithm will be exposed in a certain scene, and the construction of 5G meta-universe in the cultural tourism industry is such a long-term, involving many difficult to compare factors, that is, some factors do have obvious differences in weight, but the matters represented by the factors are extremely important, at this time, the factors will become difficult to compare. At the same time, meta-universe construction is very novel for the cultural tourism industry. There is no relevant foundation in the original construction, which is equivalent to starting from scratch. Therefore, there are a lot of factors involved, indicating that the defects of AHP algorithm will be exposed. In view of this phenomenon, it is necessary to improve the AHP algorithm. The basic idea of improvement is as follows: because the core idea of related elements under certain criteria, and then judge the quality of the scheme according to the sum of the weight, all kinds of shortcomings will be exposed in the process, indicating that as long as the optimization of factor weight calculation

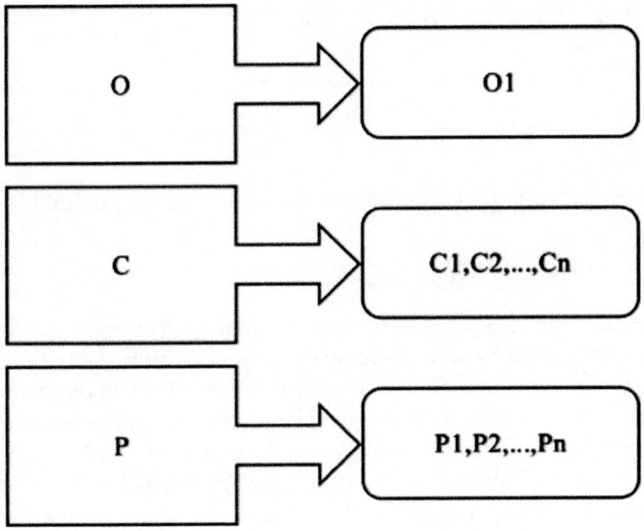

Fig. 1. Recursive classification hierarchy of AHP algorithm

Table 1. Other advantages and disadvantages of AHP algorithm

Advantages	Disadvantages
The process is simple	Comparison of factors is sometimes difficult
Comprehensive elements	If there are too many factors

method can complete the algorithm improvement. According to this idea, three weight calculation methods are selected in this paper, as shown in Table 2.

Table 2. Three methods of factor weight calculation

The name of the	The characteristics of
The normalization algorithm	The numerical values obtained are generally large, but the precision is relatively high
Geometric mean vector solution	The obtained values are generally small, but the accuracy is insufficient
Arithmetic mean algorithm	The obtained values are generally moderate and the accuracy is relatively high

Combining the above three algorithms, this paper mainly chooses arithmetic average algorithm, which is combined with the classical AHP algorithm. The expression of improved AHP algorithm is shown in Formula (1).

$$\omega_i = \frac{1}{n}\sum_{j=1}^{n}\frac{a_{ij}}{\sum_{k=1}^{n} a_{kj}} \tag{1}$$

Construction strategy of 5G meta-universe for cultural tourism industry.

2.1 Necessity Analysis of Technology Selection

It does not contain technology itself, but is just a concept. However, the products under this concept are integrated by several digital technologies, such as intelligent technology and technology. The concept product relies on advanced 5G technology to connect with the real environment, so that people can get other digital technology services in the meta-universe in the real environment. On this basis, if the cultural tourism industry wants to build 5G meta-universe, it must first do a good job of digital technology selection, which involves the necessity of technology selection, that is, 5G meta-universe does not exclude any digital technology. Therefore, on the macro level, 5G meta-universe construction includes all digital technologies and there are a lot of technology types. On the micro level, for application considerations, It is impossible for the cultural tourism industry to introduce all digital technologies into the construction of its own 5G meta-universe, but it will only choose the digital technologies that can be used for its own use. At this time, industrial organizations should analyze the necessity of digital technologies and eliminate unnecessary digital technologies.

However, in the analysis of technical necessity, many industrial organizations have encountered difficulties in determining whether a certain digital technology is necessary or not. The reason for this phenomenon lies in the technological selection thinking of industrial organizations, as shown in Fig. 2.

According to this idea, technology selection is to judge whether the relevant digital technology can meet the user needs according to the user needs. If it can meet the user needs, a high degree of matching will be produced between the demand and the technical function. If the matching degree reaches the standard, it means that the digital technology is necessary to be selected. However, in the actual situation, industrial organizations will find that there is a good match between the same user needs and a variety of digital technologies. According to the existing thinking, it means that these digital technologies need to be selected, which is obviously an unreasonable result. Therefore, industrial organizations encounter difficulties in technology selection, and the selection necessity judgment is very vague.

To solve this problem, industrial organizations can improve the existing ideas by improving the AHP algorithm. The improved ideas are shown in Fig. 3.

In other words, industrial organizations can establish multiple technology selection schemes according to the existing ideas. The specific technology items and quantity in the schemes are uncertain, which can only be played out. As long as each technology is truly related to some users' needs, it is feasible to use the improved AHP algorithm instead of manual analysis to carry out hierarchical analysis of the schemes, and compare the weight of related technologies in each level of the schemes in the same needs in pairs.

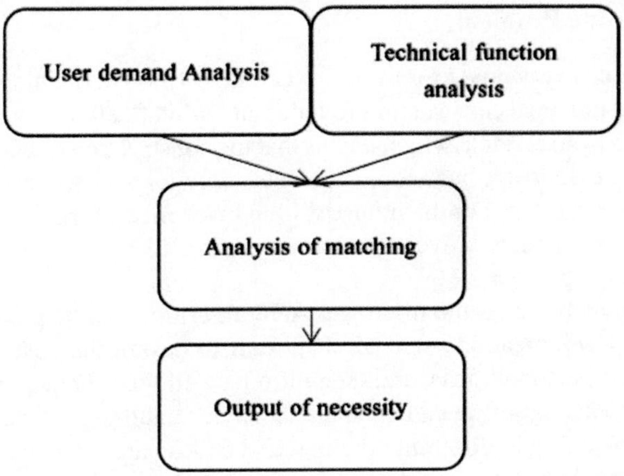

Fig. 2. Existing thinking of cultural tourism industry organization technology selection

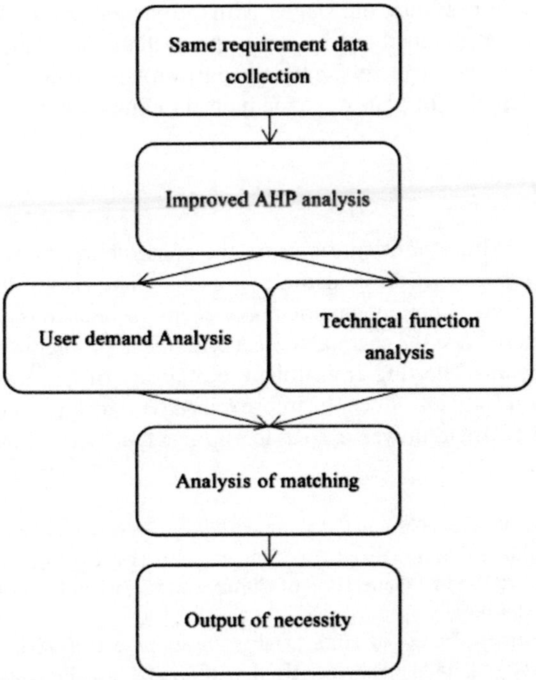

Fig. 3. Improved technology selection ideas

Finally, make an objective and accurate judgment on the quality of the scheme. At this time, the performance of the two technologies in the same needs can be judged by improving the AHP algorithm for analysis. For example, it is necessary to dig into the shopping demand of users in terms of their consumption orientation.

2.2 Cost Planning Problem

5G meta-universe is very novel for the cultural tourism industry, which means that there is a lack of meta-universe construction foundation within the industry. The construction should be started from scratch, which means that the construction of 5G meta-universe in the cultural tourism industry has a lot of contents, and a lot of costs need to be invested to complete the construction. For the industry, if an investment is made directly according to the total cost of 5G meta-universe construction, it is likely to encounter the problem of cost waste in the process.

On this basis, relying on the improved AHP algorithm can help solve the problem, that is, industrial organizations can ask their staff to design the cost planning scheme according to their personal ideas, and then introduce all the schemes into the improved AHP algorithm. The algorithm can compare the controllability of different construction stages and cost planning in different schemes, and choose the most controllable scheme. For example, Plan A divides the whole construction process into three stages, with the cost ratio of each stage being 25%, 39% and 36% respectively. After calculation, the cost controllability of each stage is 78%, 69% and 92% respectively. Plan B divides the whole construction process into four stages, with the cost ratio of each stage being 21%, 33%, 17% and 29% respectively. After calculation, the controllability of no phase is 82%, 88%, 73% and 94% respectively. By comparison, the mean controllability of Plan B is much higher than that of plan A, so the industry composition can choose Plan B.

3 Conclusion

To sum up, for the cultural tourism industry, the construction of 5G meta-universe is imperative, but the construction is also a major challenge facing the industry. There are many problems that humans cannot make accurate decisions, and the improved AHP algorithm can replace the manual to deal with these problems and make accurate and objective decisions. Starting from this, the majority of cultural tourism industry organizations should fully introduce the improved AHP algorithm and use it to promote the construction of 5G meta-universe, so as to improve their service quality and enhance economic benefits.

Acknowledgements. 1. 2022 National key support field Project of National Innovation and Entrepreneurship Training Program for College Students (202210214150): "Wonderland -- Development of Collaborative Product Generation of Cultural and Tourism Experience Content Based on Meta-universe Technology".

2. Scientific Research Project of High Quality Development of Tourism, Culture, Radio, Film and Sports Industry in Hainan Province (HNLW (ZC) 22–23): "Practical Exploration of the Concept of Metaverse and Its Impact on Culture and Tourism Development".

3. Key Laboratory of Island Tourism Resource Data Mining and Monitoring, Ministry of Culture and Tourism (KLITRDMM 2022—20).

References

1. Liu, Z.B.: Research on competitiveness of cultural tourism industry in Datong city——Analysis based on new diamond model. Territory & Natural Resources Study (2018)
2. Fei, Z., Economics, S.O.: Trade and Management. Measurement and Evaluation of Cultural Tourism Industry's Competitiveness Around Taihu Lake Area Based on Entropy-TOPSIS Model. J. Lanzhou Univ. Financ. Econ. (2018)
3. Zhi-Yuan, X.U., Chang-He, L.V.: Competitiveness evaluation of urban tourism industry in Yangtze River based on AHP model and BP neural network model. China Population, Resources and Environment (2017)
4. Li-Hui, W.U., Wang, S.L., Ma, D., et al.: Evaluating efficiency of cultural tourism in Hubei province based on DEA modeling. J. Hubei Univ. Educ. (2017)

Automatic Recognition of Civil Aviation Maintenance Records Based on Deep Learning

Jinlong Shang(✉)

Sanya Aviation and Tourism College, Sanya 572000, Hainan, China
jl_shang@126.com

Abstract. The aviation industry is not only a huge economic engine, but also one of the most important sectors that play a key role in safety and security. The analysis of aircraft maintenance records and engine maintenance records is essential for airlines, airports, aviation authorities and manufacturers to ensure that their products are properly maintained. Although this task is important, an automation solution has not been developed. This paper presents a method to automatically recognize civil aviation maintenance records from images. The main task is to identify aircraft maintenance records, engines and airframe components being processed. To solve this problem, we use the depth learning algorithm, and use ground data for pre training. We test by training the model on the image dataset and using the test dataset to compare with the classification results of human experts.

Keywords: Deep learning · Civil aviation maintenance · Automatic identification technology

1 Introduction

The development of civil aviation industry has been an important indicator of China's industrial development leve[2]l. In China, many industries and fields have carried out information construction. By introducing information construction into civil aviation maintenance, we can save maintenance costs and improve maintenance efficiency.As a result, the maintenance cost increases during the regular inspection and maintenance process or the route maintenance process because of the long logistics route [2]. Based on this, this paper studies the automatic recognition technology of civil aviation maintenance records based on deep learning.

2 Related Work

2.1 Civil Aviation Maintenance

All parts in the aircraft maintenance record only refer to parts excluding precision instruments such as engines. Whether new parts or parts being used in the aircraft maintenance record, they should be tested by professional NDT equipment. The process of nondestructive testing is the five stages of source change detection display interpretation. The

J. C. Hung et al. (Eds.): FC 2023, LNEE 1132, pp. 284–288, 2024.
https://doi.org/10.1007/978-981-99-9538-7_43

equipment generates an acoustic field/thermal field/electric field/magnetic field, which is the first step source [3]. Then the interaction between the source and the object under test causes changes, which means that the thermal field and the aircraft maintenance record parts start to interact. Then the detector on the equipment detects the changes caused by the interaction between the two, and the equipment system records and displays the signals sent by the detector, Finally, the system interprets the signal in combination with the inspected materials, and the operator can see the test results of the corresponding inspected items on the display screen of the equipment. Because the entire testing process does not damage the performance of the inspected object, it is called nondestructive testing, which is the most popular process and theoretical basis of nondestructive testing equipment.

The modern civil aviation industry has reduced the probability of aircraft maintenance record accidents caused by mechanical failures to the minimum before takeoff, regardless of the mechanical structure or technology of the aircraft maintenance record itself [4]. The main cause of modern civil aviation passenger plane crashes is still human factors. Of course, human factors here are also a very complex comprehensive concept, not only referring to the behavior of a single natural person.

2.2 Image Resolution Reconstruction Technology

The image super-resolution reconstruction technology can perform high-definition reconstruction of single frame and multi frame images of fuzzy targets in low resolution images, restore the target texture details in the images, process low resolution remote sensing images containing aircraft maintenance record targets, and compare and analyze the results before and after processing.

The CNN based method of discriminant learning has become the mainstream method of image super-resolution reconstruction because of its fast speed and end-to-end learning. Therefore, this paper mainly selects the CNN based end-to-end discriminant learning method to achieve Low to High super-resolution reconstruction.

Generation network: it adopts the structure of encoding and decoding, and consists of 17 residual modules, which are divided into three groups, including 2 sets, 3 sets and 12 sets of BatchNorm + ReLu + Conv 3 * 3 convolution operations respectively; The resolution of the input image is first reduced by 4 times by using the pooling layer, and then increased by 2 times by using the pixel shuffle layer.

Identification network: including 6 residual modules, divided into 6 groups, each group includes 2 ReLu + Conv 3 * 3 convolution operations, followed by the full connection layer, and the last two are the maximum pooling layers to improve the resolution; The residual module used is shown in Fig. 1.

The image is randomly flipped, scaled, rotated and other operations using data augmentation technology to increase the diversity of sample data. The training process lasts 200 iterations, and there are about 570000 generative network updating processes, in which the update rate ratio of the discriminant network and the generative network is 5:1; In the final stage, the identification network and generation network are updated for

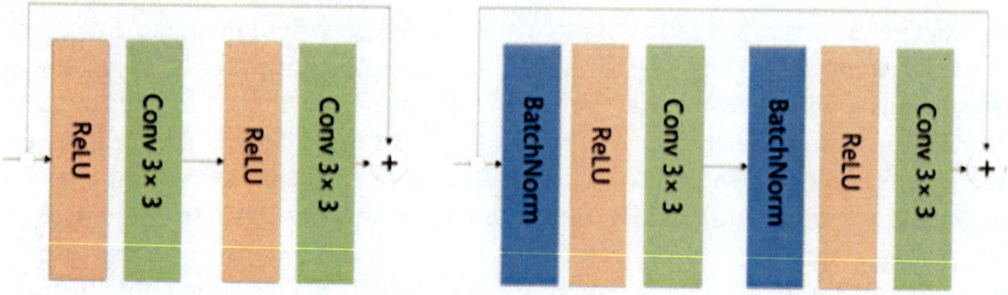

Fig. 1. Residual module

2000 times at the same time for precise adjustment.

$$J_c = \sum_{i=1}^{k} \sum_{p \in C_1} \|p - M_i\|^2 \tag{1}$$

$$AVF(x_i) = \frac{1}{m} \sum\nolimits_{f=1}^{m} f(x_{ij}) \tag{2}$$

CNN uses massive sample data for training, and constantly learns and modifies the mapping function relationship from the data, so as to enable the network model itself to obtain layer by layer abstract information until obtaining the essence of the data. According to the direction of data propagation, the training of convolutional neural network can be divided into two main processes, namely, forward propagation process and back propagation process. In the forward propagation process, the training sample data flows in from the input layer, through the processing of intermediate hidden layers such as convolution layer and downsampling layer, and finally flows out from the output layer. In the back propagation process, the error between the sample label and the actual output is calculated, and the error of each hidden layer is calculated layer by layer. The gradient descent method is used to adjust the model parameters.

The deconvolution layer is generally followed by the downsampling layer. Each feature map output from the downsampling layer is only related to the feature map of the same level of the previous layer, so the number of feature maps of the previous layer is equivalent to that of the current layer.

3 Automatic Recognition Technology of Civil Aviation Maintenance Records Based on Deep Learning

RPN network adopts a similar sliding window mode, and the convolution of the convolution kernel is regarded as the sliding process of the sliding window. For each result generated by a given convolution kernel, there is a predefined anchor box corresponding to it. Because of its convolution property, anchor has good translation invariance. Each anchor defines its scale and aspect ratio in advance, so that multi-scale and multi-scale detection can be realized in a convolution layer. Compared with the multi-scale detection method (such as image pyramid and multi convolution kernel) proposed before this method, this method is more efficient. The network structure uses 1×1 convolution, which is actually a full connection layer, but it uses the form of convolution, which is more efficient. The convolution recognition layer is shown in Fig. 2 below.

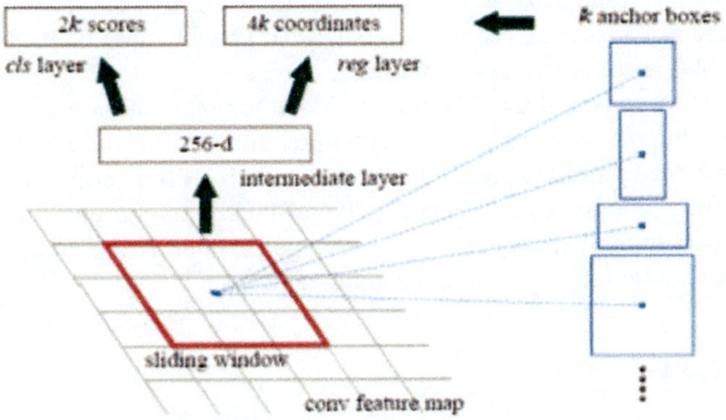

Fig. 2. Convolution recognition layer

The original YOLOv5 network has three scale detection layers: large, medium and small. These three scale detection layers output the input image after down sampling. In the research, it is found that some aircraft maintenance records in remote sensing images are very small in size and densely arranged, which adds a detection layer for small objects, is more suitable for aircraft maintenance record object detection in remote sensing images. The image data detection of aircraft maintenance record is shown in Fig. 3 below.

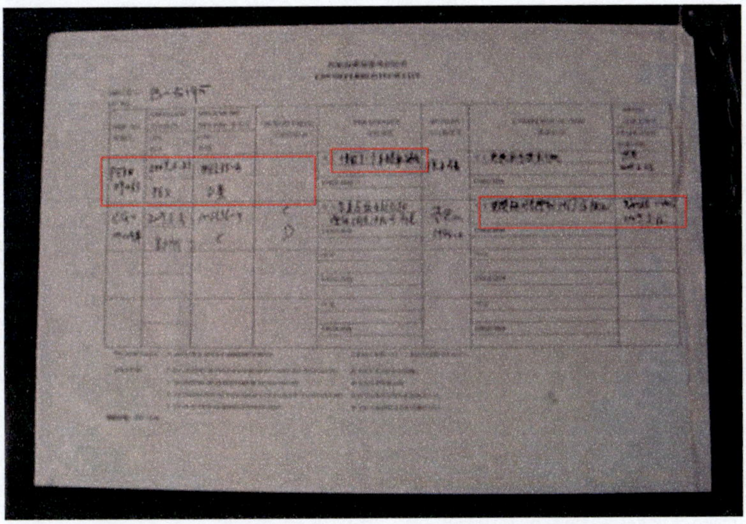

Fig. 3. Aircraft maintenance record image data detection

All our models are based on the Resnet-50 pre trained on ImageNet. In order to maintain the same level as the original Faster RCNN, we use res4f for RPN feature extraction instead of res5c, the last convolution layer of Resnet-50. For FPN, we only use Pz to Ps as the feature extraction layer instead of Po. Before image input, we will scale the image to the size of 768 on the short side. For the classified sub network, the batchsize is 2000 in training and 40 in testing.

The RPN module is still used in the candidate box generation part. The difference is that, because the network is a pyramid structure with multiple levels, there is an RPN module at each level, and the sizes of anchors in each RPN overlap but are different, we still use nine predefined sizes, but only five of them are adjacent to each level; The predefined aspect ratio of anchor remains unchanged at 0.41. In addition, each RPN will only return the gradient of the first 256 candidate boxes.

4 Conclusion

Aviation maintenance records are critical to aviation safety. Analyzing aviation maintenance records is an important task for airlines and the aviation industry. It helps predict potential failures, avoid accidents and optimize operational efficiency. Automatic recognition and maintenance of records based on deep learning is a promising research field in the past few years. However, due to human errors, equipment failures and other reasons, these important documents often need to be updated or corrected. In most cases, this process requires the participation and collaboration of multiple people to effectively find solutions. This paper presents a new method of automatic recognition of civil aviation maintenance records based on deep learning technology.

References

1. Yan, X., Wei, L., Li, J., et al.: Research on automatic recognition of pointer meter reading based on deep learning algorithm. J. Phys. Conf. Ser. **1865**(4), 042017 (2021)
2. Yang, W., Li, J., Peng, W., et al.: a rub-impact recognition method based on improved convolutional neural network. 2020(4), 283–299
3. Ren, L.: The application of internet+PBL in the civil aviation engineering maintenance practice. Experiment Science and Technology (2018)
4. Ni, X., Wang, H., Che, C., Hong, J., Sun, Z., et al.: Civil aviation safety evaluation based on deep belief network and principal component analysis - ScienceDirect. Saf. Sci. **112**, 90–95 (2019)

Research on the Training Prediction Model of Medical and Nursing Health Professionals Based on Fuzzy Neural Network Algorithm

Song Limei(✉) and Ruyong Zhang

Shandong Institute of Commerce and Technology, Jinan 250103, Shandong, China
sdszs123@163.com

Abstract. The training process of medical and nursing health professionals is complicated and difficult, so the training efficiency and quality can not be guaranteed. In order to change this situation, the related fields believe that the talent training prediction model can be constructed by fuzzy neural network algorithm, and the goal can be achieved by relying on the model. This paper mainly discusses the basic concept of fuzzy neural network algorithm, and then establishes the prediction model combined with the training process of medical and nursing health professionals, and finally introduces the specific role of the model in talent training, and carries on the demonstration.

Keywords: Fuzzy neural network algorithm · Medical and nursing health professionals training · Prediction model

1 Introduction

The medical and nursing health major is related to human health, so the quality of the professional personnel training must be guaranteed. At the same time, the personnel training has a certain time limit, so the training efficiency must also be guaranteed. And reality, past medical talent training mode to raise the health professional talents cultivation process efficiency is slow, and the quality is uneven, the overall is low, the reason is that in the past mode is more extensive, only general in talent cultivation, many problems to deal with, and makes the training efficiency and the quality is low, and not guaranteed. In the face of such situation, the medical health professionals need to be talent training mode reform, for this field thought the changes in the fuzzy neural network algorithm can provide help, using the algorithm builds the forecast model of personnel training, can prejudge the future direction and trend of training of personnel, etc., so that the relevant organization to adjust in a timely manner, to avoid or deal with the problem, promote the efficiency and quality of talent training, and to be guaranteed.

J. C. Hung et al. (Eds.): FC 2023, LNEE 1132, pp. 289–294, 2024.
https://doi.org/10.1007/978-981-99-9538-7_44

2 Concept of Fuzzy Neural Network Algorithm

Fuzzy neural network algorithm refers to a series of mathematical methods used in fuzzy neural networks to operate on different classes of fuzzy problems. Its main function is to generate specific learning logic in fuzzy neural networks. The structure of fuzzy neural network is shown in Fig. 1.

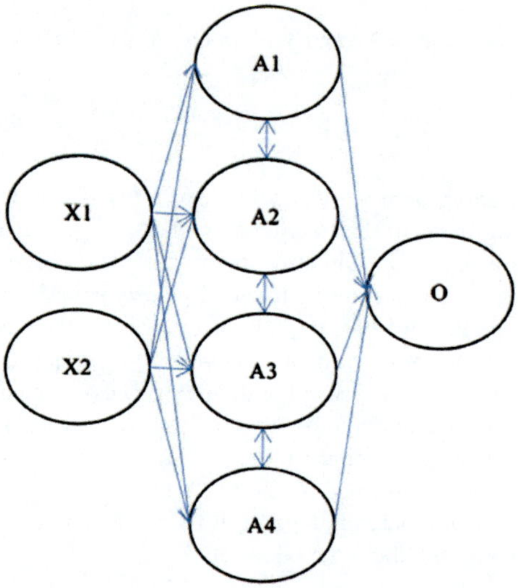

Fig. 1. Structure of fuzzy neural network

Combination can be seen in Fig. 1, as X1, X2, both were input into A hidden layer after will gradually break down for more than four hidden nodes (after decomposition, because work to base 2 (minimum number is 4), they are denoted as A1, A2, A3 and A4. There is an interactive relationship between these hidden nodes, and the interaction process represents that the model is learning and will eventually output a result O, which can be recorded by the model and become the knowledge of the model or used by the external system [1–3].

Such as in the Internet model learning involves many adjustable parameters, therefore, it is necessary to obtain answers through parameter learning, and in the process, it is necessary to choose complex algorithms that can decompose, define and integrate parameters. Common fuzzy neural network algorithms are shown in Table 1. Non-algorithm has its own advantages and disadvantages, especially in the scope of application, some algorithms can only be used to deal with a certain kind of problem, so a reasonable selection of algorithms is the key to play the role of fuzzy neural network [5–6].

According to Table 2, backpropagation algorithm has distinct characteristics and is basically the most convenient algorithm among fuzzy neural network algorithms, but its defect is also very obvious, that is, it is easy to fall into local minima when dealing with complex problems. Therefore, the selection should consider the complexity of the problem and need to be evaluated in advance.

Table 1. Common fuzzy neural network algorithms

Algorithm name	Characteristics of
Back propagation algorithm	Complete logic and simple structure
Genetic algorithm (ga)	It is suitable for optimization problem learning
Rough set algorithm	Strong learning and processing ability, can effectively reduce the dimension of knowledge space

3 Establishment of Training Prediction Model for Medical and Nursing Health Professionals

3.1 Algorithm Introduction

Rough set algorithm is a kind of all of the attributes according to the analysis of the target to establish several collections, integrating these collections based rough set again, and then according to the study of rough set, prompting machine system can understand the concept of analysis, so it can target based on the concept of combining analysis makes a fuzzy neural network to predict the current information [7–10]. At the same time, because is forecast, so it is not only accurate results, while facing different forecast results, the rough set algorithm driven by fuzzy neural network gives the probability of each prediction results and the predicted results of different coping strategies, or other decision-making information, the artificial just need to make the right choice, and perform the relevant work. According to the above principles, the basic model of rough set algorithm is shown in Eq. (1).

$$N = \begin{Bmatrix} A1, A2, ..., An \\ B1, B2, ..., Bn \\ C1, C2, ..., Cn \end{Bmatrix} \quad (1)$$

where N is the set value of the rough set of the analysis target, A, B and C are the attribute classification items of the analysis target respectively, and each classification item contains several attributes, denoted as 1,2..., n.

3.2 Model Establishment

According to the rough set algorithm, medical health professional talent training prediction model is established in this paper is divided into two steps: one target attribute extraction analysis, namely the medical keep health work has a lot of analysis of the target of talent cultivation, to use the rough set algorithm to calculate, you need to distinguish the analysis target, to extract a single mathematical analysis of the target, in order to establish the rough set. This paper takes the vocational comprehensive level of students in the training of medical and nursing health professionals as the analysis objective and extracts its attributes. See Table 2 for details. Secondly, the model is established, that is, according to the attributes given in Table 2, the relevant information of each attribute within a certain time is collected, and then the prediction model can be established according to the weight relationship between the attributes. See Eq. (2) for the model.

$$N = \begin{Bmatrix} A1, A2, ..., An \\ B1, B2, ..., Bn \\ C1, C2, ..., Cn \end{Bmatrix} \cap O = \begin{Bmatrix} D1, D2, ..., Dn \\ E1, E2, ..., En \\ F1, F2, ..., Fn \end{Bmatrix} \quad (2)$$

where O represents the analysis target, D, E and F are first-level attributes respectively, and the related second-level attributes are denoted as 1,2..., n.

Table 2. Attributes of students' vocational comprehensive level

Primary attributes	Secondary attributes
The theory of cognitive	The knowledge of cognitive
	Personal understanding
Practice	Skill levels
	The accumulation of experience
Work accomplishment	Moral accomplishment
	The spirit of accomplishment

4 Function Introduction and Demonstration

4.1 Functions

The prediction model constructed under the fuzzy neural network algorithm has two main roles in talent training, namely talent development prediction, training efficiency and quality prediction. According to the logical relationship between the prediction results of the two roles and the second-level attributes, the algorithm can understand the concept of talent training, and then output the results. The details of each role are as follows.

4.2 The Empirical

4.2.1 Empirical Plan and Preparation

In order to better implement the empirical solution, this paper carried out the preparation system is set up algorithm, namely the rough set algorithm as a kind of fuzzy neural network algorithm, and must start with a program model of embedded system, driven by system operation, or manual operation will consume a lot of time, are also susceptible to the interference of artificial subjective factors, so the need to build algorithm system. The structure of the algorithm system built in this paper is shown in Fig. 2.

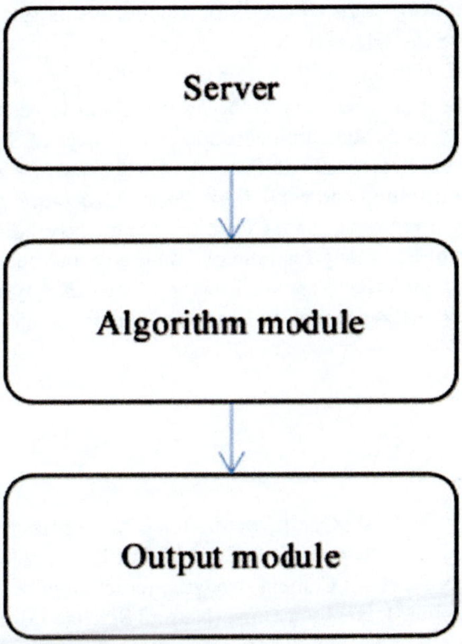

Fig. 2. Structure of the algorithm system

Combined with Fig. 2, the algorithm system is mainly based on a physical server, use of server resources to build the system framework, and then in the framework of the modular program development, select the Java development language, relying on its good realizability, realizes the algorithm module, rough set and combined with fuzzy neural network, it can be operation. The calculation results will be displayed through the output module.

4.2.2 Empirical Results

Rough set algorithm provides the trend data of the development of talent training efficiency and quality, from which it can be seen that it takes 500 h to complete the current talent training task, and the quality score is 70 points, compared with the previous data of 410 h and 86 points, indicating a downward trend. Then the data increased to 390 h and 93 min, indicating that the trend was upward, so the trend prediction result of the model was accurate. At the same time, the prediction results point out that the reasons leading to the slow decline of the development trend of training efficiency and quality may be related to the number and level of teachers. It contains the standard answer, so it also gives the decision-making scheme to improve the level of teachers. Overall, the prediction results are accurate, so the rough set algorithm is effective.

5 Conclusion

In conclusion, the correct selection of available to construct medical health prediction model of talent cultivation, the model can help people many projections for personnel training situation, including prediction, cultivating talents development efficiency and quality prediction direction, such as prediction accuracy is guaranteed, convenient for people to make accurate decisions.

Acknowledgements. The article belongs to the project about Research on the cost of running schools in major categories of public administration and services of Shandong Institute of Commerce and Technology; the project about Research on the cost of running schools in major categories of public administration and services of China Association of Higher Education.

The article belongs to the project about Pilot project of the credit bank of the medical and health professional group of Shandong Institute of Commerce and Technology; the project about Investigation and study on the effectiveness of the apprenticeship system with Chinese characteristics of 2022 Higher Education Scientific Research Planning Project of China Association of Higher Education PG0409.

References

1. Zhou, H., Bai, X., Feng, B., et al.: Quality prediction of wine based on fuzzy recurrent wavelet neural network. Computer Measurement & Control (2017)
2. Guo, Y., Chen, P., Zhao, L., et al.: Cement strength prediction of optimized bp neural network based on genetic algorithm. J. Nantong Univ. (Natural Science Edition) (2016)
3. Li, X., Ming, Y.: Research on BP neural network state prediction method for electronic system based on genetic algorithm. Electronic Measurement Technology (2016)
4. Ding, S., Wu, Q., Chang, X., et al.: Prediction of laboratory equipment support based on grey relation analysis and BP neural network. Foreign Electronic Measurement Technology (2016)

Empirical Study on the Relationship Between Renewable Energy Electricity Consumption and Carbon Emission Based on Genetic Algorithm

Yun Su(✉), Kai Zhang, Shu Liu, Shanshan Shi, and Zhaohui Wang

State Grid Shanghai Electric Power Company, Shanghai 200120, China
oppenvi@163.com

Abstract. Renewable energy is the main energy in China. It has great development potential, but it also brings severe challenges to the operation and management of power system. Firstly, this paper combs the distributed robust optimization theory, specifically including the characteristics and applicable conditions of the distributed robust optimization model, classifies them according to the different methods of fuzzy set construction of uncertain parameters, and analyzes the advantages and disadvantages of each method of fuzzy set construction, laying a theoretical foundation for the subsequent construction of the distributed robust optimization scheduling model.

Keywords: Electricity consumption · Carbon emissions · Genetic algorithm · Renewable energy

1 Introduction

In recent years, the proportion of renewable energy in the world has been continuously increasing, and more and more countries and regions have begun. The application of renewable energy not only helps to reduce dependence on traditional fossil fuels, but also effectively reduces carbon emissions and slows down the pace of global climate change. However, the application of renewable energy faces a major problem, which is the instability of its power generation, which will have a significant impact on electricity demand and supply [1]. Therefore, how to reasonably plan the utilization of renewable energy.

This article is an empirical study based on genetic algorithms, aiming carbon emissions. Genetic algorithm is an optimization method based on natural evolutionary processes, which has advantages such as efficiency, robustness, and flexibility, and has broad application prospects in energy planning. By analyzing and modeling data on renewable energy generation, electricity demand, and carbon emissions, and using genetic algorithms to solve the optimal solution, this paper first calculated the carbon emissions of 30 provinces in the past five years according to the IPCC guidelines; Then, on the basis

J. C. Hung et al. (Eds.): FC 2023, LNEE 1132, pp. 295–299, 2024.
https://doi.org/10.1007/978-981-99-9538-7_45

of STIRPAT model, the driving factors of carbon emissions are decomposed, and a two-way fixed effect model is established with static panel data, and regression analysis is carried out using stata software; Finally, combined with regression analysis, the relationship between renewable energy electricity consumption and other driving factors and carbon emissions was elucidated, providing some reference for China's energy structure adjustment [2].

2 Related Work

2.1 Promote the Consumption of Renewable Energy Power

The following are some possible effective measures:

The government formulates policies to support the development of renewable energy and encourages businesses and the public to use renewable energy electricity more. The government can introduce preferential policies for purchasing renewable energy electricity and provide incentives such as tax breaks to enterprises and the public who use renewable energy electricity.

Strengthen the promotion of the brand image and advantages of renewable energy electricity. Introduce the environmental protection characteristics and economic value of renewable energy electricity to the public through media promotion, social networks, and other channels, increase public awareness and trust in renewable energy electricity, and stimulate consumer demand.

Develop new energy service models. The use of renewable energy electricity requires a professional energy management team and service platform. Governments and enterprises can consider conducting energy management services to provide customized energy solutions and high-quality after-sales services to users.

Innovate renewable energy power products and services. Enterprises can develop various renewable energy power products tailored to different user needs, such as distributed, to meet the diverse energy needs of users [3]. At the same time, for users, we need to provide elastic, flexible, and personalized power services to create a higher quality electricity consumption experience.

In summary, promoting the consumption of renewable energy electricity requires actively implementing a series of policy measures and technological innovation, and closely connecting technology with market and social needs, continuously guiding and stimulating users' demand for renewable energy electricity.

2.2 Uncertainty Characterization of Renewable Energy Power

At present, there are mainly stochastic optimization methods, robust optimization methods and distributed robust optimization methods to characterize the uncertainty. Stochastic optimization methods are relatively mature, but they rely too much parameters, and the generated scenes often have a large scale, which leads to inefficient solution [4]; The robust optimization method does not need to rely on the probability distribution information of the uncertain parameters, as long as the fluctuation range of the uncertain parameters, and has good economy and robustness.

Stochastic optimization is prediction or statistical analysis of uncertain factors to obtain the probability distribution of uncertain parameters, and then through scenario analysis to simulate the random fluctuations of uncertain parameters, generate a scenario tree along the time axis, and add each scenario constraint to the objective function, finally forming an optimal scheduling model that introduces uncertain factors.

3 Analysis of the Relationship Between the Two

3.1 Model Construction

Among them, CO2 (carbon dioxide emission) represents per capita carbon dioxide emissions; P (population) represents the total population of each region; GDP (gross domestic product) represents the per capita gross domestic product of each region (each region is converted with 1995 as the base period);

$$\ln CO_2it = b \ln P_{it} + c \ln GDP_{it} + d \ln EI_{it} \tag{1}$$

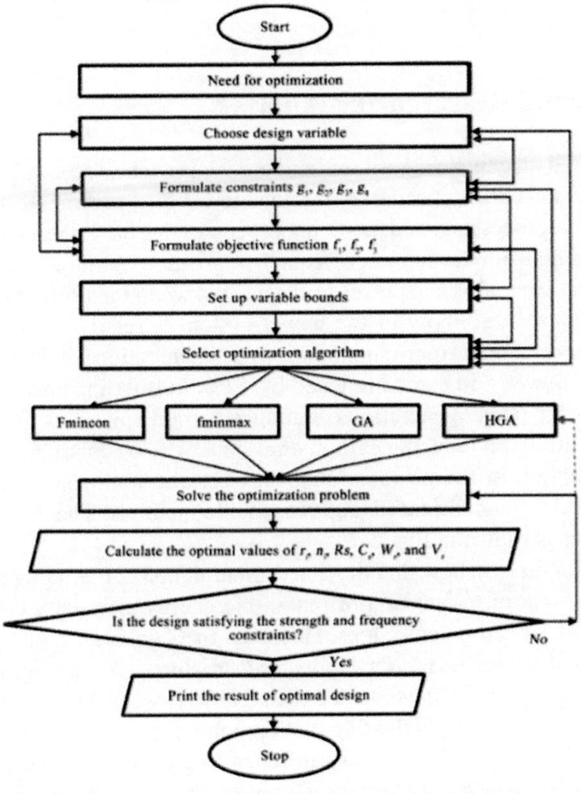

Fig. 1. Relationship between electricity consumption and carbon emissions of renewable energy based on genetic algorithm

El (energy intensity) represents energy intensity; REC (renwabl energy consumption) represents the variable. I and t represent different regional and time variables respectively; b. C, d and e are coefficients and represent the elasticity of CO emissions to population, GDP per capita, energy intensity and renewable energy consumption respectively. The based on genetic algorithm is shown in Fig. 1.

3.2 Data Source

In order to calculate and analyze CO more clearly and effectively. The panel data of the northwest region from 1995 to 2014 were collected. As mentioned above, the indicators to be used for building panel data model include: per capita carbon dioxide emissions (I), total regional population (P), regional per capita gross domestic product (GDP), energy intensity (EI) and renewable energy consumption (REC). The population data is from the statistical yearbooks of provinces and autonomous regions from 1995 to 2014; calculated by the ratio of the total GDP and the total population in the statistical yearbook of northwest provinces and regions from 1995 to 2014. Inflation and truly reflect the economic factors, the GDP of the five provinces and regions in northwest China from 1995 to 2014 was converted using the consumer price index with 1995 as the base period.

4 Experimental Research and Analysis

This study is based on genetic algorithms and uses data such as renewable energy generation, electricity demand, and carbon emissions to conduct carbon emissions. Through data collection, preprocessing, and modeling processes, we have obtained some valuable experimental results with the support of genetic algorithm optimization models.

We found that in small-scale power grid systems, when the proportion of renewable energy generation to the supply of the power system is relatively large, its effect on reducing carbon emissions is more significant. When generation reaches 40%, the carbon emissions in the power grid can be reduced by 12%. Further increasing the proportion of renewable energy power generation can reduce carbon emissions to a certain extent, but its marginal utility on reducing carbon emissions will gradually decrease.

This consumption in the power system has certain nonlinear characteristics and needs to be solved through scientific and reasonable methods. The iterative analysis of nonlinear features is shown in Fig. 2.

The regression results show that the determination 0–5721. Among the driving factors of carbon emissions, proportion of renewable energy electricity consumption can drive the reduction of carbon emissions. However, the significance of total population, per capita GDP and provincial regional industrial structure is not strong, while urbanization rate and energy intensity have passed the significance level test of 0.01, electricity consumption ratio is - 0–172, indicating that carbon emissions will decrease by 0–172 percentage points for every 1% increase in renewable energy electricity consumption ratio. On the whole, however, the driving factor that has the greatest impact on carbon emissions is regional energy intensity.

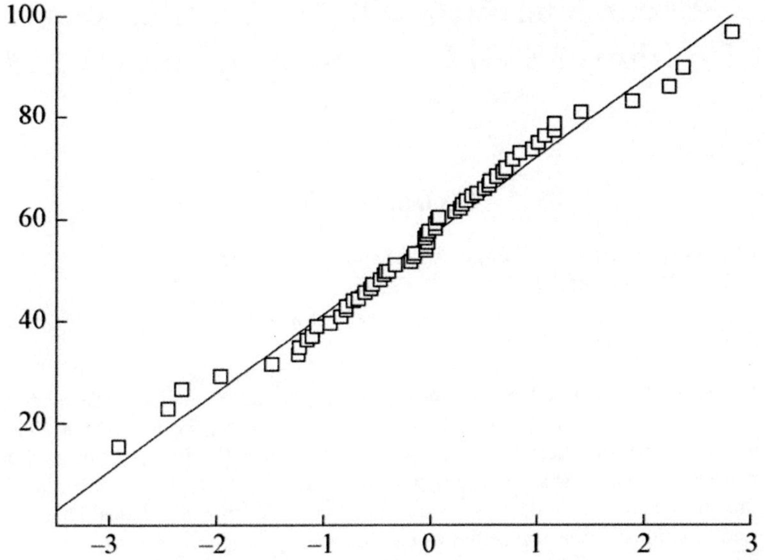

Fig. 2. Nonlinear Feature Iterative Analysis

5 Conclusion

Control carbon dioxide emissions from the consumer side, and more systematically achieve the energy conservation and emission reduction plan. Energy intensity, that is, energy consumption per unit output value, will also significantly affect the level of carbon emissions. Therefore, in the process of economic development, we should also pay attention to upgrading production processes and technologies to improve the efficiency of energy utilization, so as to effectively curb. This paper can clarify the mechanism of renewable energy electricity consumption optimization in environmental protection, provide ideas and basis for industrial transformation and upgrading, and play a reference role in decision-making for provinces to formulate targeted emission reduction plans.

References

1. Factors influencing renewable electricity consumption in China. Renewable and Sustainable Energy Rev. **55,** 687–696 (2016)
2. Ding, M.A., Shan, B., Zhu, F.: A Vision of Medium-and Long-Term Energy Demand Based on CO_2 Emission Peak Target. Electric Power (2017)
3. Zhao, X., Wei, X.: Research on the Relationship between Traditional Energy Prices and China's Carbon Emission Trading Prices: Based on the Panel Data from Seven China's Pilot Provinces and Cities of Carbon Emission Trading. Ecological Economy (2019)
4. Zhou, M., Sheng, T., Zhu, F., et al.: Dynamic relationship between carbon emission and economic growth in china's energy consumption. Safety in Coal Mines (2018)

Research on Surface Defect Classification Algorithm of Steel Plate Based on Improved BP Neural Network

Sun Maojie(✉)

School of Computer Science and Engineering, Southeast University, Nanjing 210035, Jiangsu, China
sunmj@njsteel.com.cn

Abstract. This article aims to study a steel plate surface defect classification algorithm based on an improved BP neural network. Firstly, we analyzed the classification of surface defects on steel plates and proposed an image segmentation algorithm based on shape features, which divides the steel plate surface image into multiple regions and extracts shape features from different regions. Then, we used traditional BP neural networks and improved BP neural network models to classify these shape features to determine the type of surface defects on the steel plate.

Specifically, we propose an improved BP neural network model to address the issues of low classification accuracy and slow training speed that traditional BP neural network models face in dealing with multi class problems. The model uses momentum term and learning rate annealing technology to accelerate the network training process, and uses Sigmaid function instead of the traditional step function to improve the fitting ability of BP neural network.

Through a large number of experiments, we compared and analyzed the performance of traditional BP neural network and improved BP neural network in the classification accuracy and training speed of steel plate surface defects. The results show that the steel plate surface defect classification algorithm based on improved BP neural network has significantly improved classification accuracy and training speed compared to traditional BP neural network. This algorithm has important application value for automatic recognition and classification of surface defects on steel plates.

In summary, this article studies a steel plate surface defect classification algorithm based on an improved BP neural network. In the future, we will further optimize the performance of the algorithm and improve its application scenarios to improve its practicality and universality.

Keywords: Surface defects · BP neural network · Steel plate · Classification algorithm

1 Introduction

Steel is of great significance to modern society and is widely used in bridges, ships, armor, automobiles and other fields. China is a populous country, and the construction of various infrastructure requires a large amount of steel. At the same time, China is

J. C. Hung et al. (Eds.): FC 2023, LNEE 1132, pp. 300–305, 2024.
https://doi.org/10.1007/978-981-99-9538-7_46

also a large steel producer. In 2019, the national steel output will be 990 million tons. During the production process of steel plate, due to the influence of raw materials, rolling equipment, rolling process and system control, various defects such as scratches, cracks, pits and oil stains [1]. In addition, in the production process, it will also wear and roll, resulting in production accidents such as belt breakage, accumulation and parking, which will bring incalculable losses to the production of enterprises. Therefore, online detection of great significance, and it is also the research direction of steel plate manufacturers [2].

Since the 1970s, the research on automatic steel plate surface defect detection system has been started abroad. Until the end of the 20th century, there has been a very mature surface defect detection technology. China's development is relatively late. At present, many steel plants use inefficient manual sampling methods, and some large enterprises introduce foreign advanced technology to improve production efficiency and product quality [3]. But the price of these products and equipment is too expensive, and only a few large steel plants can afford it. In order to get rid of foreign technology blockade, reduce production costs and improve production efficiency, many domestic enterprises and universities have cooperated to jointly develop their own defect detection systems, and have also made some phased achievements.

2 Related Work

2.1 Development of Steel Plate Surface Defect Detection

(1) Manual detection

In the 1950s and 1960s, manual inspection was used internationally for online inspection or open coil sampling inspection. The online detection method is to observe the steel plate surface with naked eyes during the production process of steel plate. This method has high working strength and is easy to be affected by subjective factors, such as missed detection and false detection. And the environment of the factory is very bad. High temperature, dust, noise, etc. will cause physical and mental damage to workers. In some cases, the factory will also use stroboscopic light source to detect, causing great harm to the workers' eyes[4]. Therefore, the method of online inspection gradually becomes the open coil spot check, where a certain proportion of the finished steel coils are spot checked. The main disadvantages of manual detection methods are that they cannot be quantified, are easily affected by subjective factors, cannot guarantee the detection quality, and cause great harm to workers.

(2) Machine vision inspection

The inspection method based on machine vision is to use cameras to collect steel plate images. Generally, the acquisition system is also equipped with a lighting system to improve the quality of collected images. Image processing technology is used to detect steel plate defects. There are many successful systems and cases of machine vision in surface defect detection. Which can detect surface defects of steel plates in a high-speed and stable manner, making machine vision the mainstream research direction of surface defect detection system. Westinghouse Company proposed to use linear CCD for image acquisition and high-intensity linear light source for auxiliary lighting to detect surface defects of steel plate. The EES system can collect high-quality steel plate surface defect images in real time, Have operators manually classify defects [0. There are also many colleges and universities in China cooperating with the company to jointly develop an online detection system for steel plate defects, and some achievements have been made.

2.2 Traditional BP Neural Network

As shown in Fig. 1. The number of elements in the input layer is the dimension The output layer is a (s3 × 1), w (i = 1,2,3) is the weight matrix of the main layer, and b (i = 1,2,3) is the offset vector of layer i.

$$E(w, b) = \sum_{j=0}^{n-1} (d_j - y_j)^2 \tag{1}$$

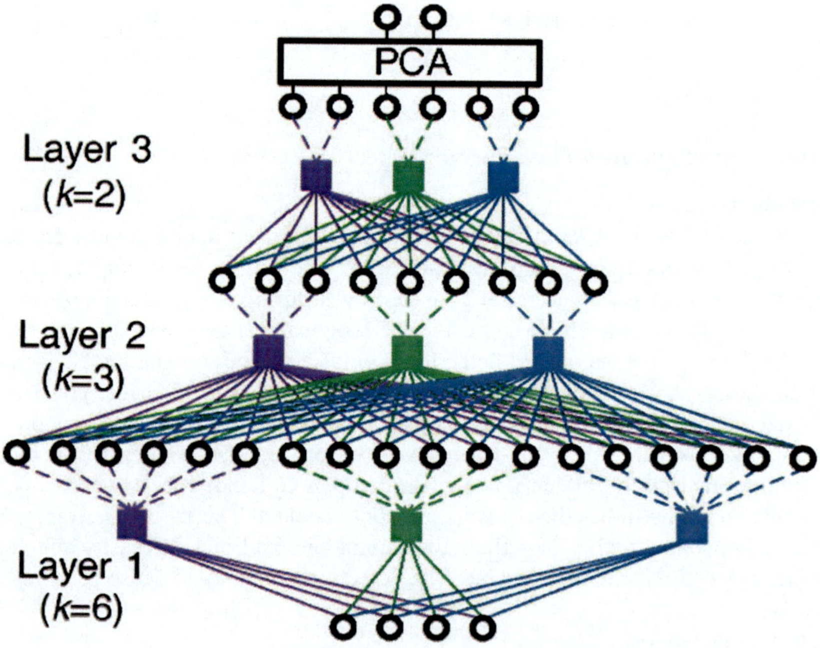

Fig. 1. BP three-layer network diagram

It can be seen from Formula (1) that BP network is mainly determined by the weight matrix and offset vector of each layer, and its coefficients are generally determined by performance learning rules, that is, the mean square error is taken as the performance objective function. BP learning algorithm has a simple structure, classification, and has also obtained good results. However, due to the shortcomings of slow training speed, convergence to local extremum and many iterations, it needs to get the optimal results through repeated experiments in the actual operation process. An effective practical technology to improve BP algorithm is discussed below.

3 Defect Screening Based on Image Gradient

A real-time online detection system needs to process huge amounts of data. If each collected image is finely processed, the hardware system needs to provide strong computing power, and the software needs to provide efficient processing. This has great requirements for the performance of the system, and its cost is also extremely expensive. These are totally unnecessary. With the development and maturity of technology, the yield of steel plate has been very high, and most of the steel plate images collected by the system are free of defects. Therefore, we should find an efficient screening method to split the steel plate images, and fine process the defect images, and directly store and display the images without defects. For the screening algorithm, it should be simple and efficient; Secondly, it should have high accuracy; Finally, the false detection rate and missed detection rate should be kept at a low level.

The common screening methods are subtraction, projection and image classification. The difference image method takes the normal steel plate image as the benchmark, and uses the collected image to make difference with it. The obtained difference image is used to screen defects. Even if it is a normal steel plate, there will be a great difference between them, so the effect of differential imaging is very poor. The projection method projects the gray image of steel plate to the horizontal axis and vertical axis respectively, and uses the variance after projection to screen defects. The projection method is better than the difference method, but it is still vulnerable to the interference of light and noise, which is difficult to meet the requirements of industrial production.

This paper analyzes a large number of defect images and normal images. It is found that the image information of normal steel plate is low frequency, while the defect image has a large amount of high frequency information. Therefore, they can be distinguished by frequency domain method. However, it is time-consuming to process the image in frequency domain, so image gradient is used instead of processing in this paper.

4 Defect Image Filtering Method

In this paper, the gradient direction projection is used for defect image filtering, and the specific operation flow chart is shown in Fig. 2.

(1) Pretreatment. The environment of the steel plant is relatively harsh, and lighting, dust, vibration, etc. can easily, and affect the subsequent processing algorithm. Therefore, the image should be preprocessed before image filtering to eliminate the impact of the environment and ensure the image quality. Filtering and grayscale transformation are common image preprocessing methods. Filtering mainly includes spatial and frequency domain processing methods to eliminate image noise. Frequency domain filtering includes Fourier transform, wavelet transform, etc. The gray level transformation of the image includes histogram equalization, gray level stretching, etc., in

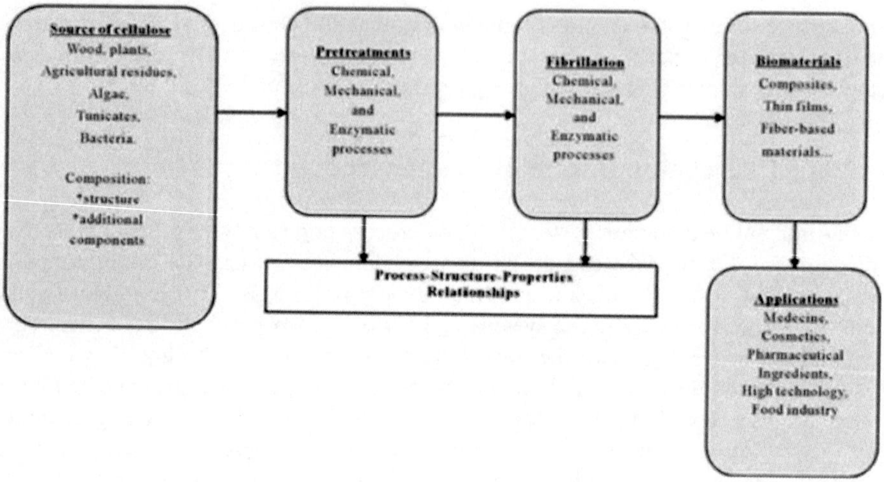

Fig. 2. Flow Chart of Defect Image Filtering

order to improve the image contrast. This section only filters the defect images, so we choose the simplest method for preprocessing. Gaussian filtering can effectively retain the information of the current pixel when removing noise, which is more suitable for the preprocessing of the current algorithm.

(2) The image gradient is calculated using Robert cross operator.
(3) Calculate gradient direction projection according to the formula.
(4) Image filtering. Through experimental simulation, observe the statistical characteristics of a large number of image mean values and standard deviations, and design appropriate thresholds for image filtering.

5 Conclusion

BP neural network adopts the neural parallel processing mode, which can quickly realize the nonlinear fitting function. However, the traditional BP neural network learning algorithm has some shortcomings, learning algorithm, the author proposes a new network learning method using the high order perturbation theory, which not only speeds up the network training speed, but also avoids the risk of falling into minima in the network learning process, When the method is applied to the classification experiment.

References

1. Shu, Y., Fan, S., Xiong, C.: Imbalances sample classification algorithm for industrial product defect recognition based on deep learning. IPPTA: Q. J. Indian Pulp and Paper Tech. Assoc. **30**(7), 255–258 (2018)
2. Wei, Y., Yang, S.: Research on fabric defect detection based on BP and SAE deep neural network. Technol. IoT & AI (2018)

3. Ge, Y.: Spark parallel optimization algorithm based on improved BP neural network. J. Phys.: Conf. Ser. **1550**(3), 032044 (4pp) (2020)
4. Lan, J.: Particle swarm optimization surface defect recognition algorithm based on radial basis neural network. Electron. Sci. Technol. (2019)

Analysis of College Students' Employment Competitiveness Based on Binary Association Rule Extraction Algorithm

Qianying Sun[1(✉)] and Yanqing Wang[2]

[1] School of Marixm, Northwestern Polytechnical University, Xian 710072, Shaanxi, China
vickysqy@mail.nwpu.edu.cn

[2] College of Arts and Sciences, National University of Defense Technology, Changsha 410015, Hunan, China

Abstract. Binary association rule extraction algorithm is a machine learning method, which is used to extract rules from text data. Text data can be in any form, such as documents, news articles, or social media posts. The rules extracted by this algorithm are called associations. These associations are then used to predict future events. This paper introduces how to use binary association rule extraction algorithm to predict students' employment competitiveness according to their performance in SAT/ACT/NAPLEX and other college entrance examinations. Use the above algorithm to extract the employment competitiveness of college students. The binary association rule extraction algorithm is applied through the following steps: Step 1: input the data into the computer and import it into the database; Step 2: Search for rules in the database and extract them; Step 3: Combine all extracted rules into a rule set; Step 4: Calculate the score of each rule set according to its accuracy, and then compare it with the scores of other rule sets.

Keywords: Association rules · Feature extraction · College students' employment competition

1 Introduction

Especially in recent years, with the gradual spread of the financial crisis in the world, the problem of difficult employment has become more and more prominent in China. In the current and future period [1].

Under the overall severe employment situation of the whole society, the employment situation of college graduates is not optimistic. College graduates are an important part of the social employment group. The employment situation of college students directly affects the overall employment situation of our society. As mentioned above, in recent years.

J. C. Hung et al. (Eds.): FC 2023, LNEE 1132, pp. 306–311, 2024.
https://doi.org/10.1007/978-981-99-9538-7_47

2 Related Work

2.1 Analysis on the Core Competition of College Students' Employment

Under the pressure of the current severe employment supply and demand contradiction and structural contradiction, improving the employment rate and employment quality of college students has become a hot and difficult issue of general concern in the whole society. For colleges and universities, improving the employment rate of graduates is not only a matter of digital improvement, but also a matter, the employment quality of students, the demand and evaluation of employers for college graduates, and so on. Many colleges and universities put improving the employment quality of graduates in the first place of employment work, and pay more and more attention to the cultivation of college students' employment competitiveness [2]. The flowchart for analyzing the core competitiveness of employment is shown in Fig. 1.

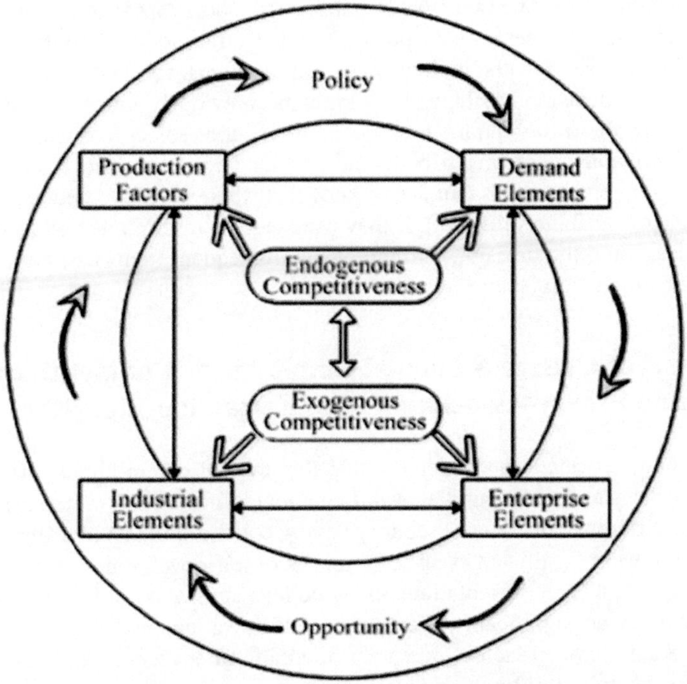

Fig. 1. Flow chart for analyzing the core competitiveness of employment

However, a large part of these colleges and universities do not realize that when the society has formed a general atmosphere of valuing college students' employment competitiveness, if college students do not seek breakthroughs to form their own core competitiveness, then this employment competitiveness will become less competitive in their future job search. For example, if an enterprise wants to survive in the fierce market competition and obtain long-term benefits, it must have its own core technology or core competitiveness and establish its own brand advantage.

Only in this way can it not be replaced by emerging enterprises and eliminated by the market. Therefore, college students must cultivate and enhance their own employment core competitiveness. Then, first of all, we must clarify the meaning, characteristics and main components of college students' employment core competitiveness.

2.2 Binary Association Rule Extraction Algorithm

Association rules were originally put forward to discover the purchase habits and purchase rules of supermarket customers. Suppose supermarket operators want to know more about the purchase habits of customers, such as what items customers usually buy at the same time, they can analyze customers' shopping baskets to explore the relationships between different goods customers put into shopping baskets. Association rules are used to analyzing the frequent occurrence of existing training sets The relationship between the itemsets of. The two most important concepts in association rules are confidence and support. Confidence is the probability that a transaction D contains an itemset X and another itemset Y when it already contains an itemset X, and support.

The value of decision attributes through some previously known attributes. First, use Aprioril 2% algorithm to mine frequent item sets, then select frequent item sets that contain both previously known attributes and classified decision attributes, and calculate whether the frequent item sets can derive known attribute values to deduce association rules of decision attribute values [3]. If they can meet the minimum confidence requirement of rules, Then the decision attribute value in frequent items can be taken as the final classification result.

3 Analysis of College Students' Employment Competitiveness Based on Binary Association Rule Extraction Algorithm

The career choice orientation indicating that the level of career choice orientation of college students is above the middle level. In demographic variables, there was a significant gender difference in material security among college students of different genders (P0.01); There were significant ethnic differences in self-development factors, material security factors and career orientation among college students of different nationalities (P0.05); There is no significant difference in self-development factor, material security factor, social demand factor and career orientation between the only child college students and the only child in self-development factor, material security factor, social demand factor and career orientation among college students of different grades (P0.01); The employment analysis of college students gives the Baseline scheme (Fig. 2), and our subsequent scheme is also based on the Baseline, and is realized through corresponding expansion and optimization.

There is no significant difference in self-development factor, material security factor, social demand factor and career orientation among college students of different majors; There is no significant difference in self-development factor, material security factor, social demand factor, and career orientation between students who are student cadres or

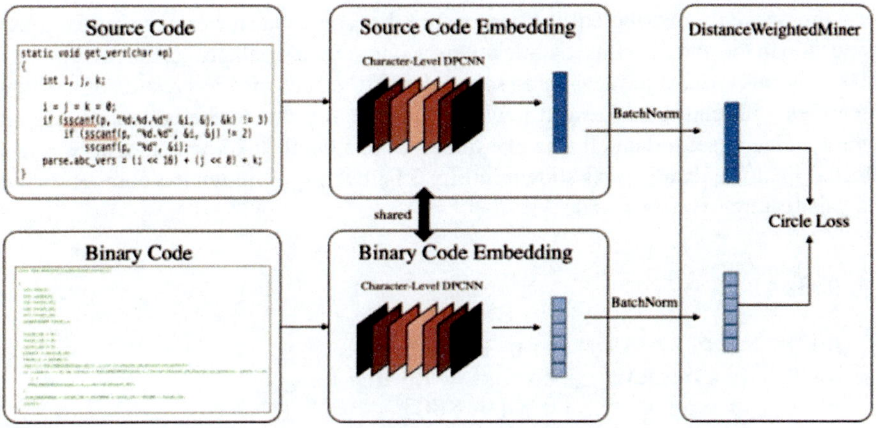

Fig. 2. Baseline Structure Chart

not; There were significant differences in self-development factor, social demand factor, and career choice orientation among college students with or without professional practice (P0.05); There is no significant difference in self-development factor, material security factor, social demand factor, and career orientation among college students who have failed in the examination; There is no significant difference in self-development factor, material security factor, social demand factor and career orientation among college students with different family monthly income; There were significant differences in family monthly income among college students of different family types in self-development factor, material security factor, social demand factor and career choice orientation (P0.05).

4 Simulation Analysis

The employment pressure and its factors of college students have extremely significant positive correlation with career choice orientation and its factors ($r = 0.131 \sim 0.229$, P0.001); The explanation rate of college students' employment pressure on career orientation is 5.2%, which indicates that college students' employment pressure has a positive predictive effect on career orientation; The occupational quality evaluation factor, employment competition environment factor, employment psychological expectation factor, lack of job search help factor, and major supply and demand contradiction factor of college students' employment pressure have significant positive correlation and positive predictive effect on self-development factor, material security factor, social demand factor, and career choice orientation, with the interpretation rates of 13.7%, 9.2%, 4.5%, and 10.7% respectively.

K-nearest neighbor is to search by giving a dataset and a manually set k value (Euclid distance is generally used as the measurement of distance between samples). The process of K-nearest neighbor to solve the classification problem is as follows: In the K-nearest neighbor, each data sample has the same number of neighbors, and different k-values in the K-nearest neighbor often result in different classification results. In real life, datasets are often unevenly distributed. It often occurs that some categories account for a large proportion in the overall dataset while others account for a small proportion in the overall dataset. In this case, if the k value is set too large, the classification results of unknown categories will tend to be skewed towards this category due to the large category, and if the k value is set too small, the classification results of unknown categories will be affected by noise data [4]. As shown in Fig. 3 below, the employment competitiveness analysis feature extraction code.

```
Rules:

Rule 1: (282/33, lift 1.3)
        checking_status = none
        ->  class 0  [0.880]

Rule 2: (641/177, lift 1.0)
        credit_history in {critical, delay, good}
        ->  class 0  [0.723]

Rule 3: (50/14, lift 2.3)
        checking_status in {<0DM, >=200DM, 0-200DM}
        credit_history in {prefect, verygood}
        ->  class 1  [0.712]
```

Fig. 3. Employment competitiveness analysis feature extraction code

K-nearest neighbor does not take into account the real data distribution.

5 Conclusion

It works by identifying the best match between two variables (i.e., the variables extracted from the dataset and the target variables). This best match is then compared with all other matches until one of them gives a better result than the others. The first step in this process is to determine which variables have been extracted from the dataset. Once these variables have been identified, we can use any method to extract new associations between these variables and our target variable (i.e. employment competitiveness).

Acknowledgements. 2022 Shaanxi Provincial Philosophy and Social Science Major Theoretical and Practical Research General Project ; Analysis and countermeasures of employment dilemma of college students(2022ND0227).

References

1. Yu, Q.H., Chen, X.L., Zhou, H.G.: The thinking of improving the family economic difficulties students' employment competitiveness based on the school-enterprise cooperation. J. Nanchang College of Educ. (2017)
2. Yang, Y.L., Sun, L.P.: On approaches to enhance college students' professional ethics education based on employment competitiveness. Heilongjiang Researches on Higher Education (2017)
3. Pan, K.: The employment competitiveness analysis of poor students in western guangdong university based on SWOT analysis. The Theory and Practice of Innovation and Entrepreneurship (2018)
4. Chao, Q.I.: Research on the construction of employment culture based on the employment competitiveness of college students. J. Jinzhou Med. Univ.(Social Science Edition) (2017)

Research on Dynamic Scheduling Method of Aerospace TT&C System Based on Chaos Genetic Algorithm

Xinghua Sun(✉)

Beijing Institute of Spacecraft Environment Engineering, Beijing 100029, China
sandflying@126.com

Abstract. The research direction is the dynamic scheduling method of aerospace surveying and mapping system based on chaotic genetic algorithm. This is a new technology to solve complex problems in aerospace systems and has been proved to be very effective in solving these problems. The dynamic scheduling algorithm, which can be used as a tool to improve aircraft performance by reducing turnaround time. This will help airlines reduce operating costs and increase profits. This paper aims to design a dynamic scheduling method for aerospace T/C system based on chaotic genetic algorithm. It also includes simulation results and performance analysis of the proposed method in different parameters (such as number of jobs, number of machines, number of processors, etc.).

Keywords: Chaotic genetic algorithm · Space TT&C system · dynamic scheduling

1 Introduction

In order to solve the problem of single result style of traditional coverage route planning algorithm and poor flexibility in adversarial environment, the surveillance coverage route with multiple styles and good adversarial performance in surveillance task execution is generated. On the basis of the artificial potential field method, the seed of the excitation potential field is coded as a gene in the form of binary string, and the diversity of seed patterns is increased through the operations of crossover, mutation, merging and other operators, so as to plan a surveillance coverage route with fewer turns, short surveillance interval, and good antagonism [1]. The task of covering route planning is to determine a path that passes through all points in the area of interest while avoiding obstacles. It is found that due to the constraints of the mission area and performance of aerospace, there are few researches on the surveillance coverage route planning, most of which are about the coverage route planning of ground robots.

They use the grid to represent the task area. The algorithm first assigns 0 to the target grid and 1 to all cells around it. Then, all unmarked grids adjacent to tag 1 are labeled 2. This process repeats until the marking face reaches the start cell. After calculating the distance conversion, the coverage path is found by not accessed from the start cell. If two or more neighbor cells have the same value, randomly select one of them.

J. C. Hung et al. (Eds.): FC 2023, LNEE 1132, pp. 312–316, 2024.
https://doi.org/10.1007/978-981-99-9538-7_48

2 Related Work

2.1 Measurement and Control Tasks

In TT&C tasks, the priorities of different satellites are different, even if the priorities of the same satellite are different in different TT&C time periods. The determination of task priorities complex work, which is related to many factors. The specific quantitative work can be determined only after repeated discussions with experts in various fields. 2. For each satellite task priority in multi satellite TT&C, we can generally use the weight coefficient of each task to reflect the relative importance of each task [2]. The more important the target, the greater the corresponding weight coefficient. From the perspective of the weight coefficient itself, it has duality. On the one hand, it should objectively reflect the relative importance of each evaluation index attribute; on the other hand, it should reflect the relative subjective importance of decision-makers to various evaluation indexes.

It is a unit that directly tracks, measures, controls, communicates and transmits data to the satellite. Its mission is to track and measure the orbital motion parameters of the spacecraft, receive and demodulate the telemetry information of the spacecraft, send remote control commands (including injected data) to the spacecraft, and communicate with the spacecraft and exchange data information under the organization [3]. The measurement and control station includes land mobile station, offshore survey ship and fixed ground station, the former two are collectively referred to as mobile ground station. The mobile ground station is only suitable for supporting short-term flight missions. Long term flight missions use fixed ground stations, such as the long-term management of satellites in orbit.

2.2 Chaotic Genetic Algorithm

The current chaos optimization algorithm mostly uses Logistic as the generation method of chaos sequence, and introduces chaos mapping into genetic algorithm. Logistic mapping equation is:

$$\mu_{r,s}(x, z) = \{\max\{\min[\mu_R(x, y), \mu_s(y, z)]\}\} \tag{1}$$

where u is the control parameter, when pu = 4, the Logistic mapping is unstable in the full interval of U (0,1), and the sequence obtained by iteration has pseudorandomness. When $3.569945972 < u \leq 4$, especially close to 4, the value generated by iteration is a pseudorandom distribution state,is shown in Fig. 1.

In the genetic algorithm, for the problem with unknown solution distribution, the initial population needs to be evenly distributed in the multi-dimensional space as much as possible, and the homogeneity of the generated population is required to be high. The sequences generated by Logistic mapping are more distributed at the boundary points and more dispersed inside. This distribution can reflect the solution domain of some problems, but for most high-dimensional optimization problems, especially when the global optimal position is unknown, uniform distribution is more universal.

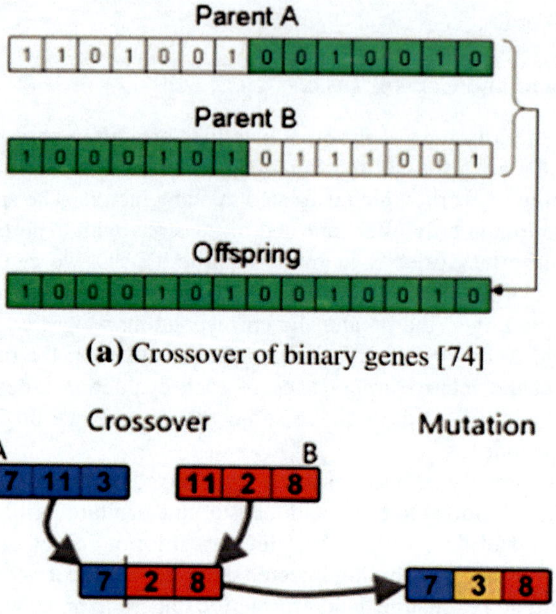

(a) Crossover of binary genes [74]

(b) Example of Mutation [74]

Fig. 1. Crossover and mutation process of genetic algorithm

3 Research on Dynamic Scheduling Method of Space TT&C System Based on Chaotic Genetic Algorithm

As shown in Fig. 2, the scheduling model of aerospace TT&C system is shown.

The does not consider the quality of task completion. For a certain task, when it is in the overlapping time window of multiple TT&C stations and satellites, the current available TT&C stations are not unique. The current status of each TT&C station is different. Due to factors such as network bandwidth and delay, the TT&C stations have different effects on completing a certain task. Here, we mainly consider the network bandwidth and delay of the TT&C stations, and introduce benefit values into the model, To analyze the efficiency of the measurement and control station to complete the task, so as to maximize the benefit value of the final completed total task.

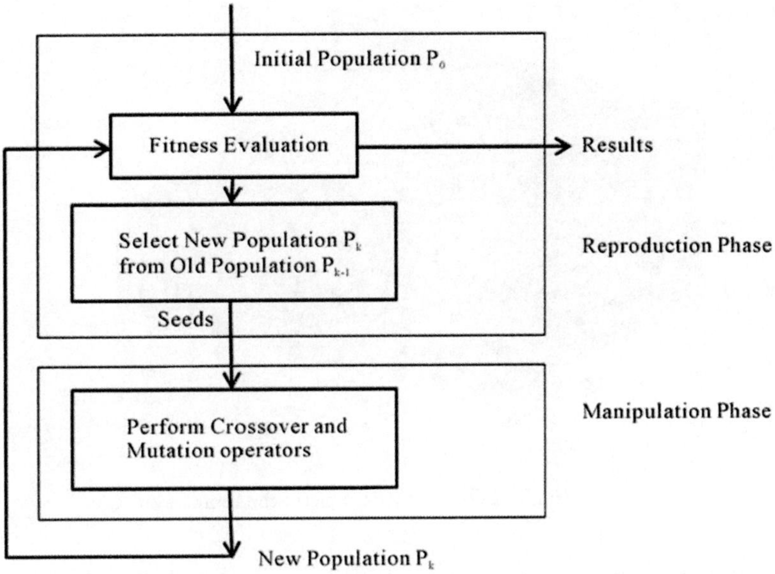

Fig. 2. Scheduling model of space tt&c system

4 Simulation Analysis

We have established the task area grid of the route. Each grid is regarded as a point in the graph G = (V, E) G = (V, E). All points form the point set V of the graph G. For a grid, we define its neighbors as eight adjacent grids, which are mapped to the graph G. That is, only when there is a neighbor relationship between points, there are edges connected.

For the initial population, we choose two typical artificial potential field seed codes of point type and line type to generate the initial population of genetic algorithm. Set that the artificial potential field of all points will increase by 0.001 every time the simulation advances, that is, p = 0.001p = 0.001. The population updating mechanism is to update according to the simulation time, with a cycle of T, that is, every time the simulation advances T time, update the population of the genetic algorithm using operators, and randomly select a seed from the population to update the current potential field, and continue to generate routes.

In the process of route generation, the performance parameters of the route are counted every 30 * 30 simulation steps, that is, the total value of the turning angle, the total value of the artificial potential field, grid potential field value and other parameters. The unpredictability in the objective function is considered to be inversely proportional to the population update cycle T, that is, the greater the T is, the worse the unpredictability is [4]. On the contrary, the better the unpredictability is. In extreme cases, When T is infinite, the unpredictability is the worst. The initial potential field of the route is shown in Fig. 3

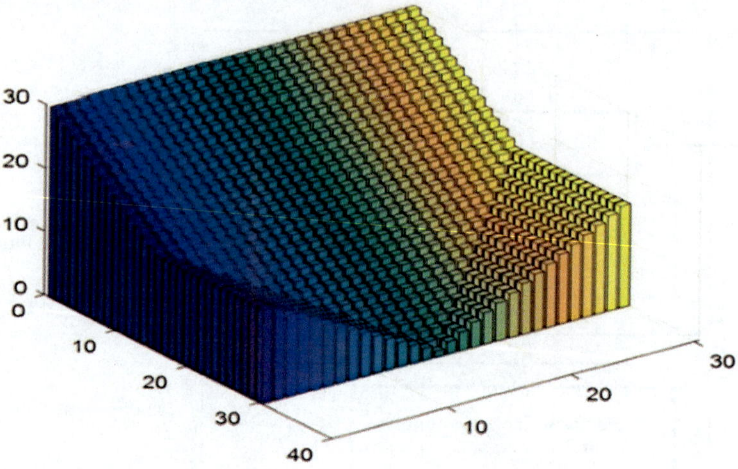

Fig. 3. Initial potential field of the route

In order to observe the statistics of the route planning algorithm more coherently and consistently, certain processing should be carried out when the potential field is updated to reach the potential field update cycle, so that the total potential of the task grid remains unchanged.

5 Conclusion

The dynamic scheduling method based on chaos genetic algorithm is a new method to solve the problem of optimal task time (TTT) in aerospace systems. The basic idea of this method is to use chaos dynamics as the model proposed, which can achieve both high-quality convergence and rapid convergence. Chaotic dynamics is used as the model of multi-objective optimization problem; Secondly, genetic algorithm is applied to solve them.

References

1. Jiang, T., Cai, Y., Yan, L., et al.: Research on method of aerospace test resource matching optimization based on improved genetic algorithm. Comput. Measur. Control (2012)
2. Yu, H.: A collaborative scheduling method of enterprise funds based on genetic algorithm (2016)
3. Zhou, W., Sun, Z., Ren, Y., Yao, F.: The application of kinematic relative positioning technology with BDS in aerospace TT&C equipment precision verification. In: Sun, J., Liu, J., Fan, S., Wang, F. (eds.) China Satellite Navigation Conference (CSNC) 2016 Proceedings: Volume I. LNEE, vol. 388, pp. 197–209. Springer, Singapore (2016). https://doi.org/10.1007/978-981-10-0934-1_19
4. Liu, Y., Ling, P.: Intelligent RGV dynamic scheduling strategy based on greedy algorithm. World Sci. Res. J. 5(9), 278–287 (2019)

Cloud Data Integrity Verification Algorithm for Accounting Informatization Under Sharing Mode

Jie Wan(✉)

Shanghai Publishing and Printing College, Shanghai 200093, China
wanjie1112@163.com

Abstract. Since the 1980s, economic globalization has gradually become the trend of world economic development. Economic globalization is gradually affecting our lives. More and more enterprises have started transnational operations to expand their business scope in order to seek broader markets and resources. With the continuous expansion of enterprises. The integrity verification algorithm of accounting information cloud data in the sharing mode is used to verify the correctness of data in the distributed system. It uses a set of rules to check the validity and consistency of the information stored in each node. This algorithm is useful when there are multiple nodes with different version information, such as when a node has been updated by another node or even after an update from another source.

Keywords: Integrity verification · Sharing mode · Accounting informatization · Cloud data

1 Introduction

Today, informatization reduces the cumbersome defects of traditional accounting methods, which not only facilitates the use of financial personnel, but also makes the accounting of financial data more accurate [1].

In the past few decades, China's economic development has risen steadily. Under this general environment, a large number of large enterprises have emerged. First, the expansion of enterprises will greatly increase costs. In the continuous expansion of the enterprise, many new branches will be established to expand the entire enterprise, and the establishment of branches will increase the cost rapidly. Second, the increasing number of branches makes the overall management of the enterprise more difficult. If each branch company has its own management system and accounting system, it will be difficult for the head office to achieve unified supervision and management, and it is easy to have problems such as lax personnel and insufficient motivation in the branches of the head office [2]. Moreover, if the enterprise spends a lot of energy on management, it is bound to have insufficient energy invested in daily production and operation activities. Third, the continuous expansion of enterprises will cause a series of

J. C. Hung et al. (Eds.): FC 2023, LNEE 1132, pp. 317–322, 2024.
https://doi.org/10.1007/978-981-99-9538-7_49

financial risks. There are countless connections between the branches of an enterprise. If a branch has financial risks during the expansion, other branches associated with it will also be involved, which is undoubtedly a fatal blow to an expanding enterprise. Fourth, the past management model cannot bring relatively sufficient financial information to enterprises, and the quality of financial information cannot be guaranteed. In the process of continuous expansion, enterprises need to constantly create new organizational forms and operating methods, so as to adapt to the changes of enterprises more quickly. As an important department in the enterprise, the financial department must change as quickly as possible to adapt to the development of the company. However, at different stages of the enterprise reform, how the financial department should change is a difficult problem [3].

Therefore, in this context, large enterprises and many small and vigorously develop financial shared service centers, which can automatically realize simple accounting work of enterprises, reduce traditional financial accounting, transform traditional accounting to management accounting, help enterprises achieve the strategic objectives formulated, reduce enterprise operating costs, and enable better allocation of enterprise resources.

2 Related Work

2.1 Domestic Research Status

"First, the development of shared services is to meet the needs of employees within the enterprise. Second, shared service centers exist independently in the enterprise, so the internal departments of the enterprise can not choose their own shared service centers, and they have the right to make external choices. Each internal unit of the enterprise uploads information to the shared service center for centralized processing, so that the internal departments of the enterprise do not need to set an information level independently They can uniformly obtain the information they need from the sharing center [4]. The shared service center can provide enterprises with financial management, personnel transfer, electronic technology support, sales and procurement services."

Liu Hanjin studied the role of shared services in his paper "Research on Decision Making, Implementation and Evaluation of Shared Services". He believed that shared services were an innovation in the internal organization of an enterprise, endowed each department of the enterprise with a market attribute, and did not control some subsidiaries.

In Financial Shared Services, Chen Hu and Dong Hao defined that the shared service center is customer-oriented, based on market prices and signed contracts, and provides corresponding services for internal and external customers of the enterprise. It places projects managed by each department of the enterprise under a unified system for centralized processing, so that enterprises can devote more energy to business development and design, It is beneficial for enterprises to create more value. Shared service center has many competitive advantages, which can reduce ensure the security and accuracy of customer information.

Cui Huimin proposed that using big data analysis technology can provide support. In the same year, Liu Zhihui proposed that management accounting should be combined with Internet +, and Sun Zhengni also proposed to attach importance to accounting reform under the background of Internet +.

2.2 Features of Financial Sharing Service Model

At present, more large companies at home and abroad. In the process of application, we can see that the characteristics are prominent, mainly in the following aspects:

(1) Centralization of financial sharing

After an enterprise has a financial sharing service center, financial personnel upload financial information and related financial, realizing centralized management of financial information. Each department of the enterprise shall make reasonable allocation, so that the financial information and various valuable resources in the financial sharing center can be used in each department. This sharing method can minimize the communication between departments, speed up the work efficiency.

The centralized management of financial sharing services enables more financial personnel to quickly obtain financial information, facilitates the formulation of financial budget, operation management and corporate strategy, and improves the accuracy of accounting decisions.

$$M_1 = \left(\Gamma_{l1} + \Gamma_{p2} - \Gamma_{p1}\right) \tag{1}$$

All accounting business processing processes one by one, and then divides the accounting business processing processes into modules. Within the scope of risk control, the most convenient and efficient way is considered, which can avoid a large amount of repetitive work. This choice of accounting methods is independent. In the process of integrating financial processing processes, the Financial Sharing Center has a set of unified standardized processing processes, which can synchronously formulate the operation details of each process, so as to ensure that all business processes are the same in the operation of different financial personnel, and ensure the stability and accuracy of the accounting quality. Model has accelerated the standardization process of financial management and financial information disclosure, including the standardization process of management such as the unified provisions of the financial system, the unified formulation of accounting subjects, the unified accounting caliber, and the unified statement consolidation caliber.

It has a more systematic processing method for the handover and review of bills, the collation of documents, and the accounting work, which can make enterprise informatization develop more rapidly.

3 Cloud Data Integrity Verification Algorithm of Accounting Informatization Under Sharing Mode

The network topology of cloud storage system is shown in Fig. 1 below.

The one user at a time. Therefore, this paper introduces the improved method proposed by Li Gangrui, but also reduce the transmission bandwidth of data required for verification between users, third-party verification organizations and cloud storage in the verification calculation. The specific cloud computing is shown in Fig. 2.

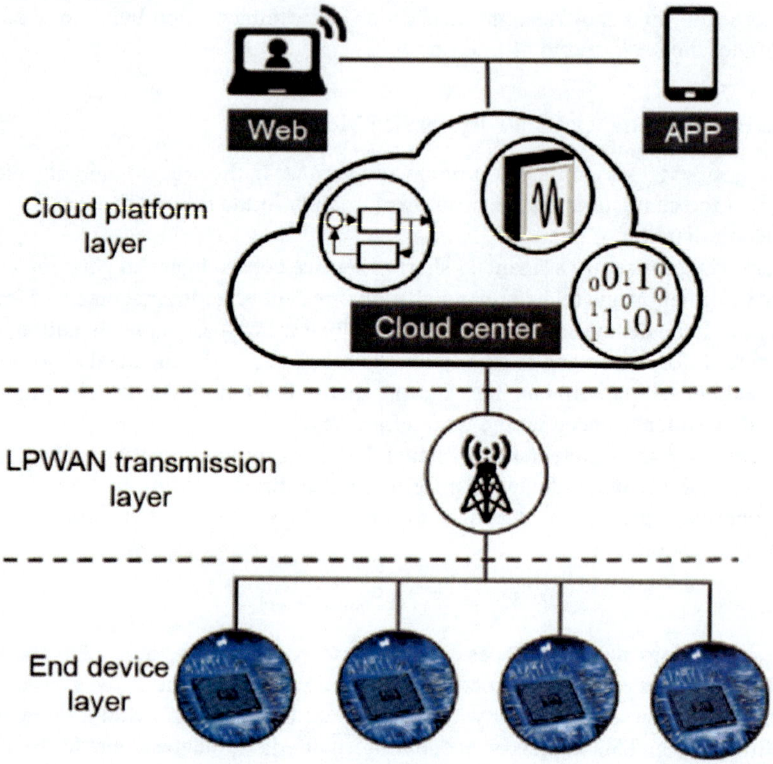

Fig. 1. Cloud storage system network topology

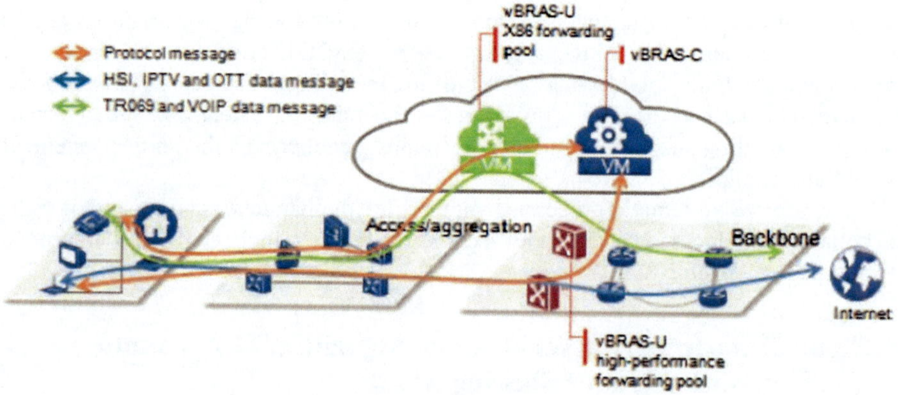

Fig. 2. Cloud computing storage security architecture

4 Experimental Results and Analysis

The experimental platform is Windows 7 operating system, the CPU is Intel Core i58400 processor, the main frequency of the CPU is 2.8 GHz, 4 GB RAM, Seagate 500 GB hard disk 7200 r/s. The development environment is Eclipse, the development language is Java, and the database is SQLServer2010.

But the average time required for each file update in the multi-user parallel verification algorithm will gradually decrease. Therefore, in the sharing mode based on cloud computing, the parallel verification algorithm has higher computing efficiency. The data performance is shown in Fig. 3.

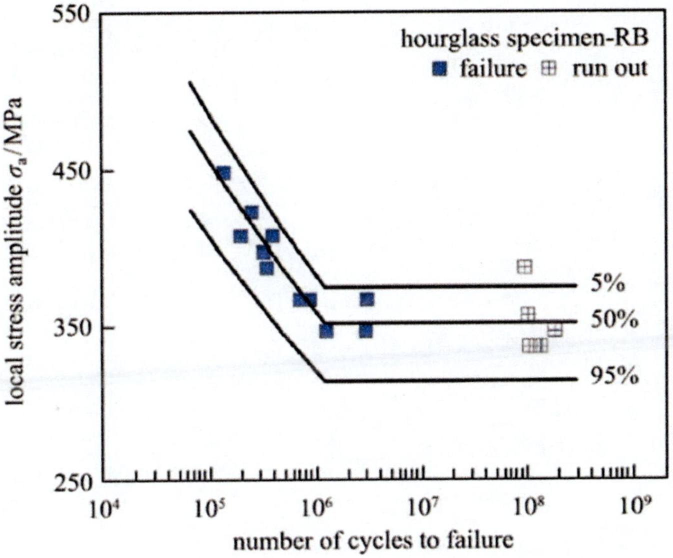

Fig. 3. Data performance

5 Conclusion

In the process of growing up of China's economy, many Chinese enterprises have also expanded their business scope to overseas markets, which has greatly changed the development direction of China's large enterprises. Therefore, the core department of enterprises, the financial department, should be the first to change with the change of enterprise development model. The emergence of financial sharing service mode has improved the management inconvenience caused by cross country and cross region operation of enterprises. This financial sharing service model in promoting the financial development of enterprises from the perspective of group management accounting informatization.

References

1. Bouvier, J.B., Nandanoori, S.P., Ornik, M., et al.: Distributed transient safety verification via robust control invariant sets: a microgrid application. arXiv e-prints (2022)
2. Fleszar, K., Slowinski, R., Artalejo, J., et al.: A branch-and-bound algorithm for the quadratic multiple knapsack problem (2022)
3. Xie, Y.: A efficient algorithm for computing k-factor. arXiv e-prints (2022)
4. Franco, S., Jacquin, L., Meller, J., et al.: Integrity verification for a software stack or part of a software stack: US20210232684A1 (2021)
5. Weintraub, G., Kadosh, A.: Integrity verification in cloud key-value stores: US20210334409A1 (2021)
6. Rorke, D.A.: Geological data integrity verification system: US20210333426A1 (2021)

Soil Moisture Prediction Method Based on Machine Learning Algorithm

Jinhua Wang(✉), Guangning Gao, Danyan, Yan Sun, Xiaoyan Wang, and Xianlong Wang

Jinan Academy of Landscape and Forestry Sciences, Jinan 250103, China
w13583162936@163.com

Abstract. Soil moisture prediction is a technique for predicting soil moisture content. It is used in agriculture and hydrology to design irrigation systems, predict crop yields, monitor the impact of weather on crops, etc. The main purpose of soil moisture prediction (SPM) based on past data and climate information. SPM uses machine learning techniques, such as neural networks, k-means clustering, support vector machines, etc., to predict soil moisture at different time points in the growing season. The purpose of this technique is to classify soils into specific categories at an accurate and high-precision rate. It provides us with information about the moisture content of various soils, which is helpful to understand their characteristics and applicability for different purposes such as agriculture, engineering and construction. We can find that there are many methods that can be used to predict soil moisture, such as mathematical models, but they cannot provide accurate results samples from different locations.

Keywords: Machine learning · Soil · Moisture prediction

1 Introduction

At present, the measurement methods for obtaining soil moisture mainly include soil drying method, neutron method, resistivity method, satellite remote sensing inversion and other methods, but these methods usually have a series of problems. For example, neutron method is a soil moisture measurement method based on the radiation principle[1]. Although it has the advantages of high accuracy and speed, it has been gradually abandoned by some countries in recent years due to its harm to human health; As the most accurate method for measuring soil moisture content, the drying method has a long history and has gradually become the comparison value of various measurement methods. However, because it needs to destroy the soil structure and consumes a lot of manpower and material resources, it has the shortcomings of wasting resources and inconvenient operation for large-scale measurement; The resistivity method has been gradually eliminated because it needs to be calibrated in advance, its operation is complex and its results change with time[2]; Satellite remote sensing technology has the ability to provide spatial information, but its scale for measuring water content is large (usually averaging the area of hundreds of square meters on the ground). For reclaimed

J. C. Hung et al. (Eds.): FC 2023, LNEE 1132, pp. 323–328, 2024.
https://doi.org/10.1007/978-981-99-9538-7_50

collapsed farmland, it cannot meet the precision and scale requirements of reclamation quality detection. Generally speaking, these methods cannot meet the requirements of rapid, nondestructive, accurate and large area measurement required in reclaimed farmland, and their costs are high[3]. Therefore, it is urgent to study a mesoscale accurate, efficient and non-destructive water content detection technology. Based on this, this paper studies the soil moisture prediction method based on machine learning algorithm..

2 Related Work

2.1 Overview of Machine Learning

Non natural heuristic algorithms and natural heuristic algorithms are two major categories of meta heuristic algorithms, adaptive search and iterative local search, many meta heuristic algorithms are naturally inspired. Figure 1 below shows the machine learning framework.

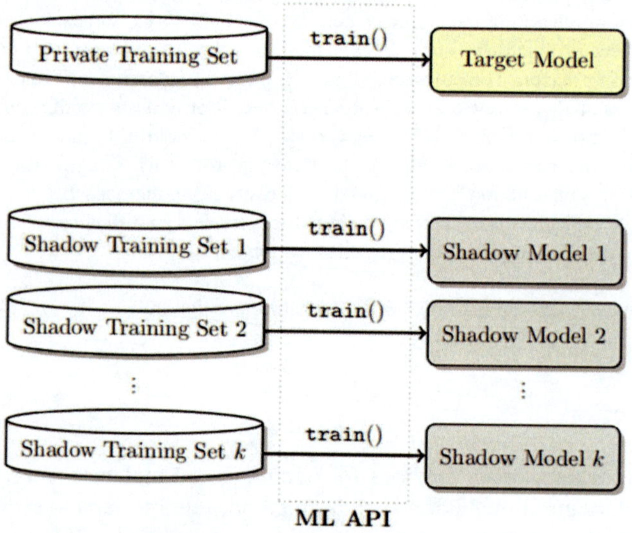

Fig. 1. Machine Learning Framework

Meta heuristics have different classification and description methods. One method is to distinguish their origins. It can be divided into two categories, unnatural heuristics and natural heuristics. As a traditional local search algorithm, the tabu strategy stores the information of the solution that has been accessed in the past through the enhancement of memory structure. When it does not allow to return to the recently accessed solution, it can promote diversification.

2.2 Prediction of Soil Moisture

In practical application, valuable soil moisture data must meet the following conditions: 1. The data must come from the same plot and multiple monitoring points. The practical experience of continuous monitoring shows that under the same agricultural plot and the same irrigation conditions, the difference of soil moisture can still reach 3% - 5%, even if the distance between the two monitoring points is only 0.5m. Therefore, valuable soil moisture data must be obtained from continuous monitoring of multiple monitoring points in the same plot with statistical significance. 2. The data must come from the monitoring of the same monitoring point and multiple soil depths. We know that there are obvious differences in soil water content at the same monitoring point and different monitoring depths; The depth distribution of active root system in different growth stages of crops is obviously different[4]. Therefore, in actual production, it is necessary to dynamically monitor the soil moisture data of the soil layer where the main water absorbing roots of crops are located, rather than the soil moisture data of the fixed depth soil layer. At the same time, we also have certain requirements for the quality of data, which must be stable and accurate. To sum up, the soil moisture data with the same location, multiple depths, stability and continuity are valuable and can guide the actual production.

3 Soil Moisture Prediction Method Based on Machine Learning Algorithm

In the existing prediction methods of site soil moisture, the original soil moisture series collected from the site are often used as experimental data. Integrated empirical mode decomposition is a signal decomposition method. After integrated empirical mode decomposition of soil moisture series, multiple modal sequences will be obtained, and the dimension of experimental data will also be improved. Compared with a single original sequence, these decomposed sequences can reflect more soil moisture changes. Convolutional neural network is one of the representative algorithms of deep learning. Because it can capture the spatial relationship of data, it is a machine learning model more suitable for processing high-dimensional data. This chapter will integrate empirical mode decomposition and convolutional neural network to predict the soil moisture of the site, and through experimental comparison has better prediction performance.

The collected soil samples are divided into three parts on average, one for soil water content data, one for soil hyperspectral data and one for soil organic matter data. The samples used for soil hyperspectral data collection are divided into two parts on average, one for wet soil spectroscopy and the other for dry soil spectroscopy. The soil sample used to determine the wet soil spectrum does not need any pretreatment, and the spectral data is obtained directly. For the soil sample used to determine the air dried spectrum, the spectral data is obtained after the soil is air dried, ground and sieved.

The median value of organic matter content in study area 1 is less than the average value, indicating that there are some soil samples with high organic matter value in the area. The variation coefficients of organic matter in study area 1 and study area 2 are 44.40% and 20.14% respectively, which belong to medium variation. The maximum value of soil moisture content in study area 1 is 311.68 g-kg-1, the minimum value is 80.49 g-kg-1, the average value is 178.60 g-kg-1, and the median value is 172.23 g-kg-1; The maximum value of soil water content in study area 2 is 272.41. The variation of soil water content in study area 1 is 22.30%, which belongs to medium variation, while the coefficient of variation of soil water content in study area 2 is 9.46%, which belongs to weak variation.

4 Simulation Analysis

From the data of each evaluation index after training, it can be seen that EEMD-CNN has the best performance in predicting soil temperature at different depths and at different time delays. This result is due to that the convolution nuclear energy used in the convolutional neural network can better obtain the hidden information in the multimodal data after the integrated empirical mode decomposition. As the shallow soil temperature is more susceptible to the impact of atmospheric environmental factors, the deep soil temperature will change more smoothly. It can be seen from the statistical characteristics of the data at each site that the variance of the data at each depth decreases with the increase of the depth, so the data of soil temperature at 5CM depth has more obvious fluctuations. When the data is processed by EEMD, the multimodal data obtained can reduce the impact of the original data fluctuations on the prediction results. Therefore, EEMD-CNN and EEMD-LSTM combined with integrated empirical mode decomposition are not vulnerable to the impact of data fluctuations when predicting low depth soil temperature, while BPNN and LSTM using the original data as input are more vulnerable to the impact of fluctuations. The full connection layer is used in BPNN, which is a common layer structure in the artificial neural network structure. This structure is neither like the convolution layer, which can obtain the spatial correlation of data, nor like the cyclic neural network node, which can obtain the sequence characteristics of data in time series. In LSTM and EEMD-LSTM using circulating neurons, the circulating neurons can connect the neurons of each time step sequentially along the time line, and the circulating neurons enable the model to inherit some previous states during the training process, so that the model can perform better in processing continuous time series data. LSTM is an improvement of recurrent neural network. LSTM adds the record of long-term memory state in the previous training process to the traditional recurrent neural network, so LSTM can be better applied to continuous time series, and has better prediction performance than BPNN. As shown in Fig. 2 below, the depth soil moisture prediction map is shown.

With the real data only in the local area, while the model with integrated. With the increase of time delay, the overall goodness of fit of each model prediction value is declining, but the decline of the model without integrated empirical mode decomposition is more obvious, especially in the area with large data standard deviation. Under each specific delay condition, each model combined with the integrated empirical mode decomposition has a better fitting degree. Among these models, EEMD Conv3D has a better fitting degree, which indicates that it has a better prediction performance in spatio-temporal soil temperature prediction. After calculation, the best evaluation index is obtained, which shows the best prediction ability of spatio-temporal soil temperature. It is the most suitable prediction method for spatio-temporal soil temperature in all models.

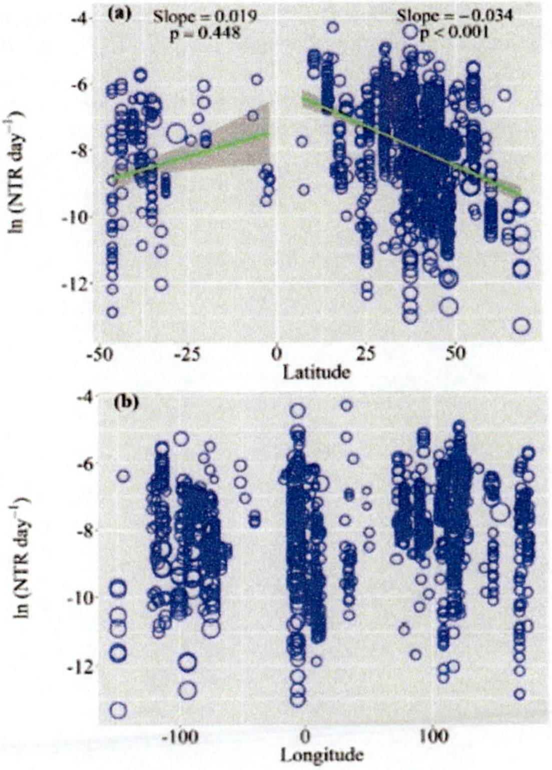

Fig. 2. Deep Soil Moisture Prediction Map

5 Conclusion

Soil moisture prediction method is a machine learning algorithm. It uses soil moisture and temperature data to predict future soil moisture. Today, it is one of the most widely used methods in many fields such as agriculture, forestry, hydrology and engineering. It uses the soil moisture observation data set to determine the best model to predict the next observation. In turn, AI models use a combination of these observations and various inputs, such as climate models, to predict future soil moisture.

References

1. Ning, X.G., Zhu, Y.L.: Power load prediction based on incremental optimization extreme learning machine. Comput. Simul. (2016)
2. Jun, Q.U., Shi L., Wang, J.: Lubricating oil moisture detection based on infrared spectrum and extreme learning machine. Lubricat. Eng. (2017)

3. Morellos, A., Pantazi, X.E., Moshou, D., et al.: Machine learning based prediction of soil total nitrogen, organic carbon and moisture content by using VIS-NIR spectroscopy. Biosyst. Eng. S1537511015304165 (2016)
4. Eyo, E.U., Abbey, S.J., Lawrence, T.T., et al.: Improved prediction of clay soil expansion using machine learning algorithms and meta-heuristic dichotomous ensemble classifiers. Geosci. Front. **13**(1), 268–284 (2022)

Design and Practice of Decision Support System for Integrated Water Resources Management

Fu Wang(✉)

Mianyang Teachers' College, Mianyang 621000, Sichuan, China
fuwang@yeah.netS

Abstract. The design and practice of based on machine learning is the process of designing a system that can make decisions using data, knowledge and rules. Decision Support System (DSS) is a computer-based tool that helps decision-makers make better decisions by providing them with information, knowledge and analysis tools. DSS is used in the field of environmental management to support decision-making on issues such as water resources development, pollution control and protection. The main objective of DSS is to provide decision-makers with a set of comprehensive data, analysis tools and models for use in the planning process.

Keywords: Water resources management · Decision making system · Machine learning algorithm

1 Introduction

As we all know, is to meet the needs of urban greening and river lake landscape. With the improvement of urban people's living standards, the external environment quality will be paid more and more attention, and the urban greening and municipal water use will show a T-level sharp increase. Due to the fact that most of the urban greening, road flushing, sanitation and other water users use water trucks to add water, and the sanitation management and maintenance work area is large, long mileage, heavy tasks, and the capacity of on-board water tanks is limited, in order to renew water, the operating vehicles have to run tens of kilometers more every day, which not only affects the efficiency of mechanized cleaning and watering operations, but also increases the traffic safety risks [1]. The construction of smart water supply points can effectively solve this outstanding problem, achieve multiple users in one well (one car, one card system), plan the route to get water nearby, insert cards to get water out, accurately measure, and also optimize water resources and enterprise management through data. To ensure the long-term management of pavement cleaning.

On this basis, integrated water resources management based on machine learning, refers to the advanced theories of domestic and foreign water conservancy experts, and combines its own mature remote acquisition control system to launch a new generation of integrated water resources management system, which is supported by GPRS, 4G technology [2], GIS technology, multimedia, wireless network and other high-tech technologies, Intelligent management of reclaimed water dispenser water sales and real-time

J. C. Hung et al. (Eds.): FC 2023, LNEE 1132, pp. 329–335, 2024.
https://doi.org/10.1007/978-981-99-9538-7_51

monitoring, comprehensive analysis and management of relevant information are carried out to change the "supply management" of water resources into "demand management", emphasize the status and role of comprehensive water resources management, and truly realize the efficient utilization and management of water resources.

2 Related Work

2.1 Requirements for Water Resource Management System

It mainly serves the users of sprinklers, such as municipal sanitation, landscaping and other units, and their management of such units Water sprinkling task, water adding route planning, water adding speed and time control, water adding vehicle scheduling, water adding situation planning, etc. are of great significance. And the background data monitoring of the system can optimize the management of the above work in real time, reduce the operating cost, maximize the rational planning of resources, increase the economic efficiency, and make the traditional management into digital intelligent management.

The equipment required for the integrated development of smart water affairs mainly has the following functions:

Monitoring function, control function and patrol function. Alarm function. And data transmission, data management, data deduction and other functions.

The intelligent water dispenser highly formulates the functions required by the above water affairs, and forms an integrated machine product by scientific and technological means. The main application directions are: flow disc, flow rate, pressure, water quality, depth, level, temperature, etc.

The design of double-sided advertising space of intelligent water dispenser box can also be used as a public service information window to popularize water knowledge. Publicize the slogan of water conservation [3].

Display function of intelligent water dispenser. Daily display of time, date and other information. When adding water, it can display the water adding data in real time, the program can also display the queuing situation of the water adding points, and the online situation of the water dispenser. Thus, the road system of water dispenser is rationally planned to reduce energy consumption.

The functions of the water dispenser are superposed, and it is not necessary to use multiple equipment in series, which is simple and convenient, and is conducive to management. System integration, a set of system management all functional equipment, without multiple interface development, interface confusion, clear data.

The "deformable and extensible product" of the water dispenser has a wide range of services. It can integrate all functions, or use the functions separately according to the needs, and adjust measures to local conditions.

The product customization space is large, such as urban advertising space, safe city monitoring point, urban convenient information display platform, smart water popularization point, etc.

The water filling points can be reasonably planned through pipelines, model cleaning, driving routes, etc., and the scheduling can be carried out in different time periods to optimize the work of the West Water Truck [4].

It mainly includes card selling part, water metering part, LCD display prompt, query part, valve control part, data remote part, data transmission part, etc. It can ensure the reliable and stable operation of the basic water selling function, and also provide a simple and convenient operation platform for users of the integrated water resource utilization management system to collect water by themselves. For multiple users, it is convenient for measurement and charging, which improves the comprehensive modern management level.

2.2 Machine Learning

The concept of machine learning is derived from the pattern recognition and computational learning theory in the field of artificial intelligence. It is committed to building specific algorithm models, learning knowledge from known data or making predictions. According to the learning mode, machine learning can be divided into supervised learning and unsupervised learning. The main difference between the two is whether the input data is labeled. Supervised learning can be subdivided into semi supervised learning, active learning and reinforcement learning. The typical supervised learning process is shown in Fig. 1, and its example is MNIST handwritten digit recognition; The input data of unsupervised learning does not contain labels. A typical model is K-means clustering. From the perspective of application task division, machine learning can be divided into regression, classification, clustering, density estimation and data dimension reduction. For example, the prediction of temperature or PM2.5 index in the next few days is a common regression problem, and the MNIST handwritten numerals mentioned above are a classification problem.

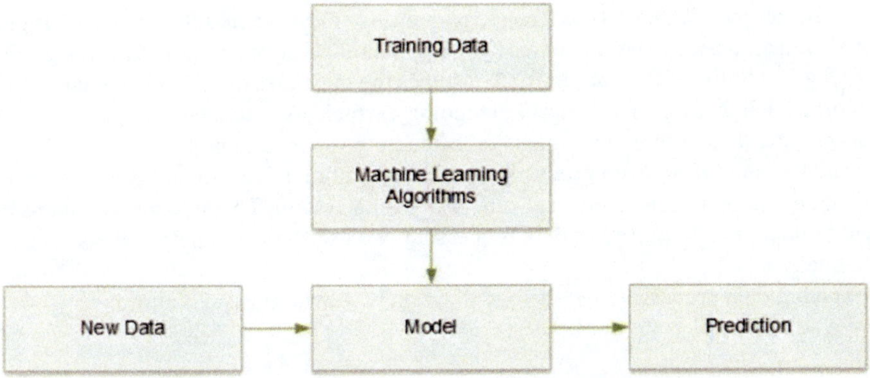

Fig. 1. Schematic Diagram of Machine Learning Process

Features are also called attributes. According to whether they are related to the current learning task, features can be divided into "relevant features", "irrelevant features", "redundant features", etc. There are two main motivations for feature selection: first, in the learning task, too many data attributes may cause the "dimension disaster" problem. If important features can be selected from the current feature set, the "dimension disaster" problem will be greatly reduced; On the other hand, removing irrelevant data features often reduces the difficulty of learning tasks. Feature selection mainly includes two processes, namely, feature subset traversal and feature subset evaluation. First of all, it is necessary to have an appropriate method to search feature subsets, because in theory, the number of candidate feature subsets is very large when there are many data features; Secondly, design a good evaluation strategy, evaluate and sort the traversed feature subset, and select the best one to get the feature selection method.

3 Design and Practice of Decision Support System for Integrated Water Resources Management Based on Machine Learning

In this project, the watershed data sent by robot fish to the server is used for real-time big data access. The robot fish collects water level, flow, water quality, weather and other information of the river sections along the watershed and sends it to the database after a series of processing. Data access is the bottom link of the project. Flume, Kafka and Storm are mainly used in this project to achieve real-time big data processing.

Flume's main advantage is that it can customize many data sources to reduce the amount of development. For Flume, the key is how to collect data and send it to Kafka. Since Flume cluster is used in this project, the configuration of Flume cluster is also critical. For Kafka, the key is how to receive data from Flume.

In this project, Kafka acts as a message queue (or message middleware), and its generated messages need to be consumed by consumers. Therefore, the key to the integration of Kafka and Storm is how our Storm consumes the messages in Kafka's message topic. In Storm, a job that we submit is called topology, which also includes the spool and bolt. In Storm, the data processing logic is reflected in the spool and bolt.

The last step of big data processing is the integration of Storm and Redis, which is the landing mode of data in our real-time processing system. That is, Storm contains the logic of data processing. After data processing, we use Redis to analyze the storage of data finally.

The big data processing technology diagram is shown in Fig. 2 below:

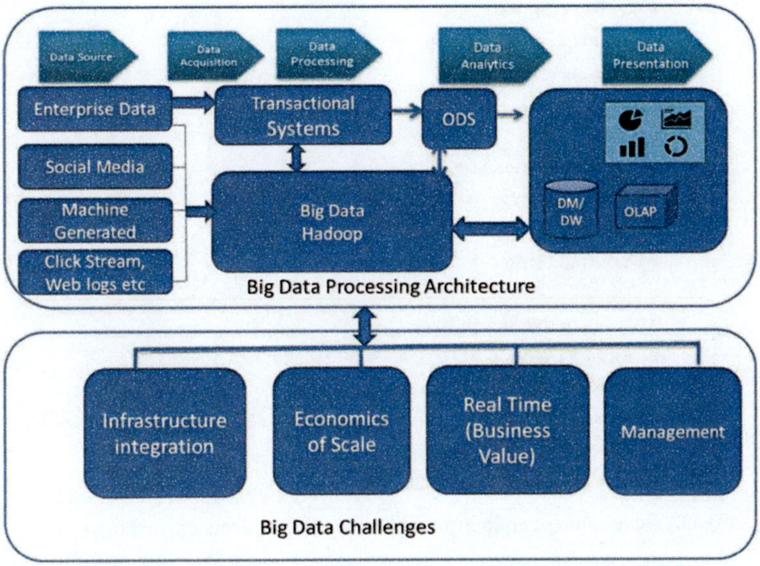

Fig. 2. Big Data Processing Technology Relationship Diagram

The main function of Flume in this project is to collect data. In the construction of the entire Flume cluster, the Flume Agent is deployed on each Flume Agent server to collect the water level, flow, water quality Weather and all kinds of information, and then use the data sinking method to send the data collected on the Flume Agent server to another Flume Consolidation Agent for summary and then to Kafka. It mainly uses Flume Avro Source to listen.

4 System Development

The technical architecture design of this system is mainly divided into several parts. The first part is big data access. The main processing mode is real-time streaming big data processing of Flume, Kafka, Storm, Zookeeper. The database mainly adopts Redis non relational database and MySQL relational database. Redis is responsible for caching data and fact data, and MySQL is responsible for storing data permanently. The backend of the system mainly uses Nodejs as the server to forward requests and access static pages, and uses the mapping from the enumeration type customized by Spring Boot to the database type. The front end mainly uses Require Js implements modular loading, develops on the basis of ArcGIS, and cooperates with front-end technologies such as HTML, CSs, JavaScrit, ECharts, etc. As shown in Fig. 3 below, it is the development code of water resources data decision-making system.

```
# configure agent
a1.sources = f1
a1.channels = c1
a1.sinks = k1

# configure the source
a1.sources.f1.type = netcat
a1.sources.f1.bind = localhost
a1.sources.f1.port = 3333

# configure the sink (kafka)
a1.sinks.k1.type = org.apache.flume.sink.kafka.KafkaSink
a1.sinks.k1.topic = sogolog
a1.sinks.k1.brokerList = hsm01:9092,hss01:9092/kafka
a1.sinks.k1.requiredAcks = 0
a1.sinks.k1.batchSize = 20
```

Fig. 3. Development code of water resources data decision-making system

All aspects of the design of the water resources management platform, including demand analysis, system business architecture, system technology architecture, and system logic architecture design. In the requirement analysis, the system introduction, functional requirements and operating environment of the system are also specifically introduced. The design of the whole system is based on the characteristics of high cohesion and low coupling, and is divided into front-end UI, presentation layer, business layer, data layer, database and big data access layer. This design idea of front end and back end separation greatly reduces the mutual influence between the page layer and the service layer, and improves the stability and scalability of the service.

5 Conclusion

The design and practice of a research project, involving the development of computer-based tools to assist decision-makers in managing water resources issues. The an effective decision support system (DSS) for use by national and international institutions involved in planning, implementing and monitoring water resource management activities. The DSS will provide an integrated approach to the collection, analysis, interpretation, presentation and dissemination of data on all aspects of water resource management.

References

1. Qiu, J.N., Shu-Yuan, L.I., Wang, H.W.: Decision Support System of Haihe River Basin Water Resources Management. Haihe Water Resources (2017)
2. Bo-Liu, X.U., Dong, Z.C., Hong, X., et al.: Database Development of Decision Support System for Water Resources Scheduling and Allocation of the Yellow River Basin. Water Resources and Power (2017)

3. Theodorakopoulos, Thomas F.: Integrated cost management system for delivering construction projects (2017)
4. Xiu, D.U.: Decision support system of water resources optimization in irrigation area. Water Resources Development and Management (2017)

Research and Development of Mobile Terminal Color Management Module Based on Android Platform

Wang Jia(✉)

Yunnan Open University, Kunming 650021, Yunnan, China
ynou_edu@163.com

Abstract. In recent years, represented by smart phones and tablets have been greatly developed. The mobile terminal color management module is a system that can be used in the field of mobile terminals. It has a function that can control the display mode, color space and other parameters of each application through its own API (application programming interface). Here are some applications: (1) Camera calibration on mobile phone screen; (2) Use external devices such as digital cameras and webcams to calibrate the screen; (3) Color correction of images taken by smart phones or tablets; (4) Image processing using image recognition technology; (5) According to different video formats.

Keywords: Mobile terminal · Android platform · Color Management

1 Introduction

As a simple communication device, become the key entrance and main innovation platform for Internet services. The way people get information has changed from traditional newspapers, TV, computers and other media to portable smart phones. At the same time, traditional paper media has entered a historical turning point [1]. Under the strong impact of new mobile media, mobile intelligent terminals are accelerating to replace personal computers as tools in people's daily work and life. With a small piece of equipment in hand, you can easily complete high-definition photography, online shopping, video games, reading and surfing the Internet, sending and receiving emails, network communication and other functions. In human history that has been widely penetrated, rapidly popularized, and has a huge impact, reaching [2]. Mobile terminal equipment is developing towards the direction of larger screen size, thinner, lighter weight and more comprehensive functions.

Consumers' positioning of mobile terminals such as smartphones and tablets is mainly based on communication functions, supplemented by shooting, reading, surfing the Internet, games and other functions. Since the advent of imaging technology, especially after the emergence of color imaging technology, clarity and color fidelity are the two most important indicators in the pursuit of image quality. As the most frequently used imaging equipment in daily life, the color rendering performance of mobile terminals has become an important indicator to evaluate their functions. The diversified

J. C. Hung et al. (Eds.): FC 2023, LNEE 1132, pp. 336–341, 2024.
https://doi.org/10.1007/978-981-99-9538-7_52

development of mobile terminals will put forward more stringent requirements on the performance of mobile terminals in some aspects. For example, the level of photosensitive components and the color reproduction performance of the screen of smart phones that are good at taking pictures will be higher than those of ordinary smart phones [3]. Some applications with high requirements for color, such as online shopping, image processing. When mobile terminal equipment is applied to printing and other industries with high color requirements, the range of color reproducible on the mobile terminal screen should basically cover the color gamut of the printing press. When the hardware conditions meet the requirements, it is necessary to control the device color to be presented in an accurate way through color management technology. For ordinary consumers, their requirements for color accuracy of mobile terminals can be summarized as "the more accurate the color displayed on the screen, the better, without increasing the screen cost" [4]. As mobile terminals gradually replace PCs to meet users' needs, users' requirements for color rendering accuracy of mobile terminals have come to a time when mobile terminal manufacturers and service providers need to provide solutions. It can be predicted, color management technology will be popularized on the operating system.

2 Related Work

2.1 Mobile Terminal

In 2007, ZTE applied for the patent.This patent adopts a color management mechanism based on ICC Profile. When using mobile terminals to display images, first call the corresponding feature file to correct the image, and then display the output image, so as to achieve the purpose of color management.

American Datacolor Company launched SpyderGallery, the world's first user defined software for correcting the color of tablet computers. This software is a professional image browser with color management function for iPad and iPad 2. The browser can not only browse photos, but also help photographers control the color output effect of the iPad display. The color management module of the software also uses ICC color management mechanism, but only provides users with two reproduction intentions (saturation and perceptible). The reproduction intention of saturation can keep the whole picture colorful, but lacks hierarchy. The perceptible reproduction intention can retain the hierarchical information of the original image to a large extent, but it will reduce the saturation of the image color. The user can select the corresponding rendering intention as required. Using this software to create a profile is simple and easy to use, which can ensure the accurate and stable reproduction of colors in the whole process.

In May 2012, Apple issued the patent Dynamic Display Adjustment Based On Ambient Conditions, which is based on the dynamic adjustment of the display effect under environmental changes. First, use the pre sensor and post sensor of the display to collect the surrounding environment information, then use the environment model to determine the positions of black points and white points, adjust the gamma curve, and modify the LUT table. In this way, the image quality under various observation environments can be reproduced as true as possible.

2.2 Feasibility of Mobile Terminal Color Management

After the above discussion, mobile terminals are in the stage of rapid development of technology, and display materials and technologies are undergoing the process of upgrading; The use environment of mobile terminals is extremely complex, which puts forward more stringent requirements for screens. Like desktop display, mobile terminal display, as a display device, is ideally capable of accurately reproducing the original color of a picture. However, due to the diversity of mobile terminal display technologies and the complex use environment, there is no unified color calibration standard that can cover all types of mobile terminals. Like the portable laptop, the portability of the mobile terminal and the user's low requirements for screen color rendering accuracy determine that the mobile terminal screen cannot provide the functions of adjusting contrast, color temperature, and single channel gain as advanced desktop displays do. All adjustments to screen color are based on the adjustment of display drive values.

At present, in the field of mobile terminals, the screen manufacturer writes the correction parameters into the display driver before leaving the factory. The products of the same batch can achieve better color consistency. Different mobile phone manufacturers use different screen materials and display technologies, which leads to obvious differences in color performance between products of different brands. For ordinary consumers, this way of color calibration before leaving the factory can meet consumers' requirements for the accuracy of color rendering of mobile terminal screens, but it is difficult to meet the corresponding requirements for industries with high requirements for color accuracy, such as photography, printing, etc.

Cai Shengyan et al. used YD/T 1607–2007, the recommended standard in the domestic communication industry, and the international standard ISO for screen proofing as the performance evaluation standards for general use and proofing use of mobile terminal displays respectively, and tested the five indicators of mainstream mobile terminal displays, including brightness, brightness uniformity, screen proofing (applicable to mobile terminals). These mobile terminal displays that meet the use requirements have the hardware basis for color management.

3 Android Platform

First, introduce the main functions of the Android platform. The Android platform provides developers with application frameworks that developers can follow to develop applications. The Dalvik virtual machine customized by Android can make the developed applications run. The Android system also integrates the browser based on the Webkit open source project, 2D and 3D graphics engines, SQLite database and other functional modules. Android system provides support for audio, video, pictures and other multimedia files, GSM calls, Bluetooth, 3G and Wi Fi. In addition, Android system also supports camera, GPS, gyroscope, accelerator and other devices (Fig. 1).

Android applications are written in Java language. After testing, developers can use packaging tools to package application data and resources together to generate an apk file. The apk file is installed on the device as the carrier of the program. Each apk file can be considered as an application.

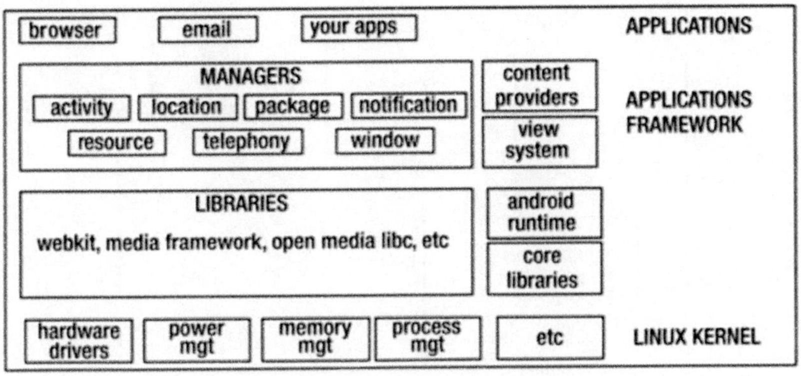

Fig. 1. Android system architecture

4 Mobile Terminal Color Management Module Based on Android Platform

The devices in the printing system often use RGB or CMYK and other device related color spaces. Due to different equipment characteristics, the same color value is on different equipment Different visual effects will be obtained when outputting. In order to obtain a consistent visual reproduction effect, it is necessary to establish corresponding color correspondence according to the characteristics of input and output devices. However, this solution is only applicable to the situation where the number of equipment is relatively small and the whole process is relatively closed. In order to achieve open color management, Apple, Agfa, Adobe and other companies jointly established the International Color Consortium (ICC) in 1993 [10]. The organization aims to solve the compatibility of color management, and its goal is to establish an open, operator independent, cross media color management system.

The idea of ICC color management is to describe the color characteristics of all devices in the system with CIE color system, and the color conversion between any devices should be conducted indirectly through CIE color space. Because CIE color system describes the color perception of people, and has nothing to do with the color presented by any device, the basic principle of ICC color management is to use color perception as the basis to uniformly describe the color presented by all devices with the color perception of people, so that all devices present the same color perception. Therefore, in the ICC color management mode, the color conversion from m devices to n devices only requires the conversion of $m + n$ device values to CIE color values. More importantly, the use of this color management method can establish a standard method for describing the color characteristics of devices, which is no longer dependent on the specific situation and relationship between two devices, so that the color management method can be standardized, and operators can complete it by themselves according to this standard method.

As shown in Fig. 2. The source device characteristics file records the conversion relationship between the source device color and the PCS color, while the label color

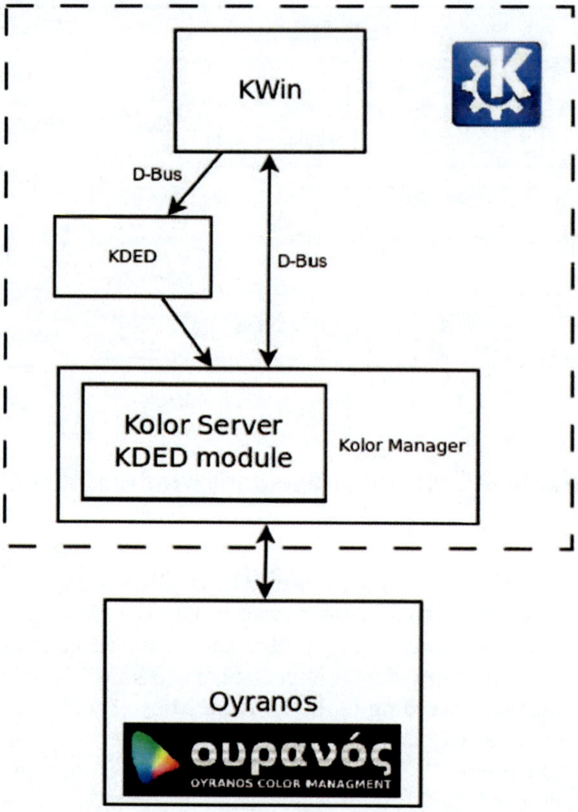

Fig. 2. ICC color management mechanism

characteristics file records the conversion relationship between the target device and the PCS color space.

When the color management software performs color conversion, it reads the characteristic files of the source device and the target device respectively, first converts the device color to PCS, and then converts. Therefore, color conversion can be realized through feature files and PCSs without establishing complex input/output correspondence for different output devices. This not only simplifies the complexity of color management, but also meets the needs of development style color management.

5 Conclusion

This paper first introduces the current status of mobile terminal display technology, pointing out that the rapid development of mobile terminals requires accurate screen color rendering. The accurate color rendering of mobile terminals cannot be separated from a good color management engine with mobile terminal characteristics. On the basis of reading a lot of color management literature and combining the characteristics of mobile terminals, this paper designs a color management engine (CMM) suitable for mobile terminals. The CMM conversion algorithm is based on the matrix/TRC model and can process ICC characteristic files based on the matrix/TRC model. According to the working principle of the color management engine, it is divided into four sub modules: interface interaction, file processing, color space definition, and color space conversion. Each module cooperates to complete color conversion in the color management process. The interface interaction module provides CMM with image color data, color space parameters and gamut mapping parameters; The file processing module realizes the functions of image color extraction, ICC file extraction, embedding and processing, and image generation; In the color space definition module, the attributes of the color space and the conversion relationship between the color space and PCS are constructed according to the color space parameters; The color space conversion module calls the above three modules to complete a complete color conversion process and feed back the conversion results to the interface interaction module for display.

References

1. Bu, J.: Research on the strategic management of local universities under the background of comprehensive reform of universities. High. Educ. Stud. **12**, 97–104 (2022)
2. Liu, R., Song, Y., Jin, M.: Design and Implementation of Buy-and-Run Delivery System Based on Android Platform. Computer Knowledge and Technology (2018)
3. Wei-Wu, H.E., Feng, H., Tan, Q.W.: Implementation of Tactical Unified Communication Client Based on Android Platform. Journal of Military Communications Technology (2016)
4. Wang, Y., Wang, Y., Zheng, Y.: Development and test of property management system based on Android. Wireless Internet Technology (2019)

Research on Key Technologies of Image-Based Virtual Restoration of Ancient Buildings

Kun Wang(✉)

School of Art Disgn, HanKou University, Wuhan 430212, Hubei, China
wk_0828007@163.com

Abstract. At studying the key technologies of virtual reality (VR) system development, construction and operation. Its main objective is to develop virtual reality technology in order to reconstruct the ancient city based on archaeological evidence, which is difficult or impossible to reconstruct by physical methods. The project will also provide a basic understanding of how virtual reality technology is applied to Archaeology and related fields. The main purpose is to carry out experimental research on three topics: 1) the development of virtual reality technology; 2) The construction method of virtual reality system; 3) Application of virtual reality system. A series of advanced methods such as 3D scanning and photogrammetry are used to study the key technologies of image-based virtual restoration of ancient buildings. The results obtained from this method will be used in our further work to improve the quality and efficiency of these technologies.

Keywords: Ancient buildings · Images · Virtual recovery · key technology

1 Introduction

With a long history of 5000 years, the world has left rich and precious cultural heritage of ancient buildings for mankind. However, due to the disrepair for a long time and the large-scale infrastructure construction and old city transformation projects carried out throughout the country, some ancient buildings have been seriously damaged, and some cities have suffered permanent losses in historical features during the transformation. The development of Tianjin's old urban area and its old buildings is extremely inconsistent with that of modern cities, which urgently needs to be scientifically reformed [1]. However, the transformation of the old city inevitably involves the contradiction between protection and development, as well as the survival of the existing traditional buildings with Tianjin's regional characteristics. How to use advanced technology to protect these precious heritage has become an urgent problem [2].

In this context, such as the aging of video tapes, the limitation of the field of view of the photographic scene, and the distortion of image reproduction [3]. The cause of cultural heritage protection has a new dawn - the emergence of three-dimensional model reconstruction technology provides a more reliable guarantee for people to study the history and original appearance of ancient buildings.

J. C. Hung et al. (Eds.): FC 2023, LNEE 1132, pp. 342–347, 2024.
https://doi.org/10.1007/978-981-99-9538-7_53

By means of virtual reality technology, a large number of photos, texts, videos and other materials collected on the spot are used to construct the whole scenic area to truly reproduce the cultural landscapes such as the Giant Buddha and the temple. It provides the most intuitive means for world heritage protection, government departments' decision-making and scenic area planning and management, and also plays an important role in many aspects such as scenic area promotion, new tourism scenic line development, tourism facility construction and scenic area ecological management [4].

2 Related Work

2.1 Research Status at Home and Abroad

Both at home and abroad attach great importance to some valuable. From the government to universities, from cultural heritage protection units to companies, and even individuals, they are doing research and design in related fields. More and more virtual ancient building systems have been designed and produced in the world [5]. The following are some pictures of domestic and foreign achievements in this regard, which also provide a good reference for the completion of this article.

(1) Digital Yuanmingyuan

Beijing Ruizhi Shengyang Culture Communication Co., Ltd. Has designed and produced the digital Summer Palace by using advanced CG technology (CG, computer graphics) after four years of efforts. The scene is magnificent, and the lifelike and meticulous restoration effect is amazing.

(2) Virtual Suzhou Classical Garden

The virtual tour system of Suzhou Classical Gardens independently developed and produced by Jiangsu Suzhou Yiwei Digital Design Co., Ltd. Uses 3ds Max to make ancient building models, and uses software to make virtual roaming in the later stage. Users can have panorama of Suzhou classical gardens through the interactive operation of the system, and can also walk freely. As shown in Fig. 1

(3) Digital Great Wall

A high-precision, three-dimensional "digital Great Wall" virtual system based on three-dimensional laser measurement technology was jointly developed by Hebei basic geographic information center and Guangxi Guineng Information Engineering Co., Ltd. Beijing Branch. According to the measurement data obtained by laser measurement technology, the system is designed by modeling. This "Great Wall" not only allows visitors to enjoy it, but also is conducive to the restoration of the Shanhaiguan ancient Great Wall.

It can be seen from the above examples that there have been many research achievements in the field of Virtual Cruise of ancient buildings at home and abroad, but there are still few reports and achievements about the application of IBMR technology to the virtual restoration of ancient buildings, which still has its research value and prospects. Therefore, this paper mainly studies the image-based virtual restoration technology of ancient buildings.

Fig. 1. Bird's-eye view of Suzhou Classical Garden virtual tour system

2.2 Image Based Rendering Technology

Image based rendering technology is to generate scene pictures of different viewpoints based on some pre generated images (or environment maps), that is, to construct new images of virtual viewpoints based on the original images of known viewpoints. This method can generally be divided into two types: one type of method does not need to recover the three-dimensional structure of the objects in the scene at all, but rather to densely sample the scene, and then organize the collected images according to a certain retrieval method. To construct the image of a new viewpoint, you only need to query the sampling set. This method requires a lot of original images and is not suitable for general virtual reality systems. The other method needs to know the structure of the object in space first, and then draw the object by using view dependent texture mapping, view interpolation and other methods according to the geometric structure and sparse sampling of the scene. Compared with traditional rendering technology, image-based rendering technology has distinct characteristics:

Each scene picture only describes the result of observing the scene from a given viewpoint along a specific line of sight. In order to get rid of the limitation of the field of view of a single frame picture, we can shoot or calculate the images along all directions at a given viewpoint and splice them into a full scene image. In order to enable users to roam in the scene, we need to establish panoramas of the scene at different positions.

3 Proposal of 3D Virtual Scene Modeling Technology

Users can observe or operate objects in the virtual environment and participate in events in the virtual environment by using human natural skills according to their own feelings. At present, virtual reality technology is very ideal for architectural performance, and there are mainly some technical problems:

(1) The establishment and real-time rendering of 3D models of complex buildings. Although there are many commercial modeling or modeling software, it is still a very heavy task to construct object models with rich details and complex structures. In particular, Chinese ancient buildings are assembled one by one in the form of mortise and tenon connection and building blocks. These wooden components are not only in a variety of shapes, but also in a large number, which is easy to cause slow real-time rendering. How to build a complex 3D virtual scene has become the primary basic problem to be studied.
(2) In terms of hardware, the speed and capacity of its display have not been well solved. Most of the special equipment for virtual reality systems are not only expensive, but also not very convenient to use, and the effect is relatively limited, which can not meet the ideal effect of virtual reality.
(3) Solution it is often unbearable for users to wait a long time when the overall speed of the existing network is slow. How to optimize the complex model has also become a key technology to be studied in modeling.

4 Key Technologies of Image-Based Virtual Restoration of Ancient Buildings

4.1 Overall Solution Design for Virtual Restoration of Ancient Buildings

(1) The contact measurement technology can be used to collect the field data of the ancient buildings, record the surveying and mapping data in detail, and draw the architectural surveying and mapping map. This process is to make an accurate model of the cultural heritage of ancient buildings and restore the original appearance. Among them, the key point of measurement is the plane data of ancient buildings, the elevation data, and even the section data. Then, the three-dimensional cultural relics and buildings, and then the restoration plan of ancient buildings is made into a specific, accurate and vivid three-dimensional restoration system of historical relics and historic sites. Using this method, we can clearly browse the overall layout of ancient buildings in a three-dimensional environment, and also enter the interior of the building to understand its internal structure. However, this method is difficult to implement and requires a lot of architectural knowledge.
(2) If ancient buildings exist and are well preserved, but there are some places that cannot be reached by surveyors, or places with great risks, non-contact measurement technology can be used, which can be completed by visual inspection or estimation. However, it is inevitable that data estimation errors will occur. In order to improve the accuracy of measurement, multiple three-dimensional laser scanning measurement technology can be used. With this method, high-density point data acquisition

can be carried out on the surface of the measurement target. This measurement method greatly improves the measurement standard and measurement efficiency of ancient building data, also very high. Using these data and 3DS MAX software for three-dimensional modeling greatly improves the realism of the restoration effect of ancient building cultural relics. But this method requires expensive investment.

4.2 Design of Image-Based Virtual Restoration System for Ancient Buildings

The two-dimensional information in the practical problem of ancient building modeling and virtual restoration, that is, three-dimensional reconstruction and virtual restoration based on photos. This system is designed and developed for the restoration of the original appearance of ancient buildings that have been damaged or no longer exist.

The overall architecture of the image-based virtual restoration system of ancient buildings developed in this paper is shown in Fig. 2

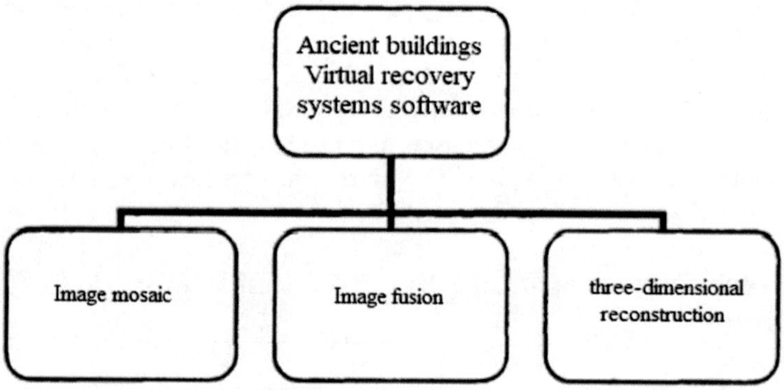

Fig. 2. Overall architecture of virtual restoration system for ancient buildings

5 Conclusion

The research of this subject involves knowledge of many disciplines: ancient architecture, three-dimensional modeling, virtual reality, etc. due to the limitation of time and professional ability, it is impossible to conduct in-depth research on all ancient building models. At present, there are still many aspects to be further studied: first, there is a certain gap between the understanding of ancient buildings and professionals, and further efforts are needed; At present, the research mainly focuses on the ancient architecture of Nanputuo temple in Xiamen. Although it is representative to a certain extent, it is only to throw a brick to attract jade; At present, the research of virtual simulation system based on Internet has just begun, and the visualization of the model has been realized. Further research is needed on the realism of materials and the interactivity of users.

Acknowledgements. 2019 School-level Curriculum Ideological and Political Special Teaching and Research Project (No.: 2019sz05).

References

1. Yan, S.: Research on Key Technologies of Intelligent Protection for Marine Power System. Clausius Scientific Press (2021)
2. Tan, L., Wan, R., Wan, G., et al.: Research on Key Technologies of Real Time Shadow in Global Virtual Geographic Environment. Journal of System Simulation (2017)
3. Qin, X.J.: Research on key technologies of ship navigation system based on hybrid tracking registration. Ship Science and Technology (2017)
4. Jia, J., Zhou, C., Jiang, N., et al.: Research on 3D Modeling Key Technologies of Complex Ore Body Based on Geological Database (2016)
5. Yongjin, L.U., Bo, W.U., Lin, R., et al.: Key Technologies of the Three-Dimensional Engine Database Foundation based on CATIA V6 Software. Machinery (2016)

Application of Weighted Least Squares Algorithm in Machine Vision System

Liping Wang(✉) and Zhongliang Wang

Tongling University, Tongling 244000, Anhui, China
wanglip_tl@163.com

Abstract. The relationship between camera position and reconstruction accuracy is analyzed, and the weight value is estimated on this basis. The application of weighted least squares algorithm in machine vision system is a method for estimating unknown model parameters. The weight is calculated by using some known data, and then applied to unknown data for estimation. This method is also called supervised learning or regression based on training set. Here, we use all the features in the image and train them with available data sets containing preprocessed and normalized training sets so that they can be easily compared with other images in the same category.

Keywords: Machine vision · Weighted least squares algorithm

1 Introduction

The vision system is generally based on pinhole model, and the projection matrix equation is used to reconstruct feature points. Due to the error, the least square method is often used to obtain the feature points in practice. The error analysis of machine vision system is complex. In order to including the influence of the distance from the camera to the feature point and the rotation matrix of the camera on precision. The second is to establish an error model and use the error model to correct the data containing errors; the third is to design a nonlinear optimization algorithm and use the optimization algorithm to correct the deviation between the actual value and the theoretical value [1].

In the process of factory production, in order to ensure the high quality of production products, it is necessary to continuously improve the production process, define, measure, analyze, improve and control from the perspective of DMAIC's Six Sigma. Among them, the data obtained through measurement is an important prerequisite to ensure the effective implementation of analysis and improvement. In order to ensure that the obtained measurement data are referential, it is necessary to evaluate the ability of the measurement system for product measurement, that is, use statistical methods to identify the influencing factors affecting the measurement results and their degree of influence, and evaluate the ability of the measurement system, It is also concluded whether the measurement system meets the requirements for use. "The higher the capacity of the measurement system, the better the reliability of obtaining measurement data [2]. The

J. C. Hung et al. (Eds.): FC 2023, LNEE 1132, pp. 348–353, 2024.
https://doi.org/10.1007/978-981-99-9538-7_54

QS9000 quality system requires enterprises to test the reliability of the measurement system before using the measurement system to obtain data, so as to ensure that the measurement system is qualified and meets the requirements for use.

2 Related Work

2.1 Research Status at Home and Abroad

Set the threshold value according to the product processing surface, and when the surface wear is greater than the threshold value, replace the tool, which is more accurate than the human eye observation [3]. Liu Ze and others used a machine vision system to detect rail surface defects, calibrate the location of rail surface defects, and count defect characteristics, thus realizing rapid rail detection. In order to accurately detect the wear status of the tools used in the production process, Qin Guohua et al. used the machine vision system to take pictures of the integrated circuit board, automatically selected the optimal threshold value using the bimodal method, and determined the edge of the wire bonding position according to the threshold value to determine the accuracy of the wire bonding position. Sun Xuechen et al. designed an online automatic detection system based on machine vision. For the obtained camshaft surface image data, the neighborhood weighted segmentation method was used to separately analyze the value of each pixel in the image and perform threshold segmentation, extract camshaft surface defects, and judge whether the product is qualified. In the field of food inspection, Zhao Jun and other scholars summarized in the inspection of potato quality by domestic and foreign scholars. They took photos of potatoes through the machine vision system, and graded potatoes according to their appearance, shape, surface defects and other factors. Tong Zhao and others analyzed the application status of agricultural product quality detection. It saturation of rice, removing impurities in tobacco leaves, and grading apples according to the color, shape, size and surface defects extracted from image data [4]. In the medical field, Gan Lan used the the image of gastric mucosal tumor cells, and proposed an image recognition method based on quantum self-organizing feature mapping neural network, which can effectively classify gastric mucosal tumor cells into three categories: cancer cells, proliferative cells and normal cells. The machine vision system can also be used for the detection of ceramics. Weng Zhengkui and other people took photos of ceramics through the machine vision system. By traversing the edge contour of ceramics, they obtained the details of their edge structure and realized the non-destructive classification of ancient ceramics' shape, structure, glaze color and decorative patterns.

2.2 Overview of Machine Vision System

The machine vision system uses a machine to obtain information instead of the human eye. It uses image sensor devices (CCD, CMOS, CID, etc.) to obtain scene information by charge injection, transfer, and readout, mechanical control execution module, and intelligent judgment and decision-making module. The Robotics Branch of the American Manufacturing Engineering Association believes. After acquiring the image, the machine vision system transmits the image to the processing unit [5]. Through the image processing technology, it judges the size, shape, color, etc. of the product according to the pixel distribution, brightness, color and other information of the image data, and then makes the action of production improvement according to the discrimination result. The composition of the machine vision system is shown in Fig. 1. Now, in industry, machine vision system is mainly used for assembly positioning, product identification, dimension measurement and quality inspection.

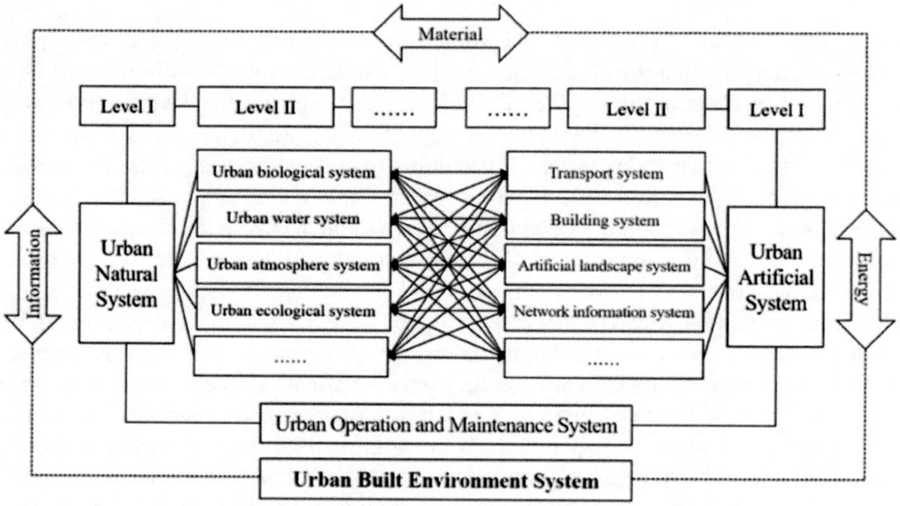

Fig. 1. Composition of machine vision system

3 Application of Machine Vision System

When the machine product detection, the detection device has no direct contact with the tested object. Therefore, in many working environments where human eyes cannot perform, the machine vision system can be used instead. The machine vision measurement system can detect products quickly, with high accuracy and stability, and can eliminate the differences between individuals and reduce the recognition loss. Applied in industry, agriculture, textiles and other fields.

The industrial field. In the industrial inspection field, it has realized the functions of component identification and online rapid detection, which are used to detect the size information of products, tool wear status, product surface defect degree, etc., greatly improving the quality and reliability of products. In addition, the machine vision system can solve the problems of low efficiency and hard to guarantee quality caused by manual sorting operations, and realize efficient and high-quality automatic sorting operations.

In the agricultural field, the machine vision system can not only judge whether the crop fruit is qualified according to the color and shape of the mature fruit, and grade it. In addition, it can also be used to monitor the leaf area, leaf circumference, stem and diameter, petiole angle, damage and other parameters of the plant, understand the growth of the plant, and judge the maturity of the fruit and the lack of fertilizer and water of the crop according to the color and size of the fruit surface.

In the medical field, the machine vision system can be used for the detection of medical devices. It can detect and verify the safety and precision of medical device materials, components and medical device products. In addition to the detection of medical devices and other hardware facilities, the machine vision system can also identify cancer cells, proliferative cells and normal cells according to the obtained cell images, which is of great significance in the medical field.

Machine vision system can be used to detect fabric defects in the textile industry. It can classify fabric defects and evaluate fabric grades according to the extracted defect features. It can quickly and accurately obtain fabric defect information, and plays a very positive role in monitoring the quality of fabric products.

Machine vision system can replace human beings to carry out various high-precision operations under various dangerous and harsh environments, liberating people from heavy and boring physical labor. In addition, it also has the advantages of improving labor efficiency, ensuring product quality, and fast response.

4 Weighted Least Squares Algorithm for Machine Vision

4.1 Parameter Equation

The model of machine vision is shown in Fig. 2. After the camera is calibrated, its internal and external parameters can be known. It can be seen from the figure that O0P = O0O1 + O1P. The left superscript represents the coordinate system. In the camera coordinate system, O0O1 = (0,0, f), O1P = (x, y, 0), point on the image plane. If u and v represent pixel values and superscript represents coordinate system.

$$O_oP = \begin{bmatrix} x \\ y \\ f \end{bmatrix} = \begin{bmatrix} dx & o & -u_o dx \\ 0 & dy & -v_o dx \\ 0 & 0 & f \end{bmatrix} \tag{1}$$

4.2 Weighted Least Squares

Before analyzing the problem of weight estimation, it is assumed that the cameras in the machine vision system are the same, but the stations are different, or the same camera shoots at different stations, and the image point errors on the image plane conform to the normal distribution.

First, describe the results used many times in this paper without proof:

$$\text{Of which}: \ d(x_{il}, x_{jl}) = \begin{cases} 0, & x_{il} \neq x_{jl} \\ 1, & x_{il} \neq x_{jl} \end{cases} \tag{2}$$

First, consider a single camera vision system, such as the camera on the left in Fig. 2. Under the image plane coordinate system, the projection of space points.

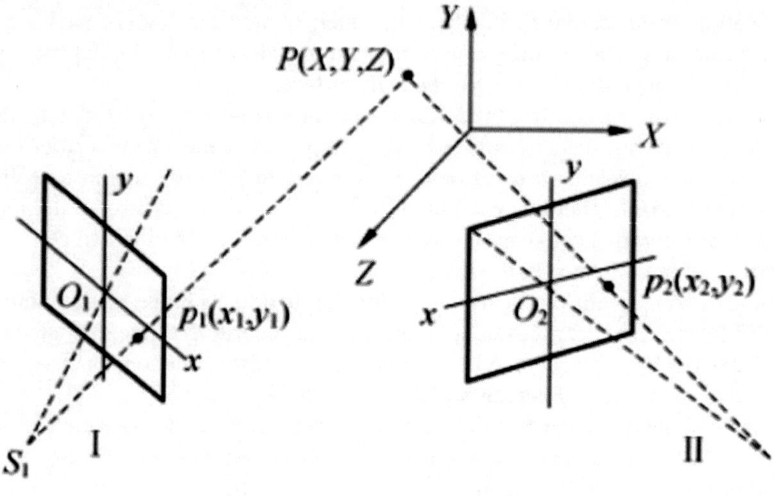

Fig. 2. Algorithm schematic diagram

The measuring system obtained in this paper has low capacity. In this paper, after obtaining image data, the chip area is segmented, and number of pixels C in the chip edge area are taken as the quality characteristics of the product. Analyze the influence of various influencing factors on product quality characteristics, and judge whether the machine vision system meets the use requirements. In Shi's paper, after acquiring the chip image data, the chip area is fitted with an external rectangle. Because the chip edge is irregular, the chip edge error is reduced after rectangle fitting. Therefore, the error between chip image data is greatly reduced. This on the original image data of the chip. After noise reduction and grayscale processing, it analyzes the image data of the chip with irregular edges, solves the chip error reduced in Shi's paper, and evaluates the machine vision measurement system more realistically and effectively.

5 Conclusion

The machine vision system is re examined from the perspective of parameter equation, and the relationship between camera position and reconstruction accuracy is analyzed. Based on this, the weighted least squares algorithm is applied to the 3D reconstruction algorithm. The key of the algorithm is to estimate the weight, and the formula for calculating the weight is given. The simulation and experimental results show that the weighted least squares method is better than the ordinary least squares method, because it highlights the role of cameras with better layout. It has better accuracy when the number of photos taken is less than 30. In the measurement of distances of 37.0310, 24.9704, 26.0153 cm, compared with the ordinary least squares algorithm, the precision of the weighted least squares algorithm is improved by 0.4 cm. When the number of photos is more than 40, The accuracy of the two algorithms is close.

Acknowledgements. Anhui University Scientific Research Project:: Research on intelligent detection and marking system of flexible PCB surface defects based on machine vision (KJ2020A0703).

References

1. Xiaofeng, D.: Application of deep learning and artificial intelligence algorithm in multimedia music teaching (Retraction of Vol 38, Pg 7241, 2020). J. Intell. Fuzzy Syst. Appl. Eng. Technol. **5**, 41 (2021)
2. Shupei, M.O., Guoliang, L.I.: Design of real-time positioning system for underground personnel by using piecewise weighted least squares. Mining Saf. Environ. Protect. (2019)
3. Qing, M.Y., Chuang, L.S., Bo, W., et al.: Weighted total least squares for the visual localization of a planetary rover. Photogram. Eng. Remote. Sens. **84**(10), 605–618 (2018)
4. Abedi, M.: AIRRLS: an augmented iteratively re-weighted and refined least squares algorithm for inverse modeling of Magnetometry data. J. Geol. Res. **1**(3), 12 (2019)
5. Liang, K., Clay, M.J.: Iterative re-weighted least squares algorithm for l-minimization with tight frame and 0. Linear Algebra Appl. **581**, 413–434 (2019)

Application of AI Technology in Internet Finance and Analysis of Security Risks

Ou Wang[1], Huan Ye[2], and Runfa Li[3](✉)

[1] Guangzhou Huali Science and Technology Vocational College, Zengcheng, Guangzhou 511325, China
[2] Guangzhou International Economics College, Guangzhou 510540, China
[3] Guangzhou College of Technology and Business, Guangzhou 510850, China
lirunfa@gzgs.edu.cn

Abstract. Internet finance is a new financial market formed by the interaction of traditional financial markets and emerging financial markets. In this article, we will focus on how to use AI technology to analyze security risks. Artificial intelligence (AI) is an intelligent technology that can make decisions based on certain knowledge or rules without human intervention. AI can help us reduce human error. The application of artificial intelligence technology in internet finance is becoming increasingly widespread, including risk control, customer service, intelligent investment advisory, and other fields. Among them, artificial intelligence algorithms can help financial institutions determine whether there are risks based on users' personal information and behavior patterns, improve risk management level, and improve service quality. At the same time, artificial intelligence technology can also provide personalized investment advice to customers, helping them achieve better returns. However, the application of artificial intelligence technology in internet finance also brings some security risks. For example, hackers can exploit algorithmic vulnerabilities, attack artificial intelligence systems, and steal users' personal information and funds. In addition, artificial intelligence systems are susceptible to human manipulation and erroneous indications, resulting in misjudgment and bias. This requires financial institutions to adopt strict security measures, improve their technological level, ensure the safety and privacy of customers, and ensure the sustainable and healthy development of artificial intelligence technology.

Keywords: finance · artificial intelligence · internet · security risk

1 Introduction

Various financial business models such as P2P financing, cash lending, crowdfunding, digital currency, online payment, and online banking have been derived. In the process of operation and operation of Internet financial products based on cash loan mode, various artificial intelligence resources for intelligent data processing and analysis, such as: the risk coefficient that customers can bear, the customer's good credit, the risk of fraudulent loans and bad debts [1].

J. C. Hung et al. (Eds.): FC 2023, LNEE 1132, pp. 354–359, 2024.
https://doi.org/10.1007/978-981-99-9538-7_55

More and more Internet financial enterprises have been applied to risk control management based on AI technology. The use of AI helps to continuously improve the risk control system and establish a lasting risk control system. Compared with traditional risk control technology, AI risk control adds more dimensions and relevance analysis. Traditional risk control judges whether a person has loan qualification, reviews deposits, income, collateral, family conditions, etc., and sets medical access threshold [2]. If it fails to meet the requirements, it cannot apply for loans. AI risk control can find data from other aspects, not only financial data, but also social data, payment data and life service data of borrowers.

These personal data information are highly sensitive and of great value. The security construction of artificial intelligence platform has become the top priority of information security construction and risk management of Internet financial enterprises.

2 Related Work

2.1 Application and Research Status of AI Technology in Internet Finance

At present, the domestic research on AI and Internet finance mainly includes two aspects: the research on Internet financial security risk and regulatory mechanism: Zhang Chenghui (2016) proposed that the Internet financial activities should be targeted according to the exposed problems, the differentiated regulatory and risk control system should be adopted for the Internet financial platform, the cross-sectoral coordination and supervision should be strengthened, and the behavior supervision of formal financial institutions should be strengthened [3], We will intensify efforts to crack down on illegal fund-raising and strengthen risk education for investors. Hou Jianqiang et al. (2016) believed that payment innovation information behavior of traditional finance [4]. The diversification of information sources processing technology have brought new technologies and resources to Internet financial risk management.

Hu Chen (2017), starting from the characteristics of P2B industry, demonstrated the structural path of P2B Internet financial risk by means of questionnaire survey and analysis, and thus proposed seven principles of controllable P2B Internet financial risk. Analyzed the possible impact of AI application, and put forward countermeasures on this basis. Cheng Dongliang (2016) introduced the application status of AI in the financial field, analyzed the risks faced in its development, and based on this, put forward suggestions such as strengthening access control and identity authentication, introducing audit measures and monitoring measures.

At present, there are few documents and materials to discuss the issues of Internet finance in the context of AI. The existing research on the security risks and monitoring mechanism of Internet finance is still based on the past technical standards.

2.2 Concept and Characteristics of Internet Finance

Internet finance, that is, the "Internet plus finance" model. In 2012, Professor Xie Ping first proposed the concept of "Internet finance". His main report, "Research on Internet Finance Models", details the definition of Internet finance and its three core parts: payment method, information processing and resource allocation.

The data security capability maturity model was jointly drafted by Alibaba (Beijing) Software Service Co., Ltd., China Information Security Evaluation Center, Tsinghua University, the Institute of Software of the Chinese Academy of Sciences, the National Information Security Engineering and Technology Research Center, Huawei Technologies Co., Ltd., and 3600 Technology Co., Ltd, It is issued by the National Technical Committee for Information Security Standardization (SAC/TC 260).

3 Information Security Risk Analysis and Protection Strategy Research of Internet Financial Enterprises

3.1 Make Full Use of Data Security Capability Maturity Model

The data security Capability Maturity Model is a way to evaluate the organization's data security capabilities, which can help organizations understand their own security capabilities, find weaknesses in security management and improve measures, so as to improve the level of data security assurance. In terms of making full use of the data security Capability Maturity Model, organizations should start from the following points:

Determine the standard model: There are a variety of data security Capability Maturity Model to choose from, and organizations need to choose their own standard models according to the actual situation. For example, organizations can choose a maturity model based on NIST CSF or customize it according to other standards.

Conduct security assessment: Organizations need to evaluate their data security capabilities based on the selected standard model, understand the current status of security capabilities, and identify security vulnerabilities.

Develop an improvement plan: After the evaluation is completed, the organization needs to develop a data security improvement plan, specifying improvement goals and measures.

Implementation of improvement measures: Organizations should gradually improve their safety capabilities in accordance with the established improvement plan, such as strengthening risk assessment and safety awareness training.

Continuous improvement: Organizations need to continuously improve the maturity of data security capabilities through regular security assessments, and further enhance the level of security assurance.

By making full use of the data security Capability Maturity Model, organizations can fully understand their own data security situation, find shortcomings, develop corresponding improvement measures, and constantly improve the level of data security assurance.

Figure 1 above is the data security capability maturity capability building framework. It is designed according to the capability maturity level 3 (fully defined level). Each module is designed to meet the following requirements:

Compliance and business requirements: For any organization, before carrying out the process of data security capacity building, it must be based on the premise of meeting national laws and regulations. It can ensure business development and promote business production through data security construction.

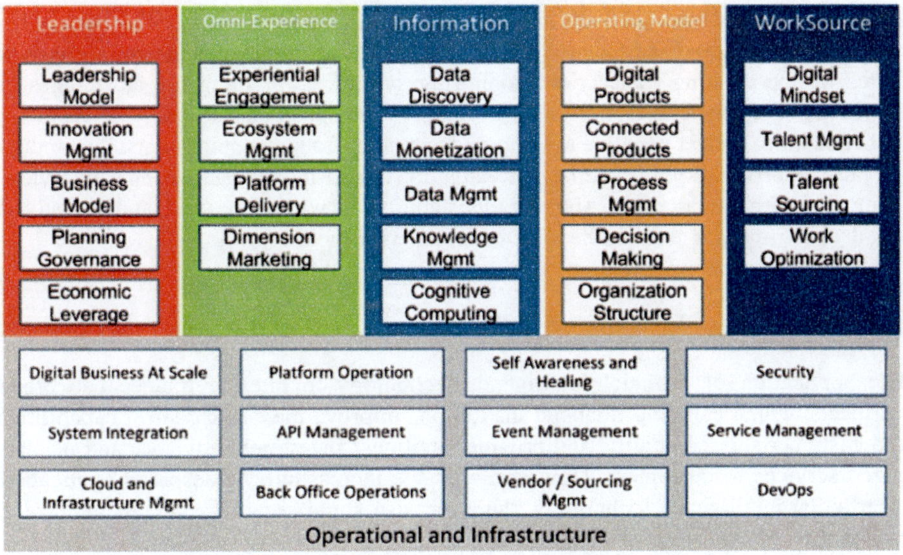

Fig. 1. Data security mature capacity building framework

Organization construction: data security construction can not be completed by one or two people. It needs a special data security organization. The data security organization should be established and its responsibilities should be defined.

System process: establish data security management system and promote its implementation, including data security system charter and management specifications.

Technical tools: technologies and tools supporting the data security system, promoting effective implementation, and security technology platform tools at all stages of the data life cycle.

Personnel capabilities: the capabilities required to ensure data security organization construction, system process, technical tool construction and implementation. The core capabilities include data security management, data security operation, data security technology and other security capabilities.

3.2 Use Machine Learning and Neural Network to Assist Decision-Making and Early Warning

The field of Internet finance is basically a pure data field. AI can serve as an online intelligent financial adviser, provide appropriate financial investment plans for users according to the calculation and analysis results and their personal investment experience, and calculate risks for users for reference. The application of AI in Internet financial decision-making mainly refers to the provision of algorithmic online investment advisory and asset management services for customers, often referred to as intelligent investment advisory.

For example, making full use of cognitive computing, intelligent robot process automation, identity analysis, network analysis, machine learning and other advanced analysis functions can speed up due diligence and help Internet financial enterprises effectively understand and manage a large number of anti-money laundering alerts generated by the existing transaction monitoring system. Combining the advantages of AI technology, financial institutions can improve the speed and accuracy of customer authentication, collect negative news to understand customer requirements, thus reducing false reports and speeding up the investigation of anti-money laundering alert review. Compared with human work, AI has higher stability and will not be tired, and its analysis and decision-making will not be affected by external factors.

4 Application Structure Research

The application of artificial intelligence technology in internet finance has broad prospects, which can help financial institutions improve their risk control capabilities and customer service quality, and provide intelligent investment advisory and personalized services to customers. However, with the increasingly widespread application of artificial intelligence technology, there are also some security risks, including the exploitation of algorithm vulnerabilities, hacker intrusion into systems to steal customer personal information and funds, artificial intelligence systems being manipulated, and so on.

In order to address these security risks, financial institutions need to take a series of measures. Firstly, strengthen data encryption and identity recognition technologies to enhance data security capabilities; At the same time, organizations should pay attention to the following issues when using the data security Capability Maturity Model:

Model selection: Different standard models have their own advantages and disadvantages for organizations, so when choosing, organizations should consider their own situation comprehensively and customize appropriate models to truly leverage the advantages of the models.

Evaluation object: The organization should clearly define the scope of evaluation objects, such as customer data, financial data, etc. The evaluation scope varies, and the evaluation results also vary. Therefore, all data that needs to be protected should be considered.

Data dynamics: the data security Capability Maturity Model can only reflect the current status of data security at a certain time, and the organization should regularly evaluate it to maintain dynamic monitoring of data security status.

Employee training: Organizations should strengthen employee training, enhance security awareness and skills, enable employees to understand the importance of data security, and enhance their sense of responsibility for data security.

To sum up, the data security Capability Maturity Model is a good tool for evaluating the organization's data security capabilities, which can help organizations fully understand their own data security situation, find and improve weak links, and constantly improve the level of data security assurance. However, attention should be paid to the above issues to ensure that effective implementation methods are followed.

5 Conclusion

When making decisions and implementing artificial intelligence systems, people do not want their actions to violate the basic ethics and moral principles of human society. Therefore, it is necessary to consider this issue at all times during the design and development stages of artificial intelligence systems. Therefore, it is necessary to establish ethical norms and technological research and development norms for artificial intelligence. Researchers of AI systems need to conduct ethical and moral hazard assessments of products to ensure that AI systems will not engage in anti human and anti-social behaviour.

References

1. Khanagar, S.B., Vishwanathaiah, S., Naik, S., et al.: Application and performance of artificial intelligence technology in forensic odontology - a systematic review. Legal Med. (48), 48 (2021)
2. Filipiak, D., Stróyna, M., Wcel, K., et al.: Application of AI and in-memory computing for extracting vessel movement patterns from historical data. In: 14th NATO Operations Research and Analysis (OR&A) Conference: Emerging and Disruptive Technology: Meeting Proceedings, North Atlantic Treaty Organization (2021)
3. Liu, A., Zhang, G., Zou, Q., et al.: New Application of Horizon Flattening Technology with Seismic Data for Exploration and Development in Fula Sub-basin, Muglad Basin, Sudan (2021)
4. Buffon, E., Mendona, F.: Application of RPAs to disaster risk reduction in Brazil: application in the analysis of urban floods. J. Unmanned Veh. Syst. (2021)

Check for updates

Traditional Culture Utilization and Protection System of Health Care Tourism Destination Based on Android Platform Design and Implementation

Xiuxia Wang(✉) and Mengmeng Sun

Shandong Institute of Commerce and Technology, Jinan 250103, Shandong, China
sdszyl2022@126.com

Abstract. In recent years, with the significant improvement of China's economic development level, the tourism industry has ushered in a period of rapid development, and health care tourism has gradually become one of the hot spots of the people's attention. In the process of carrying out health tourism, the utilization and protection level of the traditional culture of health tourism destinations can be further improved. Based on this, the article is based on the Android platform, and the paper deeply discusses the design and realization of the traditional culture utilization and protection system of health tourism destination, hoping to help the protection and development of the excellent traditional culture of health tourism destination.

Keywords: Android platform · health care tourism · traditional culture · utilization · protection

1 Introduction

Since the reform and opening up, the development speed of China's tourism and cultural industry is getting faster and faster, health care tourism is health care tourism, more focused on the comprehensive integration of the natural ecological environment and cultural environment in the tourism, to achieve physical fitness and self-cultivation. With the full implementation of the rural revitalization strategy, many rural positive development kang tourism projects, in order to better realize the health tourism destination the use of traditional culture and protection, the article based on Android platform into traditional culture utilization and protection system design planning and function implementation, in order to improve the level of health tourism projects in our country.

2 The Implementation Significance of the Utilization and Protection of the Traditional Culture in the Health Care Tourism Destination

Since the 21st century, China's social and economy is developing faster and faster. In the national economic system, tourism industry accounts for a heavier proportion. Relying on China's unique rural industry and rural revitalization strategy, rural tourism

J. C. Hung et al. (Eds.): FC 2023, LNEE 1132, pp. 360–366, 2024.
https://doi.org/10.1007/978-981-99-9538-7_56

has gradually become one of the main directions of tourism development in the future. Health care tourism focuses more on the good natural ecology of tourism destinations. In the process of tourism, people should further realize the goals of physical fitness and self-cultivation in the process of tourism, and at the same time, multiple concept systems such as health care tourism, health tourism, elderly recuperation tourism and so on are derived. But investigate its fundamental, health tourism often can bring people good state, physical and psychological aspects, realize the harmonious unity of man and nature in the natural ecological environment, in the process of increasing the number of passengers, will inevitably cause negative effects on the traditional culture of the tourist destination, so we need to further strengthen the use and protection of traditional culture. For example, the specific development mode of health care tourism destination. For example, health care tourism projects focusing on ecological health care need to fully protect the current quality natural resources and effectively recuperate and recuperate the bodies of tourists; cultural health care requires special types of cultural resources, such as Confucianism, Buddhism culture, TCM culture, longevity culture, food culture, folk culture, and relying on the existing traditional cultural elements. It can be seen that further strengthening the utilization and protection of the traditional culture of health tourism destinations is conducive to improving the existing tourism service system, highlighting the core attraction and brand competitiveness of health tourism projects, and helping the positive development of the health tourism industry [1].

3 Design of the Traditional Culture Utilization and Protection System of Health Care Tourism Destination Based on Android Platform

3.1 System Requirements Analysis

Based on the Android platform, the traditional culture of the health care tourism destination is specially utilized and protected. In terms of the system design, we first need to fully understand the needs of the system design. Related users and participate in health tourism tourists can through the existing system anytime and anywhere to understand the destination specific information, especially in terms of traditional culture, but also can apply the system to strengthen the special utilization and protection of traditional culture, to further improve the protection awareness of health tourist crowd, improve the development process of digital and health tourism destination protection. Thus, in the system design stage, it is not only necessary to reflect the value of the protection and utilization of traditional culture, but also to improve the convenience and rationality of the services provided by the system. In the system planning and design stage, need to fully consider the system interaction interface clear, smooth and reasonable, to ensure that the users to provide the corresponding service content, users can according to their own needs, in the system to retrieve the traditional culture elements, location, development history, and many other information, clear the value of traditional culture protection and related measures, to reflect the convenience.

In the overall planning, Android client interface to reflect beautiful and generous, logical structure to clear, further stimulate the use of health tourists and other users, in the traditional culture use and protection of content design planning to ensure the rationality of the information presentation, can set classification navigation, to ensure that the user needs guide clear reasonable, help users to quickly retrieve or understand the relevant content. The existing system client needs to provide users with navigation services and resource management functions, regularly update the content and resources related to the traditional culture of health care tourism destinations, and make personalized push combined with the number Settings and knowledge collection of different users. If there are derivative products or cultural products in the relevant traditional culture, orders can also be ordered and processed in the system.

3.2 System Architecture Design

Design the traditional culture utilization and protection system of health care tourism project destination based on Android platform, which mainly adopts C/S architecture. The server applies the JavaEE hierarchical structure to set up different functional modules at different levels of the control layer, the business layer, and the persistence layer, while operating around the overall system core. As shown in Fig. 1, the Web client needs to operate based on the existing traditional culture database of health care tourism destinations, and carry out data communication in an orderly manner according to the HTTP protocol to realize the effective interaction of the Android application client. In the overall system architecture, it can be divided into two major elements: service port and client port. Users can directly interact with the server through Android systems such as smart phones or computers, and directly obtain resources and information in the system according to different needs. In the existing system function structure need to enrich the traditional culture utilization and protection resources database, which covers text, image, sound, video and other types of data information, form the resource management module and business interface module, user interface module, and other modules, ensure the server, the client needs to show the existing modules and the functional interaction [2].

Fig. 1. System Architecture

3.3 System Module Planning

3.3.1 Knowledge Resource Management Module

For health tourism destination traditional culture utilization and protection system is crucial module of knowledge resource management module, in order to make more tourists through the existing system and related knowledge resources, to clear tourist destination cultural embodiment and cultural types, protection measures, etc., targeted selection routes and tour activities, so in the knowledge resource management module need to existing health tourism destination, all kinds of traditional cultural knowledge resources, to carry out the basic management. For example, Fig. 2, traditional handicraft culture, such as ancient oil pressing, weaving and dyeing techniques, ceramic techniques, water wheel, etc., traditional folk culture, such as folk festivals, markets, traditional clan culture, such as families, ancestral temples, temples, and folk performing arts, such as minority songs and dances, folk art, and cultural performances. Such diversified traditional culture needs to collect knowledge resources in pictures, text, audio and video and other forms, and submit them to the existing knowledge resource management module. Managers can modify and delete the existing modules, and users can also realize the interaction with the database in the knowledge resource search.

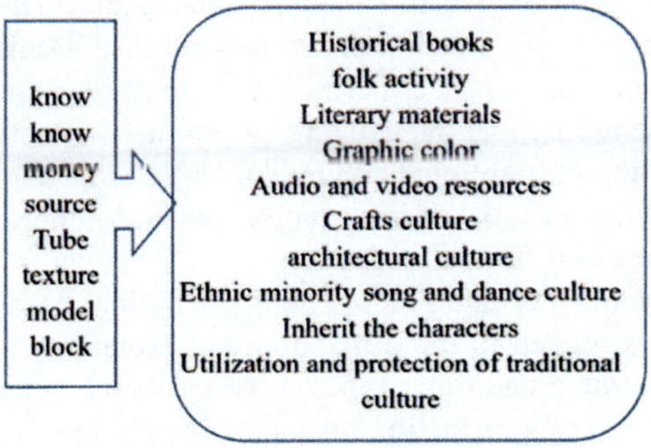

Fig. 2. Organization form of the knowledge Resource management module

3.3.2 Communication Interface Module

In this system, the communication between the client and the server side needs to be carried out in the form of HTTP, so the communication interface modules need to be designed for data and information interaction. It mainly covers user registration, login and exit interfaces, and other module information interaction interfaces, etc. Different module interfaces and classification interfaces need to ensure the smooth flow of information, and clarify the interface elements in different items such as request mode, request parameters, parameter format, and return value.

3.3.3 Traditional Cultural Commodity Exhibition and Sales Module

Many health tourism projects from the existing traditional culture, make full use of traditional culture for commodity research and development, such as the three gorges region in health tourism project by the three gorges people eat fish fishing traditional way of life and cultural customs, the fish bone production for handicrafts sales, such traditional culture tourism kang goods pay more attention to the design. Therefore, in the existing utilization and protection system of traditional culture, the cultural and creative products of traditional culture can be designed and planned as commodities. The system manager selects the cultural and creative products to release the commodity information in time, and the users need to track the order consumption in the existing Android platform. Therefore, the access and operation rights of merchants, managers and users need to be added in this module. In addition, combined with the current commodity logistics needs, we can also link Taobao, JD and other commodity sales platforms in the existing modules to improve the comprehensive utilization value of the system [3].

3.4 Database and Client Adaptation

Database system developers comprehensive collection kang tourism destination traditional cultural elements, such as traditional architecture, rural scenery, traditional customs, religious beliefs, traditional festivals, etc., will be related to the historical allusions, development, cultural elements, clear different class purpose ID and classification name, serial number and so on. Managers need permission to modify, create, and update in existing classes. It is worth noting that the existing system needs to effectively enhance the utilization and protection of traditional culture content. SQlite database can be used for data collection and collation, and data information can be configured combined with the functional requirements of different modules.

Secondly, in terms of information and client adaptation, because there are various contents and resources related to the utilization and protection of traditional culture, such as text, pictures, video and other types, it is necessary to uniformly process and deconstruct such diversified information resources. Apply the file transcoder to effectively convert the existing resources to form the standard audio and video or text files, and improve the adaptation of the information content in the Android platform.

4 Function Realization Analysis of the Traditional Culture Utilization and Protection System of Health Care Tourism Destination Based on Android Platform

4.1 User Registration and Login Function

When users application smartphone or tablet login software, need according to their own information needs, timely view the information related to health tourism destination content, the traditional culture use and protection content is the top priority, the user can register in the system, then commodity purchase, information query, or participate in the traditional culture protection related activities, which can apply SSL single sign-on scheme, users can login in the existing system or open permission can access to related application system, such as taobao, etc. [4].

4.2 Information Search Function

The core of the utilization and protection of traditional culture lies in the tourists to truly understand the essence and embodiment of various traditional cultures in health tourism destinations, further stimulate the desire for protection, fully participate in traditional culture protection activities, and stimulate creativity with existing traditional culture to buy cultural and creative commodities. Therefore, users can browse the points and related information in the system according to their own needs, and they can also apply big data and engine algorithms to realize the personalized recommendation of traditional culture utilization and information protection. From the perspective of users and tourists, many people do not have a clear main intention of applying the information search function of the system, which can be unified recommended, recommend the use and protection of the traditional culture of typical audio, video, pictures and text information on the home page, and effectively carry out information output.

4.3 Commodity Order Management Function

Administrators can upload cultural and creative products related to traditional culture to the system orders and commodity sales modules, or directly link to cultural and creative product sales platforms such as Taobao and JD. In the existing order management stage, it is necessary to meet the series of functions of order generation, modification, query, submission, deletion and deletion. At the same time, the maintenance data of related products can be effectively processed according to the updated content in the database to realize the connection between products and information. Businesses can provide goods for tourists, release information in time, and carry out order tracking, but we should pay attention to the operation authority of businesses.

5 Conclusion

To sum up, in the future development process of China's tourism industry, health tourism is one of the main trends, which is closely related to the development trend of population aging and the sub-health state of people in China. It is of great significance and value to further strengthen the traditional cultural utilization and protection of health tourism destinations. Article based on Android platform for kang tourism destination traditional culture utilization and protection system design and function implementation comprehensive analysis, respectively demonstrated the system requirements, system architecture, system module planning, database and client adaptation, etc., hope to be able to strengthen the health tourism destination of traditional culture utilization and protection, promote the comprehensive development of in the tourism industry in our country.

References

1. Wang, W.: Design and implementation of Tujia Brocade Culture mobile knowledge service system Based on Android Platform. Central China Normal University, Hubei (2017)

2. Lu, S.: Research on the traditional longevity culture of Hechi under the Forest Health Care area. Sino-foreign Exchange **23**, 64–65 (2020)
3. Peng, J., Wang, L.: Research on the development mode of health care tourism at home and abroad. J. Taiyuan City Vocat. Tech. Coll. **6**, 55–57 (2022)
4. Su, P., Pang, S., Zou, F., et al.: The design of the rural tourism system based on Android. Coop. Econ. Technol. **5**, 94–95 (2018)

Visual Communication Design Based on Temporal and Airspace Filtering

Yan Wang(✉)

Hunan First Normal College, Changsha 410006, Hunan, China
wyan_0919@163.com

Abstract. Weak target detection in the process of target scale, background noise, causing weak target image visual transmission effect is unsatisfactory, to improve the weak target image visual transmission effect, now propose a weak target image visual transmission design scheme: using guide filter technology to enhance image characteristics, using partial differential equation to obtain image background baseline, to achieve better weak target image background suppression effect;

Histogram equalization algorithm is used to improve image quality and obtain good image enhancement effect, and image feature extraction technology and particle swarm algorithm are used to complete the tracking and visual transmission of weak target images. Experimental analysis proves that the proposed design scheme in this study can effectively reduce the influence of interference factors of weak target detection, produce a good visual transmission effect of weak target images, and meet the actual needs of weak target detection and tracking.

Keywords: temporal and airspace filtering · visual communication · weak target · background suppression

1 Introduction

Based on the continuous improvement of science and technology level, the performance of the detection system supported by science and technology has been greatly improved. It has the advantages of strong concealment, high flexibility, good anti-interference ability, wide detection surface, strong practicability, and clear image display, and has been widely used in many fields. In the practical application process of the detection system, the remote detection phenomenon is often encountered. Because the farther apart the detection target is from the detection system, the smaller the size in the image, so the imaging scale of the detection target is small, and the target signal is weak, and such images are called the weak target image. In addition, in the process of long distance detection, the transmission distance, climate conditions, atmospheric environment, detection target background, detection system imaging equipment and other factors, lead to weak target image lack of obvious pixel gray characteristics, geometric characteristics, and there is serious background noise, cannot be weak target information comprehensive, clear, accurate communication, affect the detection system application role. Therefore, it is imperative to improve the visual transmission effect of weak target images. In this

J. C. Hung et al. (Eds.): FC 2023, LNEE 1132, pp. 367–372, 2024.
https://doi.org/10.1007/978-981-99-9538-7_57

study, we design a visual communication method of weak target image based on time and airspace filtering, hoping to reduce the influence of background noise, enhance the quality of weak target image, and improve the visual transmission effect of weak target image.

2 Design Analysis

2.1 Background Inhibition

Spatial domain algorithm is a widely used algorithm for weak target image background modeling. Compared with other algorithms, its data features are more intuitive, rich and acquired, which can meet the needs of most detection systems for weak target image background modeling. In this regard, this study design uses the spatial domain background modeling algorithm to perform background suppression and reduce the influence of background and noise on the visual transmission of weak target images. At present, there are three common algorithms for spatial domain background modeling: filtering, visual physiological mechanism, and low-rank matrix approximation. The filtering algorithm enhances the image effective features and increases the difference between target image and non-target image [1]. The algorithm has good image enhancement, but it will smooth and damage the target information to some extent. Visual physiological mechanism algorithm specifically strengthens the weak target to effectively distinguish the weak target from the background to achieve the background inhibition effect. The algorithm is greatly affected by the specific characteristics of the weak target, but when the contrast of the weak target cannot meet the actual needs, it will reduce the background suppression effect, thus reducing the visual transmission effect of the weak target image. The low-rank matrix approximation algorithm distinguishes the weak target from the background through the data component analysis of the weak target image. The scene applicability is strong, and the background suppression effect is good, but the algorithm is highly complex, and the visual communication timeliness is poor. Considering the requirements of the detection system for the visual transmission of weak target images and the advantages and disadvantages of different algorithms, this design determines to choose the filtering algorithm for background suppression. In order to avoid the smooth damage of the image edges, the relevant parameters are set reasonably, and the control of the image details is strengthened. In order to obtain a better background suppression effect, partial differential equation is used to obtain the background baseline of weak target image sequence, so as to accurately grasp the gray change of weak target image and realize the effective suppression of the background and noise of weak target image.

2.2 Image Enhancement

After the suppression of the weak target image airspace and the weak target image, the design of histogram equalization algorithm is used to reduce the proportion of background and noise in the image, increase the proportion of the gray space of the weak target in the image, enhance the effective feature extraction advantage of the weak target, and improve the quality of the target image. First, the probability of any gray value of the weak target image is calculated by the following formula: P (αj) = ejIn e (j = 1, 2, 3, 4, 5…… n), in P (αj) Represents the chance of the j th gray scale value, e represents the gray scale series of the weak target image, j represents the number of pixels, ejRepresents the j th number of gray-scale pixels [2].

Secondly, the gray scale images are balanced to obtain the gray scale level (T) after the image enhancement processing of each pixelj), The formula is as follows: Tj = 4095, T if the weak target image is 20%j = 4095 × 20% = 819. It is worth noting that the background saturation phenomenon needs to be considered during the image enhancement processing process, which can give an appropriate constraint during the threshold value setting process. It is found that at the threshold value = 100, the image enhancement processing effect is relatively ideal.

2.3 Detection of Weak Targets

The is found that after the background suppression and the enhancement of the weak target image, the position of the target position in the image and the position of the target in the display may have a certain deviation in the image, which requires target information repair processing to ensure the accurate and effective detection information of the weak target. Particle filter detection technology is a widely used object detection and tracking technology, which consists of the initial state simulation stage, prediction stage, correction stage, resampling and its processing stage. Through prediction and correction, the particle state can be adjusted for prediction, so that the particles can maximize out equally close to the real state. Therefore, the study scheme is designed to introduce particle filter detection technology to solve the target position offset problem.

2.4 Image Feature Extraction

The effective extraction of weak target image features has a very important impact on improving the visual transmission effect of weak target image. Therefore, after background suppression, image enhancement, weak target detection. At this stage, the wavelet domain transform algorithm and gray algorithm of weak target image. The effective application of the wavelet domain transform algorithm can highlight some specific information in the weak target image to some extent, effectively suppress the non-target information, and further separate the weak target from the background and noise [3]. The effective application of the gray-scale projection algorithm can realize the accurate and fast extraction of the target image features. Under the application of wavelet domain transformation algorithm and gray scale projection algorithm, the target value () can be expressed by the formula, in the formula, it is weighted factor, representing image gray scale, image detail features and image motion features respectively.

$$\zeta_{(x)}\zeta_{x} = \phi f(\rho_{x}) + \varpi f(Q_{X}) + kf(\theta_{x})\phi\ \varpi\ kf(\rho_{X})f(Q_{x})f(\theta_{x})$$

2.5 Image Visual Transmission

In order to obtain the better visual transmission effect of the weak target image, the particle swarm optimization algorithm is introduced. The algorithm has the advantages of few parameters, low complexity, fast calculation speed and wide application range, which can ensure the visual transmission effectiveness of weak target images and improve the visual transmission timeliness of weak target images. Particle swarm optimization algorithm process is usually "start particle swarm parameter initialization particle position and speed random initialization condition judgment (meet the conditional output optimal solution) does not meet the conditional particle speed and position update particle adaptation value calculation particle individual history optimal adaptation value and position update particle population history optimal adaptation value and position update parameter update condition judgment (meet the conditional output optimal solution)". This study scheme design separates the target features from the background features, and the particle swarm optimization algorithm is mainly used to weight the weak target feature components.

Assuming that the weak target image feature vector is y1, y2, y3, then the image feature vector optimal solution is W ' = [y1, y2, y3]. W ' acquisition strictly follows the subgroup optimization algorithm process. Assuming that there are n particles and each particle is in the same space as the optimal solution, each particle corresponds to an optimal solution. If the optimal solution dimension is g, each particle orientation can be expressed as a after h-step iteration l(h) = (al1, al2, al3......als), The local optimal solution for each particle is the Wl(h) = (Wl1, Wl2, Wl3......Wln,W1G) [4].

3 Experimental Analysis

3.1 Experimental Environment

In order to verify the application effect of the weak target image visual communication design scheme based on temporal and airspace filtering, the experimental analysis is carried out. Experimental analysis mainly has two parts: first, build the serious background noise sky scene (1), background noise serious forest scene (2), gradually halo strong not smooth sky scene (3), gradually halo strong not stable forest scene (4), different scene simulation simulation experiment, scene basic information shown in Table 1. Second, the design scheme and the previous design scheme (such as infrared spectrum, hierarchical analysis, virtual reality image visual communication scheme of the image visual transmission effect of weak targets) for comparative analysis. The experiment was refined in the windows10 operating system of the central processor, the Intel ® Core ™ i5–8500 (9M cache, 4.10GHz).

3.2 Experimental Results

Overall and analysis of the experiment information, obtain the following results: (1) different scenarios of weak target image enhancement effect: from Table 2, in 1,2,3,4, the original image contrast mean is 5.35,8.98,3.34,4.71, respectively in temporal airspace filtering technology application, weak target image contrast are greatly improved, mean bag 33.76,31.42,50.88,71.19, compared with the original image, the maximum difference is more than 66. It shows that this research scheme can effectively improve the image quality of weak targets and produce better image enhancement effect. This is

Table 1. Basic information table of the experimental scenario

scene	Total frame number	resolution ratio	noise-signal ratio (dB)		
			maximum	minimum	mean
No.1	468	250 × 180	0.57	7.38	3.58
No.2	480	250 × 180	0.65	8.20	1.84
No.3	524	250 × 250	-14.14	6.50	1.23
No.4	353	250 × 250	-3.04	5.14	2.08

related to the fact that the background noise is effectively suppressed, the target features are prominent, and the non-target energy and the target signal energy gap widen.(2) Visual transmission effect of weak target images under different design schemes: when the image frames are 200 frames, 300 frames, 400 frames, 500 frames, 600 frames, 800 frames and 900 frames, The success rate of visual communication in this study protocol was 97%, 98%, 96%, 98%, 96%, 99%, and 97%, respectively;

The success rate of visual transmission based on infrared spectroscopy is 89%, 85%, 88%, 90%, 86%, 87% and 89%, respectively; The success rate of visual communication based on hierarchical analysis was 64%, 65%, 67%, 66%, 66%, 67% and 69%, respectively; The success rate of visual transmission based on VR was 77%, 75%, 74%, 80%, 82%, 84%, and 85%, respectively. Compared with the four schemes, the design scheme has the highest visual communication success rate. The four visual transmission time increased by image size, but under the same image size, the shortest visual transmission time, 500500 pixels, for example, 20.25 s, 50.01 s for infrared spectrum, visual transmission time based on hierarchical analysis, 60.12 s, and 54.48 s based on virtual reality. This is not very complicated with the application algorithm of this study, which can meet the needs of rapid transmission of different images.

Table 2. Comparison of the image enhancement effects of weak target images in different scenarios

scene	Original image contrast		because ofImage contrast of time-time and airspace filtering	
	scope	mean	scope	mean
No.1	2.05–9.34	5.35	8.21–54.90	33.76
No.2	3.65–14.50	8.98	14.93–48.44	31.42
No.3	1.72–7.65	3.34	11.49–108.74	50.88
No.4	2.32–11.42	4.71	31.11–114.15	71.19

4 Conclusion

Improving the visual transmission effect of weak target image has become the objective requirement of detection system application. According to the imaging situation of the weak target, the factors affecting the visual transmission of the weak target image are the lack of available target features and the serious background noise of the target. According to these influencing factors, this study proposes a visual communication design scheme of weak target images based on temporal and airspace filtering. Through the rational use of temporal and airspace filtering technology and image enhancement technology, the background noise is effectively suppressed, the image available features increase, the algorithm scene adaptability is enhanced, and the visual transmission effect of the weak target image is improved. It can provide guidance for the visual communication inquiry and practice of weak target images to a certain extent. However, it is worth noting that the visual communication design of the weak target image needs to obtain innovation and development through continuous exploration, so as to achieve multiple reinforcement of the image visual communication speed, quality, form and experience.

Acknowledgements. 《Research on the Development and Inheritance of Hunan Embroidery Art from the Perspective of New Economy and Society》 (18YBA106).

References

1. Yongqiang, W., Xiaopo, L.: Low-resolution laser image enhancement methods based on visual communication techniques. Laser Mag. **43**(06), 121–125 (2022)
2. Xiaoyuan, L.: Visual communication method of multi-frame laser scanning. Laser Mag. **43**(05), 161–165 (2022)
3. Ganaixiong, Q.L.: Research on weak light image enhancement based on visual communication. Laser Mag. **42**(09), 114–118 (2021)
4. Mingyong, G.X., Fan, Y.: Infrared weak object detection algorithm based on time-airspace and high-order moments. Foreign Electron. Meas. Technol. **40**(12), 1–6 (2021)

Research on Hyperspectral Image Target Detection Algorithm Based on Depth Learning

Zhiping Wang[1](✉), Cundong Tang[2], Li Chen[2], and Yi Wang[2]

[1] School of Digital Media and Art Design, Nanyang Institute of Technology, Nanyang 473306, Henan, China
3021054@nyist.edu.cn

[2] College of Information Science and Technology, Northwest University, Nanyang 473004, Henan, China

Abstract. This method includes two steps: (1) The first step is to use neural network to extract depth from the collected hyperspectral images, and then use it to classify the hyperspectral images. (2) The second step is to use the classification result obtained in step 1 to detect the target in the collected hyperspectral image. we need more algorithms, such as classification algorithm and data mining algorithm, which can help us achieve our goals. In this technology, the training data is divided into two parts: one part contains the original images captured with different spectral bands; The other part contains the same original image after processing with various algorithms. The final HIC model consists of a set of convolutional neural networks, which are trained to predict.

Keywords: Deep learning · Spectral images · testing

1 Introduction

HSI---Hyperspectral Image. The nonlinear relationship between hyperspectral data objects makes it impossible for traditional methods to classify accurately [1]. As a powerful feature extractor, depth learning method is used in hyperspectral image classification task. 1. Summarize the shortcomings methods used in HSIC, and then understand the advantages of deep learning methods to solve these problems. 2. The latest depth learning framework is divided into spectral characteristics, spatial characteristics and spatial spectral characteristics. 3. How to efficiently produce annotation datasets.

Therefore, it is difficult to build a classification and recognition model with sufficient accuracy depending on a single character information for some special needs scenarios. Hyperspectral imaging technology, as a spectrum integrated nondestructive testing technology, integrates the advantages of spectrum and machine vision, and can obtain spectral and spatial information I4 at the same time, thus improving the accuracy of the detection model. In recent years, the application of nondestructive testing technology represented by hyperspectral image technology in the field of target recognition has been widely reported [2]. However, the following challenges still exist in target purity detection based on hyperspectral image technology. Some target imperfect grains that

J. C. Hung et al. (Eds.): FC 2023, LNEE 1132, pp. 373–378, 2024.
https://doi.org/10.1007/978-981-99-9538-7_58

are damaged by external force or biological invasion and do not have growth ability need to be removed before target storage or sowing. At present, the imperfect particle detection model is mainly divided into multiple categories based on the causes of imperfect particles, and the multi classification model is built together with the normal target.

In addition, hyperspectral image data has rich information, but also has a lot of information redundancy, so it is necessary to extract features from the original data according to the requirements of recognition. The proper feature extraction method can fully guarantee the performance of the recognition model. The current hyperspectral image recognition technology mostly focuses on the extraction of spectral features that are easy to obtain.

2 Spectral Depth Learning Model

HSI analysis is mainly divided into: dimension reduction operation, spectral decomposition, channel detection and classification, feature learning for classification, repair and denoising, and resolution improvement. This paper focuses on HSI classification. List a step framework for applying deep learning to HSI classification problems: learn spectral and spatial features separately; learn spectral and spatial features; learn how to improve generalization performance considering limited training samples; The mathematical form is: the HSI cube can be expressed as: here, all frequency bands belonging to the class are represented, and each frequency band contains the total number of samples, and its corresponding class is. The traditional manual feature methods mainly include two categories: global feature information: color histogram, GIST (global scale invariant transformation), texture descriptor local feature information: HOG (directional gradient histogram), SIFT (scale invariant feature transformation). These manual features are mainly used to build the visual bag model (BOVW) and the model based on HOG features [3]. B. One of the main problems deep learning challenges is the lack of labeled HSI data. Other problems: high internal variability----uncertainty change of reflection value caused by concentrated environmental factors; Data degradation caused by noise; Redundant bandwidth----calculation complexity of HSI equipment impact model; Spectral mixing----spatial resolution related to HSI. 4. Representation of HSI HSI data is represented by a cube: among them, the spectral segment information and spatial information contained in a sample are all spectral segments, and are spatial parts, as shown in Fig. 1 below.

A. The spectral segment indicates that in the above representation, It is not easy to distinguish different types of data only by spectral segment information. Because of the influence of spectral segment mixing, the same material may display the same spectral segment information, or different materials may display the same spectral segment information. B [4]. The space represents the pixel information in the bandwidth of each spectral segment, that is, the spatial information, which is expressed in the form of a matrix: the spatial correlation is very high, adjacent pixels are likely to belong to the same category, and the adjacent relationship of pixels can determine the use of pixel cores and pixel center windows. C. Spectral segment space represents the joint use of spectral segment and spatial information.

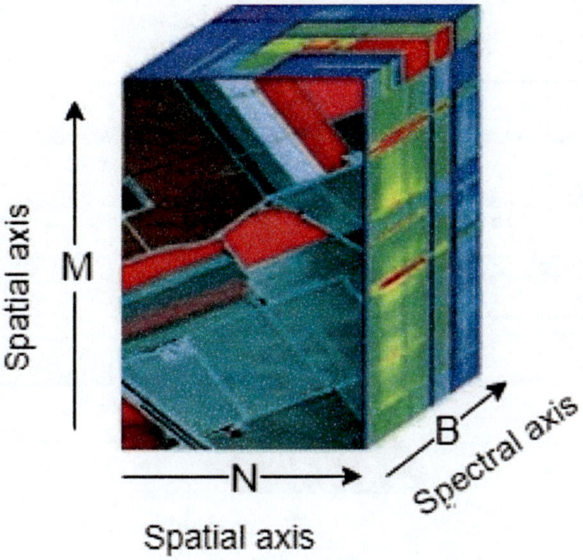

Fig. 1. HSI Data Cube

3 Research on Hyperspectral Image Target Detection Algorithm Based on Depth Learning

Target detection realizes target classification and positioning in the picture, and outputs the classification results of interested targets in the picture by classification. The target is output by positioning the position of the rectangular box. Different operation actions have different occlusion conditions on the target, and the degree of darkness in the experimental environment also affects the target detection results.. As shown in Fig. 2 below, the highlight spectrum detection algorithm framework.

This can force the network to capture different aspects of objects, and at the same time enhance the network's sensitivity to inaccurate positioning. Adaptive max pooling is ROI max pooling; The latter uses FCN for target segmentation, splicing the feature map of the last layer with the feature map generated by the former as the final feature map. In order to accurately locate, three sample frame correction methods are used, namely, Bbox expression, Iterative localization, and Bounding box voting. Bbox region: after the whole picture passes through the last layer of the network, it is connected to a Bbox region layer. Different from RPN, the Region layer here is two layers of FC and one layer of prediction layer. In order to prevent the frames obtained by the Selective Search from being too close to the objects to frame the objects well, the candidate boxes are expanded to 1.3 times the original ones. Iterative localization: The initial box is the box obtained by Selective Search, and then the box is evaluated using the existing classification model. The boxes below the given threshold are filtered out. The remaining boxes are resized using the Bbox regression method and filtered iteratively. Bounding box voting: First, apply NMS to the boxes processed by Iterative localization, IOU =

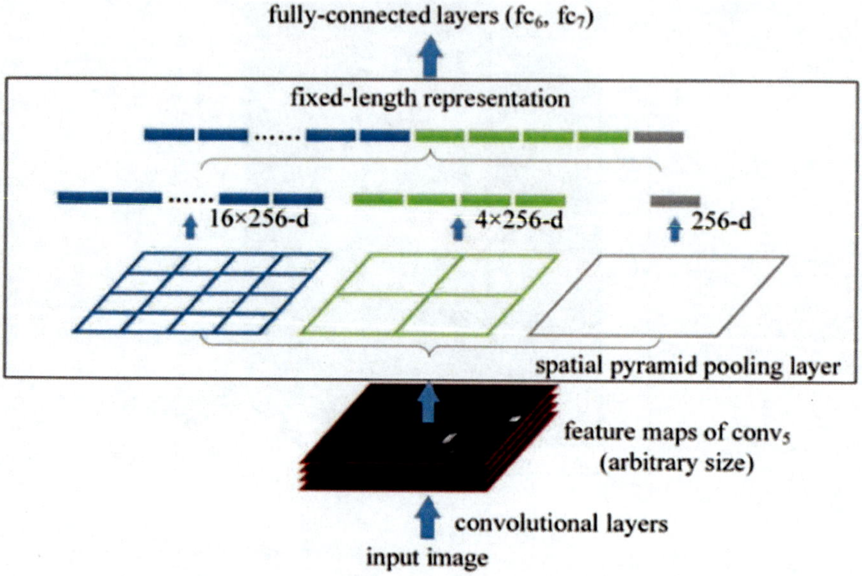

Fig. 2. Framework of high light spectrum detection algorithm

0.3, and get the detection results. Then, for each box, use the weighted coordinates of each box of the same type and IOU > 0.5 to get the final target sample box.

In Fig. 3 below, we further list the horizontal visual comparison of the performance of various detectors. It can be seen from the figure that the two-stage detector has the characteristics of high detection accuracy but slow detection speed. However, with the further, the detection speed of the single-stage detector is fast and the detection accuracy is constantly improving and approaching the traditional two-stage detector.

It is difficult to judge whether a point is abnormal by analyzing only its pixel values Therefore, in most image anomaly detection tasks, it is necessary to jointly analyze the image background and surrounding pixel information for classification, and the detected exceptions are mostly context or pattern exceptions Of course, there is no very strict boundary between these three exception types For example, some methods extract various features of an image and compare them with the features of a normal image to determine whether they are abnormal, which translates the detection of pattern anomalies in the original image space to the detection of point anomalies in the feature space [5] The image anomaly detection task can be divided into two categories: qualitative anomaly classification and quantitative anomaly location The classification of qualitative anomalies is similar to the image classification task in traditional image recognition tasks, that is, it gives an overall judgment on whether the anomaly is abnormal, without the need to accurately locate the location of the anomaly.

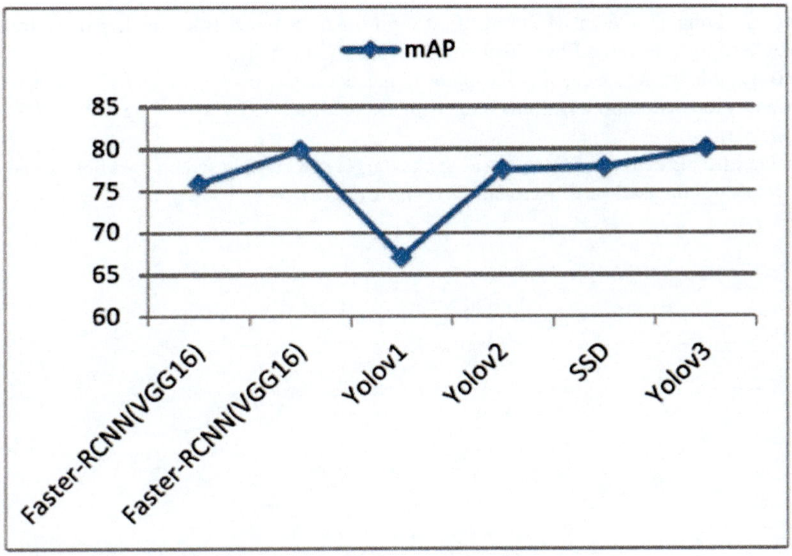

Fig. 3. Comparison of Multiple Detectors mAP

4 Conclusion

There are difficulties and challenges in applying depth learning methods to target vision detection, such as the imperfect depth learning theory and the lack of large-scale and diverse data sets. In order to solve these problems, we think we can use parallel vision to research. Online optimization of visual system through "parallel execution" object vision detection.

Acknowledgements. This work was supported in part by the National Natural Science Foundation of China under Grants 61402517, in part by National key research and development projects(subject) under Grant 2020YFC1523301, and 2018YFC1504705, and 42027806, in part by Shaanxi Key Research and Development Program Project under Grant 2019ZDLSF07-02, and 2018JM6029, and 2019ZDLGY10-01, in part by Shaanxi Province Industrial Innovation Chain Project of China under Grant 2017ZDCXL-GY-03-01-01, in part by Xi'an major scientific and technological achievements transformation and industrialization projects under Grant 20GXSF0005, in part by Henan's scientific and technological breakthrough projects 222102110045 and 222300420582.

References

1. Zhang, N., Zhao, R., Yu, B., et al.: Research on Hyperspectral Target Detection Algorithm Based on Ensemble Learning Constraint Energy Minimization. Aerospace Shanghai (2018)
2. Zhang, X., Pan, Z., Hu, B., et al.: Target detection of hyperspectral image based on spectral saliency. Image Process. IET **13**(2), 316–322 (2019)

3. Yang, S., Song, Z., Yuan, H., et al.: Fast high-order matched filter for hyperspectral image target detection. Infrared Phys. Technol. **94**, 151–155 (2018)
4. Zhao, C., Meng, M., Yan, Y.: Sub-pixel target detection on hyperspectral image based on dictionary reconstruction. Harbin Gongcheng Daxue Xuebao/J. Harbin Eng. Univ. **39**(9), 1582–1588 (2018)
5. Liu, D., Li, J., Electronic SO: Spectral-spatial target detection based on data field modeling for hyperspectral data. Chin. J. Aeronaut. **31**(4), 11 (2018)

Research on Reliability of Automatic Control of Intelligent Robot

Wang Zihan(✉)

Robot Engineering, School of Electrical Information, Southwest Petroleum University, Chengdu 610000, Sichuan, China
H1570950910@126.com

Abstract. The research on automatic control reliability of intelligent robots is a research project carried out by the Institute of Automatic Control, Chinese Academy of Sciences in recent years. The main purpose is to study the reliability and robustness of the intelligent robot control system. In this paper, we will focus on the design and analysis of reliable and robust controllers for intelligent robots. We also discussed some practical applications related to our work. In this paper, aiming at the problems of frequent maintenance and low work efficiency caused by the accuracy of automatic robots in the actual working environment, the calculation method of automatic control reliability of intelligent robots is studied. Starting from the use of tools for mechanical accuracy reliability analysis, analyze the calculation process of mechanical motion reliability, establish a virtual prototype, and discuss the variables. The experimental results show that the reliability calculation method of robot automation designed in this paper is effective.

Keywords: Automatic control · Intelligent robot · reliability

1 Introduction

Industry is the foundation of a country's development, and robot automation is the lifeblood of industrial development. Strengthening the automation of industrial robots is conducive to the rapid development of industry. Although China has increased investment in the field of industrial robot automation control, there are still many problems in the application of industrial robots in the field of automation control, which not only hinder the development of industrial robot automation control, but also cause the stagnation of industrial automation production, greatly limiting the development of industry [1]. As the saying goes, "If you want to do a good job, you must first sharpen your tools." Strengthening the automatic control of industrial robots can maximize the development of China's industry. This paper analyzes the practical application of industrial robots in the field of automation control.

Robot is one of the greatest inventions since the new era of mankind [2]. The automatic operation of the machine frees people's hands from labor, so as to conduct more valuable technical and theoretical research. The automation level of a country can be reflected in the degree of industrial mechanization of the country. Today, robots have

J. C. Hung et al. (Eds.): FC 2023, LNEE 1132, pp. 379–384, 2024.
https://doi.org/10.1007/978-981-99-9538-7_59

been used in industry, agriculture, commerce, even tourism and other third-party industries, ranging from maintaining national security to transporting goods. The efficiency of automation work is one of the factors affecting China's progress towards a manufacturing power. The accuracy of a robot is a very important function of the machine in the process of work and operation, and also an important indicator to judge whether the robot can perform this work. In some work requiring high accuracy, such as assembly, welding, measurement, etc., the accuracy requirements of robots are more stringent [3]. Therefore, it is imperative to study the accuracy and reliability of industrial robot automation control.

2 Related Work

2.1 Analyze the Development Status of Industrial Robots in China

In the practical application of the automatic control system of industrial robots, first of all, it can be taught and reproduced. The staff can use this function to input information into the robot program, and the robot will automatically save the input information in the corresponding location. When the staff need relevant information, they will automatically extract it and explain it accordingly, so as to give full play to the role of teaching and reproduction. Secondly, the industrial robot has the function of using control [4]. It can use corresponding methods to enable the robot to have different motion speeds and motion patterns. Thirdly, the industrial robot can automatically calculate the surrounding information and work out a reasonable operation route according to the relevant calculation results. This function is also the most critical part of the entire control system. Finally, industrial robots can provide automatic feedback. When the robot is running, no matter its state or position will change constantly, and industrial robots can self feedback according to their own state and position.

At present, there is still a big gap between China and foreign countries in terms of industrial robots. This is mainly because China's industrial robots started late. Although China's industrial robots have made certain achievements in technology, they have not mastered some core technologies, and there are not many industrial robots with real intellectual property rights in China. In a word, At present, the technical level of industrial robots in China is not very high, specifically because of manufacturing and assembly. Because there must be a gap between the precision of parts in China and that in the world's advanced countries. Although some achievements have been made in key technologies and assembly, there is still a big gap between these breakthroughs and those in foreign countries, which needs attention.

2.2 Functions and Characteristics of Industrial Automation Control

(1) Industrial automation control function

The robot automatic control can realize the reappearance teaching function. The robot can complete the corresponding teaching process through the command. First, the code of position, speed and action is input into the robot control system. The command is stored in the memory. The control system transmits the code to the robot operating system through the command, thus realizing the reappearance of teaching. Secondly, the function of controlling motion refers to the function of controlling the speed and motion form of the robot through the operating system.

(2) Characteristics of industrial automation control

Industrial automation robot can liberate human labor force and complete some dangerous and difficult tasks that human beings cannot complete. Robots are closely related to mechanics and structure. To describe their state, we can use coordinates, and control the robot's action by changing the corresponding coordinates. Secondly, industrial automation robots need to be endowed with intelligent tasks, so that they can act according to human consciousness, change the parameters of industrial automation robots, and their states and actions will change accordingly. The robot can be operated, managed and decided through the information base.

3 Reliability Analysis of Intelligent Robot Automation Control

3.1 Mechanical Accuracy Reliability Analysis Tool

The simulation analysis tool used in this paper is a reliability analysis tool of motion accuracy based on the automatic analysis theory of mechanical dynamics. Mechanical dynamic automatic analysis is a simulation analysis software that integrates modeling, calculation, conclusion and three-dimensional visualization. The automatic analysis of mechanical dynamics is a very powerful software. It can not only be used for the analysis of virtual prototype, but also for the static and dynamic motion analysis of virtual mechanical digital platform. It can also be used as a tool to develop virtual operation products. The basic steps are shown in Fig. 1.

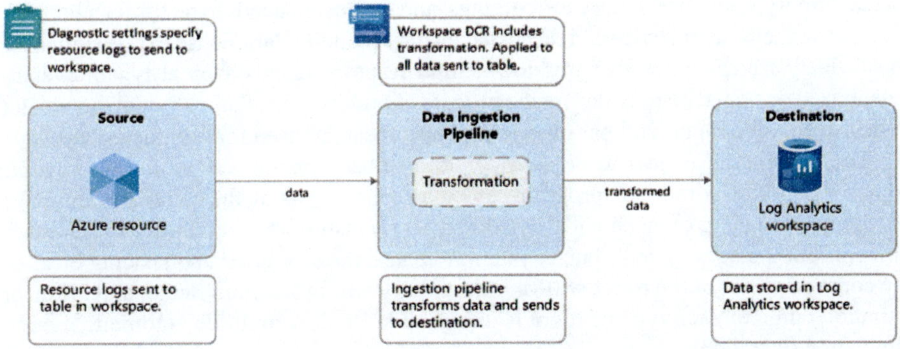

Fig. 1. Basic step diagram

Users can automatically set menu options and dialog boxes according to their usage habits, and even enter code commands to automatically repeat mechanized detection and calculation.

3.2 Mechanical Motion Reliability Analysis Method

The key objective of machine reliability analysis is to establish the relationship between the output parameters of machine performance and the random variables that affect the changes of these output parameters, and establish the relevant functional model, as shown in Fig. 2.

$$\gamma_{mi} = -\left[\frac{\partial L(y_i, F(x_i))}{\partial f(x_i)}\right] \quad (1)$$

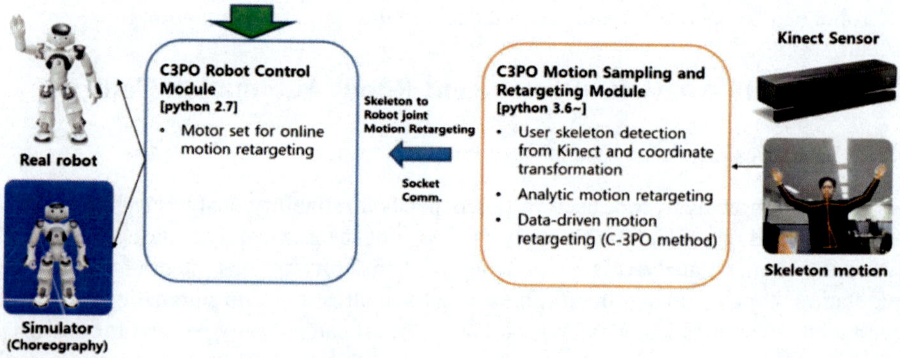

Fig. 2. Reliability Calculation of Robot Automation

After establishing the virtual model, through data analysis, it can be found that the virtual prototype still has many shortcomings, and the design needs to be further changed. Parametric method is provided in the visual display module: Parametric expression is the most widely used parametric expression method in automatic mechanical dynamics analysis. Based on visual display during the process of building simulation model, parameter values need to be input, and parametric expression can be used for parameterization.

The influencing factors to be considered have been parameterized in the previous paper. The motion simulation program can be compiled by using the command language of automatic analysis of mechanical dynamics. As for institutional reliability, the research object is not a single system, but a system. Because the size error and spacing error of the connecting rod of the robot will vary irregularly within a certain range, and the error variation range of each type of robot is different, it requires multiple automatic motion simulation operations.

4 Application Fields of Industrial Robots

At first, industrial robots were applied in the automobile manufacturing industry. Social economy and science and technology were developing rapidly. Industrial robot technology was also developing and progressing constantly, greatly improving the function of industrial robots, and the scale of application fields was constantly expanding. Industrial robots can replace workers' work in conventional environments. Industrial robots can give full play to their role in high-temperature, high-risk, toxic and high heat environments. Some work environments are relatively harsh. The effective use of industrial robots can replace manual work for heavy and repetitive operations, which can not only ensure the personal safety of workers, but also effectively improve the production efficiency of enterprises. Industrial robots have many advantages. They have been widely used in the field of automatic production. Industrial robots can give full play to their advantages in special working occasions, harsh working environments, high-risk work and other related aspects.

In the process of automatic production, the industrial robots often used include inspection, welding, painting, handling, painting and assembly robots. In the process of automatic production, the inspection robot mainly determines the size of the parts, classifies the parts and controls the quality of the parts; The welding robot is used to weld the load-bearing girder and the automobile structure; The material handling robot is used for loading and unloading, unloading, stacking and directional grasping of parts; Spraying and painting robots are mainly used for spraying and painting operations. Industrial robots can do this operation with no less than five degrees of freedom. In order to facilitate the operation of industrial robots, industrial robots can work on rails; Assembly work is the most complex part of all production links. Assembly robots, whether visual sensors and auditory sensors, proximity sensors and tactile sensors and other related sensors, must be equipped to enable industrial robots to successfully carry out assembly work.

5 Conclusion

This paper mainly uses virtual prototype technology to calculate and analyze the motion reliability of industrial robots, and uses mechanical dynamics automatic analysis software to establish a parametric simulation model of robots, which provides an effective method for users to simulate robots of different sizes, making it possible to use digital technology to calculate the motion reliability of mechanisms. At the same time, the influence of the size error and spacing error of the connecting rod is analyzed and calculated through experiments, and the calculation results are compared with the field experimental data, which finally proves the effectiveness of the method. Based on this, an effective calculation method is designed for the reliability analysis of complex robot automation operation.

References

1. Yang, F., Hu, C., Feng, Z., et al.: Research on connector assembly system and control strategy for large-scale panel. IOP Publishing Ltd (2022)
2. Zou, D.: Research on the Project Intelligent Financial Expenditure Control of R&D Department of High-Tech Enterprises under the Guidance of Performance. Francis Academic Press 2021(12)

3. Jw, A., Mt, B., Hao, W.C.: Research on the design of intelligent recycling system for cosmetics based on extenics. Proc. Comput. Sci. **199**, 937–945 (2022)
4. Park Y . Intelligent Washing Machine and Control Method of the Same, US20210324560A1[P] (2021)

Research on the Digitization of Dunhuang Architectural Paintings Under the Background of Big Data

Meng Weng(✉)

Xi'an Academy of Fine Arts, Xi'an 710065, Shaanxi, China
wemg2004@sina.com

Abstract. Dunhuang Mogao Grottoes is an art museum in the desert, and the architectural paintings on the murals are a treasure house of architectural history research. The development of big data in recent years has greatly promoted the research of architectural history. The launch of the Dunhuang Digital Museum has shown its advantages. The mathematical study of Dunhuang architectural paintings can also blaze a trail for the study of architectural history and the promotion and development of traditional culture. In 2017, the State Council issued the "Opinions on the Implementation of the Inheritance and Development Project of Chinese Excellent Traditional Culture", emphasizing the inheritance and development of Chinese excellent traditional culture. Through digital research on architectural paintings, big data and digitization provide technical support for the "protection, inheritance, dissemination and exchange" of traditional culture. Through the digitalization of architectural paintings, the blueprint of architectural historical research can be protected, and architectural paintings can be excavated on the basis of in-depth development, laying the foundation for cultural inheritance and modern communication. This paper conducts a special research on architectural paintings in Dunhuang murals, combs the development status of architectural painting digitization, and the technical process of realization to provide specific directions for later research and development.

Keywords: Big Data · Dunhuang Architectural Paintings · Digitization

1 Introduction

In recent years, "big data" and the Internet+ have profoundly changed people's lives. Since the emergence and development of data science in the 1960s, we have introduced a new term-"big data". Big data is a global quantitative information asset, which has profoundly affected social production, life, politics and economy in many aspects [1]. The use of new, large-capacity acquisition, storage, management and analysis processing modes can provide a wealth of data collections for various fields.

The new technologies and new methods derived from the analysis of these "data collections" also have a huge impact on architectural design and architectural history

J. C. Hung et al. (Eds.): FC 2023, LNEE 1132, pp. 385–391, 2024.
https://doi.org/10.1007/978-981-99-9538-7_60

research. Therefore, the research methods for architectural history are gradually changing in this torrent. The acquisition, storage, analysis and processing of big data need to be realized through computer terminals. The computer digitizes the content, and the Internet will form a network of data. Through digitization and networking, "big data" has formed a new breakthrough in the study of many problems in the architectural paintings on the Dunhuang murals. For example, the research on the development of tower-like buildings, how to gradually emerge from the cladding-style tower into a loft-style tower, and the comparison of the towers in architectural paintings, are of great significance to the study of the evolution of the towers. Further research can also be done on the architectural image, construction technology, and building structure before the Tang Dynasty.

2 The Technical Characteristics of Big Data and Its Application in the Study of Dunhuang Architectural Paintings

Big data provides a data basis for scientific analysis of the digitalized building image, and lays a solid foundation for its subsequent technical means, communication channels, interactive experience, commercial value, and cultural communication. Big data presents the characteristics of large capacity, diverse data types, fast acquisition speed, authenticity and high post-derived value. These have a strong correlation with the study of architectural images in architectural paintings.

Large-capacity characteristics of big data [2]:

Big data is characterized by large capacity, and data units currently have B, KB, MB, GB, TB, PB, EB, ZB, YB, and BB levels. And big data is more valuable and potential development power because of the large amount of data. Therefore, a large-capacity database has a great basic support for the information collection and storage of architectural paintings and later analysis.

3 The Status Quo of the Digital Development and Protection of Dunhuang Architectural Paintings

The study of Dunhuang has a wide range of studies, and the study of architecture is closely related to the sculpture murals. The architectural paintings in the Dunhuang murals reflect the rich architectural image. The digitization of architectural paintings is undoubtedly the inheritance and protection of the Dunhuang Mogao Grottoes and the architectural history. The digital work practice of Dunhuang murals thus started in the 1990s. Dunhuang architectural paintings are included in the Dunhuang murals [3], There are few examples of wooden architecture before the Tang Dynasty in our country, and the architectural images of the Northern Wei, Northern Zhou, Sui and Tang Dynasties are preserved in the cave murals. For example, in Cave 445 of the Sheng Tang Dynasty, the demolition diagram in the Maitreya Sutra accurately depicts various architectural components, which has important reference value for the study of wooden architecture before the Tang Dynasty.

The traditional method of mural protection is to combine fine art copying with photography and video. In 1993, the combination of computer storage and management of Dunhuang murals was the first attempt to realize the big data and informatization of Dunhuang murals.

The scholars mastered effective architectural painting information by visiting the virtual caves, and then analyzed and studied architectural graphics.

Data source: Dunhuang Academy Yearbook 2010–2016 and digital Dunhuang website.

4 Technical Process for the Realization of Digitalization of Dunhuang Architectural Paintings

The digitization of Dunhuang architectural paintings is realized through computer and digital image technology. With the maturity of digital photography technology, the protection of murals has changed from copying and traditional film shooting to digital shooting, and the form of protection has also shifted from physical protection to both physical and virtual spaces [4]. In March 2010, Microsoft Asia Research Institute cooperated with Dunhuang Research Institute. The billion-pixel digital camera of "Flying" can generate high-quality digital images. The specific digitization process is: First, use a computer for digital analysis of objects, use professional digital cameras for digital collection of objects, collect and classify massive amounts of original resources, establish an artificial intelligence platform, perform digital storage, conversion, and finally achieve digital output. The output results are reflected in the three aspects of virtual reality, augmented reality and interactive reality (Fig. 1). At present, the digitization of Dunhuang architectural paintings is mainly embodied in virtual reality and interactive reality. Users can visit more than 30 caves by clicking on the digital Dunhuang website with special VR glasses, and can intuitively see the content of the architectural paintings on the cave paintings. For example, there are large-scale buildings depicted in the Wuliangshou Jingbian on the north wall of the main room of Cave 217. The image of the building on the mural can be obtained through Internet technology, and a clearer architectural painting can be seen. However, due to the limitation of the current working

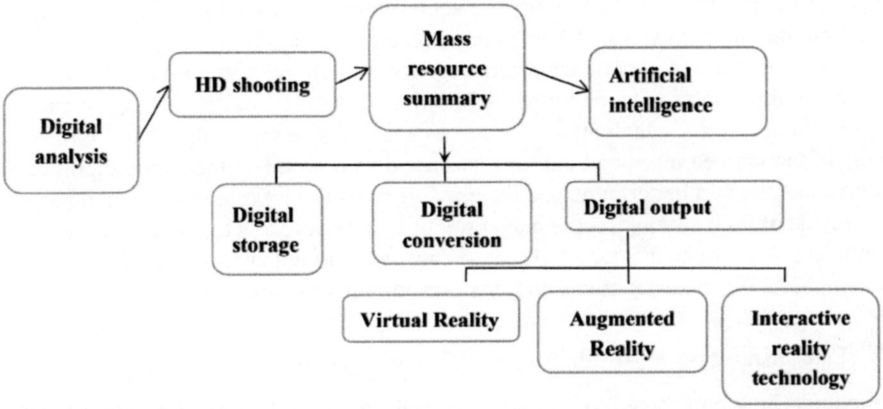

Fig. 1. The digital technology flow chart of Dunhuang architectural paintings (self-painted)

depth, the number of architectural paintings that can be seen is less, and the typicality is also weak. The interactive experience is poor, and it still stays above the latitude of the two-dimensional plane.

5 Development and Research on the Digitalization of Dunhuang Architectural Paintings

Although the digitization of Dunhuang architectural paintings, with the "digital Dunhuang" project, there has been a great development. The digitization of Dunhuang architectural paintings, but even if the architectural image in the Dunhuang murals is perfectly restored, it only stops at the production of two-dimensional images and only takes the most basic step. The history of architectural development, architectural color composition, architectural technology, structural characteristics of ancient buildings, structural composition, etc. reflected in architectural paintings cannot be presented [5]. The study of architectural paintings only stays on the surface of the drawing. To do in-depth digital development of architectural paintings, this article attempts to analyze and explore the possible development of digitalization of Dunhuang architectural paintings from the following points:

5.1 In-Depth Digital Development of Architectural Paintings

In-depth analysis of architectural paintings with the help of 3D modeling technology and VR technology.

With the maturity of VR technology, if the digitization of Dunhuang architectural paintings and the digitization of Dunhuang murals are basically the same in the early stages and technical paths, then the nature of the production of cave VR and the production of architectural painting VR is completely different. The digitalization of architectural painting requires a profession closer to the architecture itself. The main purpose of the cave is to restore and the latter is to recreate. Digital achievements can restore history, reflect the building wisdom of the ancients, and research and promote digital foundations. The production process of 3D modeling and the production of architectural VR can be a concrete way to embody the achievements of digital paintings.

Accurate digital analysis of Dunhuang architectural paintings.

The digitization of general cave paintings focuses on the restoration of cave shapes and the restoration of cave paintings, while the digitization of architectural paintings requires more details, which places higher requirements on the accuracy of digitization. Many of the warped maps and bonsai maps are drawn based on the architectural background, and the number is large, and the drawing reflects a lot of architectural details. In the process of digitization, it is the main direction of the research to capture the details of architectural paintings in a targeted manner, in-depth digital analysis in the early stage, and emphasis on characteristics and transformation in the later stage.

5.2 Establish a Special Dunhuang Architectural Drawing Database

Ming and Qing, and there are rich architectural types. Architectural painting is combined with Buddhist architectural art, and there are houses, palaces, pagodas, pavilions, temples, post stations, and so on. It can be used as a branch of special research for special summary. A database of architectural drawings is formed to provide a wealth of information for the study of architectural history theory and architectural design practice. A clustering algorithm is required to form a building map database. Clustering is a process of assigning organizations to different categories based on the attributes of objects. Objects in one cluster have the same or similarity, but are different or dissimilar to objects in another cluster. Among many clustering algorithms, Randomly set k cluster centers, and calculate the similarity of cluster centers for other objects, and assign them to the class of the cluster center closest to it. For each new cluster, find the mean value of all objects in it, and use it as the cluster center of the new cluster. Through the clustering method, we can obtain new clusters with the following characteristics: high cohesion and low coupling, that is, the elements within each cluster are close, and the connection between each cluster is small. The advantage of this algorithm is that it can efficiently classify large data sets, but once a local optimal solution is obtained, it will end, and the global optimal solution cannot be obtained. Many researchers have proposed improvements to the algorithm based on their research. Spectral clustering algorithm is another effective clustering method. The idea of spectral clustering algorithm is to regard samples or pixels as vertices, and the similarity between samples or pixels as weighted ones. how to define the vertex similarity and how to retain important edges, so that the constructed similarity matrix more truly reflects the similarity relationship between data points. The Gaussian similarity function is a method of calculating the similarity between samples in the classic spectral clustering algorithm, but the selection of the Gaussian scale causes the function to have certain limitations. In order to avoid the problem of parameter selection, many scholars have studied the process of constructing the similarity matrix. The most common method that does not use Gaussian similarity function is to use the neighborhood algorithm to connect each point with the closest point. The similarity matrix is obtained by solving the quadratic programming problem, which reduces the spectral clustering algorithms impact on The dependence of the parameters makes the algorithm more stable. The formulas used in the clustering algorithm are:

$$D = \sum\nolimits_{i=1}^{n} \left[\min_{r=1,\cdots,k} d(x_i, c_r)\right] \tag{1}$$

$$SNR = \frac{\left|\int_{-w}^{w} G(-x)f(x)dx\right.}{\sigma\sqrt{\int_{-w}^{w} f^2(x)dx}} \tag{2}$$

$$L = \pi\left\{\frac{\int_{-\infty}^{\infty} f^2(x)dx}{\sqrt{\int_{-\infty}^{\infty} f(x)dx}}\right\}^{-\frac{1}{2}} \tag{3}$$

5.3 Research and Develop a New Type of Dunhuang Architectural Interactive Experience Software to Make a New Attempt for the Promotion of Dunhuang Culture

At present, there is a problem of poor interactivity after the digitalization of Dunhuang architectural paintings. It can be said that there is no specific interactive experience, and the experience of virtual reality is not intuitive enough. The development of a new type of Dunhuang architectural interactive experience software (see Fig. 2) is undoubtedly a good way to promote traditional culture. Through interactive experience, you can restore the historical architectural scenes to bring people a deeper experience, and you can learn more about the building itself immersively, and it is also a better inheritance of cultural history.

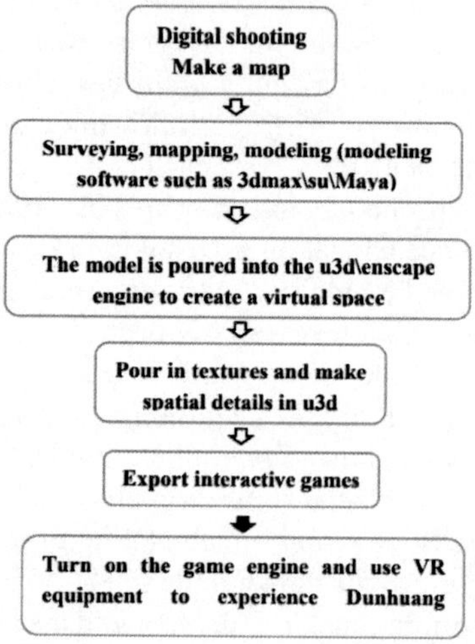

Fig. 2. Digital interactive game production process

6 Conclusion

In the context of the development of global big data, digital photography, 3D modeling, and VR technology have greatly developed. The digital research project of Dunhuang architectural paintings has solid technical support, and the progress of the research on the subject appears practical and urgent. The realization of the digitalization of architectural paintings and the in-depth study of digitalization have established a new type of resource acquisition platform for scientific researchers, art lovers and general audiences. The traditional architectural culture embodied in the Dunhuang murals can be popularized through education, protection, inheritance, innovation, development, and communication through another dimension.

References

1. Liu, G., Lu, D.: Digitization of Dunhuang Murals. Dunhuang Studies, 2003, (No. 4) (2003)
2. Liu, G., Zhang, J., Diao, C.: Research on 3D digitization technology of Dunhuang Mogao Grottoes. Dunhuang Stud. No. 4, 107–108 (2005)

3. Fu, X., Ma, X., Sun, Z.: Research on Digital Restoration of Damaged Murals——Taking Dunhuang Murals as an Example. Decoration, (1st issue) (2019)
4. Edited by Dunhuang Academy [M]. "Dunhuang Academy Yearbook"
5. Sun, Y., Sun, R., Fan, J.: Interpretation of Dunhuang Medieval Architectural Paintings. East China Normal University Press, Shanghai (2016)

Refined Identification of Distribution Network Planning Survey Based on Improved Convolutional Neural Network Algorithm

Wu Guoyue(✉), Zhang Chenxi, and Lin Lixiang

Guangzhou Power Supply Bureau of Guangdong Power Grid Co., Ltd., Guangzhou 510000, Guangdong, China
wuguoy126@126.com

Abstract. Distribution network planning is an important guarantee for power grid construction and transformation, which can ensure the reliable, stable, economic and flexible development of the system. With the increasing capacity of distribution network, the amount of data to be processed and analyzed in distribution network planning has increased dramatically. Especially for the high voltage, medium voltage and low voltage superior power supply, distribution network structure and operation status, the workload is huge and the complexity is high. If the planning process completely depends on the planners to analyze and calculate, it is easy to have calculation errors, incomplete analysis or other uncertainty errors. Based on the improved convolutional neural network (CNN) algorithm, a new identification algorithm is proposed to improve the measurement accuracy of distribution network planning. The proposed method uses CNN to extract the first k important variables, and then combines them with the previous two methods, one uses nonlinear regression, the other uses linear regression. In addition, we propose a new metric method. In order to evaluate our results, this article uses a large number of real data sets.

Keywords: Distribution network planning · Convolution neural network · Fine identification

1 Introduction

With the continuous improvement of customer requirements and the concern of power grid companies on the reliability and economy of distribution networks, distribution network planning has become an important project for the transformation and construction of distribution networks. As a guiding text, it determines the development direction and mode of the planned area. If the design is wrong, it will bring huge cost waste to power supply enterprises and poor quality electric energy service products to residents. The superior company also gradually noticed that the limited resource allocation of power supply enterprises [1]. Therefore, more detailed requirements have been put forward in terms of management means and work content. The distribution network involves a large area, many voltage levels, and a complex interconnected structure. In

J. C. Hung et al. (Eds.): FC 2023, LNEE 1132, pp. 392–397, 2024.
https://doi.org/10.1007/978-981-99-9538-7_61

economic construction, and residential power consumption, flexibility of power supply of medium and low voltage distribution networks, and reduce power grid losses, It also brings problems such as distorted communication between superiors and subordinates, and disjointed monitoring. Therefore, the introduction of a good communication and management mechanism is a strong guarantee [2].

In the previous planning of distribution network, the lack of comprehensive standards and guidance makes the investment and construction not suitable for the development of economy and demand, and also increases the cost of power grid construction and operation. At present, the network loss rate of the country is about 10%. The general distribution network planning is to configure and transform according to the line overload, heavy load or other voltage reasons, replace the cable with large conductor section or increase the line and capacity, rather than a complete and comprehensive construction of a reliable and reasonable network structure to adapt to the construction requirements of urbanization [3]. The state pays more and more attention to the construction of distribution network, invests more and more money, and has made many achievements. However, the foundation of distribution network is relatively weak, mainly because the equipment is old but the load is heavy, the automation level is low, and the wiring is unreasonable. In particular, the medium and low voltage distribution network urgently needs comprehensive guidance standards.

2 Related Work

2.1 Domestic Distribution Network Planning Status

The "Guidelines for the Design of Urban Power Networks" issued by the Ministry of Energy in 1993. This document has unified provisions from seven aspects, including planning requirements, load forecasting and design principles, and selection of power supply facilities, and has become an important basis for modern, the project staff should be able to take into account the impact of all aspects when planning based on the past planning experience and certain models. However, experience and subjective judgment alone cannot make full use of the collected data and information to make scientific and accurate judgments, which will consume more manpower and material resources in the actual planning, and have little guiding significance for the development of the distribution network [4].

Some large cities in China have focused on the development of distribution networks for many years. Shanghai, Yantai, Guangzhou and other cities have built a better distribution network automation and network structure, which can achieve fault isolation and dispatching automation and other functions, and even add information collection and low-voltage reactive power compensation control functions.

The distribution network planning can get better development in the above cities, which can not be separated from information technology and computer aided analysis. The computer technology that is widely used in China is mainly GIS technology. Many experts have designed some software packages and codes for the application of distribution network planning, but there is still a long way to go from comprehensive guidance and perfect software. This paper makes an in-depth study on the transformer location and line routing of the 10 kV distribution network, puts forward a preliminary research scheme, and automatically solves simple problems with the help of language development tools. Using MapX and VB as the development language and taking advantage of the high efficiency of GIS technology, the fault data of power system can be well processed. The development of artificial intelligence technology has injected new impetus into the development of distribution network. At some stages of planning, the combination of geographical data and artificial intelligence technology is used to solve the current problems, so that the automation solution can wait for a certain development. Through heuristic rules and expert system, different planning methods are proposed for different stages to solve the problem of distribution and coordination of substation planning.

Some new genetic algorithms have improved the idea of solving nonlinear problems in distribution network planning. The scientific conclusion is greatly improved by using genetic algorithms to optimize problems. In general, some cities in China gradually introduce computer aided decision-making system in planning to increase the rationality of distribution network transformation. GIS has also been greatly promoted in some control planning, but there is still much room for improvement.

2.2 Research Status of Distribution Network Planning Software

With the continuous improvement of customer requirements and the concern of power grid companies on the reliability and economy of distribution networks, distribution network planning has become an important project for the transformation and construction of distribution networks. Distribution network planning requires a large amount of data to comprehensively analyze the superior power supply, distribution network structure and operation status of the distribution network. The workload is large and the complexity is high. If the planning process completely relies on the planners to analyze and calculate, it is easy to have calculation errors, incomplete analysis and other problems.

Some front-line employees reported that many of the data collected each year were collected last year or before, and the design unit did not sufficiently inherit the data, resulting in heavy workload and low efficiency. Because the investigation takes up a lot of time of the front-line infrastructure department staff, if we can quickly get all the analysis indicators and perfect planning reports after obtaining the original data, and deal with the problems found in the correction of errors and the status quo analysis reduce the interference to the normal work of front-line staff. Guangdong Power Grid Corporation has a detailed guidance process for the revision of the entire distribution network planning. However, due to unexpected problems such as wrong capital collection and re revision, the design unit often has to make important changes or make major changes when handling the revision task of the distribution network planning in the previous year. In serious cases, the task cannot be completed within the specified time, resulting in the rescheduling of the subsequent schedule, which wastes human and material resources. The provincial power grid corporation also expressed dissatisfaction with this. In addition, the design process of different design units is different, and the final report may be different in data format and table text format, which is not standardized and neat enough.

Static voltage stability analysis and security risk assessment, and the database is used to uniformly manage the data required by each module. However, in the planning process, only N-1 criteria and part of N-2 security verification are verified, lacking other indicators that comprehensively reflect the current situation of the network.

3 Distribution Network Planning Based on Improved Convolutional Neural Network Algorithm

3.1 Graph Convolution Neural Network

At present, convolution based on spatial method and convolution based on spectral method are two main research directions of graph neural network. Graph convolution neural network is a representative model based on spectral method convolution. With its excellent processing ability for non Euclidean spatial data, it has rapidly become a research hotspot in the field of artificial intelligence. Graph neural network based on spectral method is widely concerned because of its simpler model structure and lower computational complexity. The convolution method used in this paper is graph convolution technology based on spectral method, which is introduced below.

In 2009, Graph Neural Network (GNN) was proposed. It uses the structure of traditional neural networks for reference to directly process data in non Euclidean space, which is the enlightenment work of graph neural networks and lays in the future. Laplace operator and Fourier transform are used to solve the problem of convolution calculation of graph data, and the first graph convolution neural network is proposed. The GCN model is proposed by using the idea of first-order adjacent order aggregation, which is used as the basic GCN model by most subsequent studies.

Before using Laplacian matrix, graph convolution neural network needs to use two tricks to improve its performance: (1) add a self loop to all nodes to ensure that nodes themselves and their neighbors participate in reasoning together, that is, $A \leftarrow A + I$ (it is still recorded as A in this paper); (2) Normalization is carried out to prevent gradient explosion or disappearance, as shown in Fig. 1.

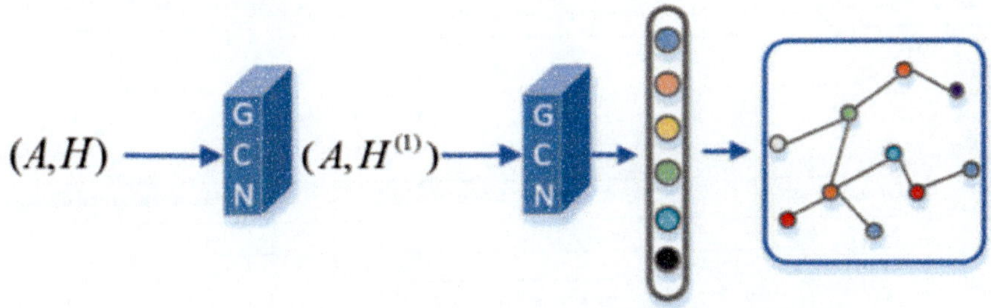

Fig. 1 Structural model of graph convolution neural network

3.2 Transformation Method of Distribution Network

The reliability of power supply and distribution is increasingly concerned by the whole society and power supply enterprises. At present, the transformation method of the current distribution network mainly depends on the established technical standards and the experience of the staff.

First, determine the construction in progress and planned scheme of high-voltage power supply in the distribution network, analyze the transformation of substation planning and medium and low-voltage distribution network planning in the previous year, investigate the load adjustment of medium and heavy load, overload lines and distribution transformers in the overall and all areas, and improve the problem index value; Secondly, according to the situation of enterprises and large customers using telegraphy, combined with the city's industrial layout and development, social natural growth rate, and the service factor of telegraphy, load forecasting is carried out from top to bottom and from bottom to top, so as to obtain the power demand of the whole city and districts; Before formulating the project transformation scheme, it is necessary to have a comprehensive understanding of the power balance and planning and design principles, including the electricity demand analysis of power balance and step-down capacity, the detailed discussion on the design principles of the capacity load ratio, grid structure and guide section of high-voltage distribution network, and the technical principles of the structure, switch stations, distribution lines and distribution devices of medium and low-voltage networks, taking into account their actual application; Then, carry out a comprehensive design of high-voltage, medium and low-voltage distribution network transformation scheme. The transformation scheme of high-voltage distribution network mainly considers the current situation and development prospects of its high-voltage distribution network connection mode, and plans the distribution points and network structure of the substation, and carries out sensitivity analysis and risk check on the implementation

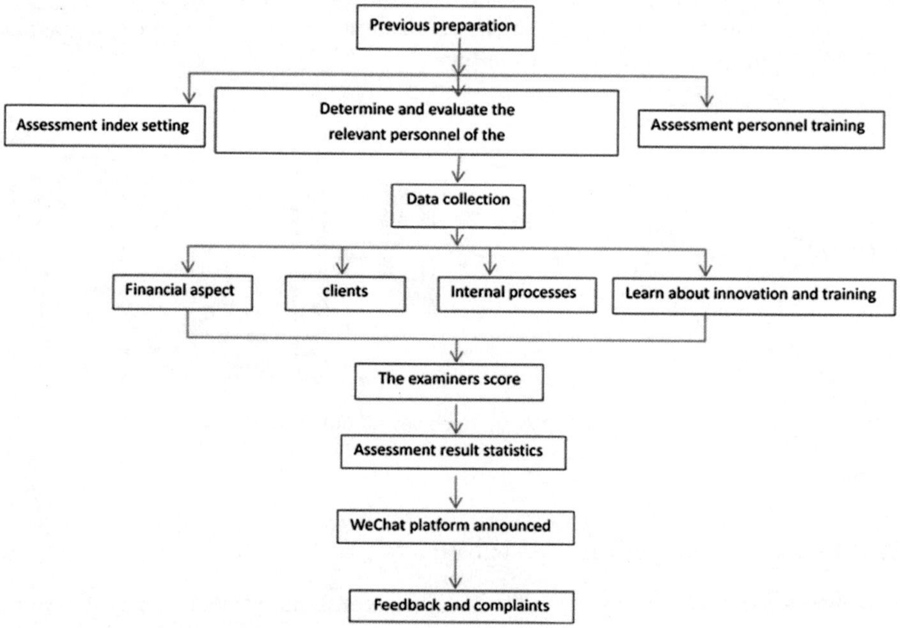

Fig. 2 Flow Chart of Distribution Network Transformation Scheme

scheme. For the medium and low-voltage distribution network, according to the planning of high-voltage substation and power supply scope, the new outgoing lines Medium voltage grid transformation and low-voltage grid improvement projects, and analyze the rationality of its high, medium and low schemes. The final transformation plan also needs to carry out investment estimation and economic evaluation, which are detailed to the estimation of the main network, distribution network and the whole network; In addition, it also evaluates the effectiveness of the whole planning project, such as the degree of solution to the problems existing in high-voltage and low-voltage distribution networks and the reliability of power supply, and analyzes the line loss rate and investment efficiency rate after the implementation of the scheme. The process is shown in Fig. 2.

4 Conclusion

With the rapid development of distribution network reconstruction in China, distribution network planning has become a task with huge amount of data and various processing indicators. It is urgent to change from the previous experience oriented planning to an efficient and accurate planning method combining theory and experience. It is an urgent goal to develop a convenient and practical intelligent network planning computer aided system. The intelligent auxiliary analysis system for distribution network planning introduced in this paper is guided by the technical regulations for distribution network planning, which is based on Access 2003 database and VB NET language programming to realize the interaction between the computer and the planners, and realize the automatic calculation and analysis of different index values. Combined with the characteristics of various forms and types of capital collection data, the rationality judgment for digital information and the specification rules for text information are formulated. With this function, the data can be kept in a reasonable range and accuracy. The text data specification can be correctly invoked by the system.

References

1. Wang, J., Huang, C., Zhao, L., et al.: Lightweight identification of retail products based on improved convolutional neural network. Multimedia Tools Appl. **81**(22), 31313–31328 (2022)
2. Li, S., Yin, S., Deng, H.: Lipid droplet recognition based on watershed algorithm and convolutional neural network. In: ICMIP 2021: 2021 6th International Conference on Multimedia and Image Processing (2021)
3. Qu, M., Wang, T., Li, F., et al.: Distribution network electrical topology identification algorithm based on deep learning. IOP Conf. Ser. Earth Environ. Sci. **645**, 012076 (2021)
4. Ponciroli, R., Rovinelli, A., Ibarra, L.: A convolutional neural network-based approach to field reconstruction (2021)

Localization Technology of Small Current Ground Fault Section of a Distribution Network Based on Multi-terminal Synchronous Waveform

Jinlian Wu(✉), Bo Wu, Feng Liu, Bo Wang, and Yongting Lu

Dingxi Power Supply Company of State Grid Gansu Electric Power Company, Dingxi 744300, China

15095460248@163.com

Abstract. The role of fault section location in the grounding diagnosis is very important, but there is a problem of low diagnostic accuracy of small current. The GPS positioning method does not solve the problem of fault section location in the grounding diagnosis of the distribution network, and the small current diagnosis ability is low. Therefore, a multi-terminal synchronous waveform method is proposed to analyze the grounding diagnosis. Firstly, the synchronous theory is used to judge, and the fault section positioning standards are divided according to the fault section positioning standard to reduce the fault section positioning Disturbing factors. Then, the synchronization theory forms grounding diagnostic standards and synthesizes the fault section positioning standards OK. The fault segment positioning accuracy and fault section location time of the multi-terminal synchronous waveform method is superior to the GPS positioning method.

Keywords: synchronization theory · multi-terminal synchronous waveform method · Distribution network grounding diagnostics · Voltage · Current

1 Introduction

Fault diagnosis capability is one of the important contents of power grid fault section location [1], which is highly significant for distribution network grounding diagnosis. However, in the actual fault section location process, there is a problem of poor diagnosis ability of small current [2]. Some scholars believe that the application of the multi-terminal synchronous waveform method to the grounding diagnosis of distribution network can accurately analyze the ground fault problem Segment targeting is supported [3]. On this basis, a multi-terminal synchronous waveform method is proposed to optimize the location of fault segments and verify the model's effectiveness.

J. C. Hung et al. (Eds.): FC 2023, LNEE 1132, pp. 398–404, 2024.
https://doi.org/10.1007/978-981-99-9538-7_62

2 Related Concepts

2.1 Mathematical Description of Multi-terminal Synchronous Waveform Method

The multi-terminal synchronous waveform method uses the positioning theory to diagnose minor current faults [4] and finds the location of ground faults according to diagnostic indicators.

Hypothesis 1: The fault standard is y_i, the threshold for small current identification i s_i s, the fault segment positioning accuracy is x_i, and the fault segment positioning function is $P[x_i \in (0 \sim 10)]$ as shown in Eq. (1).

$$P(x_i) = \sum x_i \subseteq \oint y_i \mapsto \xi \tag{1}$$

2.2 Selection of Small Current Diagnosis Strategy

Hypothesis 2: The small current diagnostic function is $T(x_i)$ and the weight factor is w_i, then the small current fault diagnosis strategy is shown in Eq. (2).

$$T(x_i) = z_i \cap \prod P(x_i, y_i) \vee w_i^2 \tag{2}$$

2.3 Determination of Fault Segment Location

Before the multi-terminal synchronous waveform method analysis, the fault segment positioning strategy should be analyzed in multiple dimensions, and the fault segment positioning criteria should be mapped to the two-dimensional sample library to eliminate abnormal small current signals. First, the grounding diagnosis of the distribution network is comprehensively analyzed, and the threshold and index weight of the small current signal is set to ensure the accuracy of the multi-terminal synchronous waveform method [5]. Distribution network grounding diagnosis is a system test of small current signals, and a small current diagnosis is required. If the distribution network grounding diagnostics are distributed nonnormally, their small current signals will suffer, reducing the overall accuracy of fault section location. In multi-terminal synchronous waveform method and improve the fault section positioning level, the small current signal should be selected, and the specific information selection is shown in Fig. 1.

The survey results show that the small current grounding information presents a multi-dimensional distribution that aligns with the objective facts. The GPS positioning method has no directionality, indicating that the small current fault information has strong randomness, so multi-terminal synchronous waveform analysis should be performed. The small current grounding information conforms to the normal standard. It can be adjusted by the multi-terminal synchronous waveform method to remove repeated and irrelevant fault signals and supplement the default fault signals so that the dynamic correlation of locating faults in the entire fault section is strong.

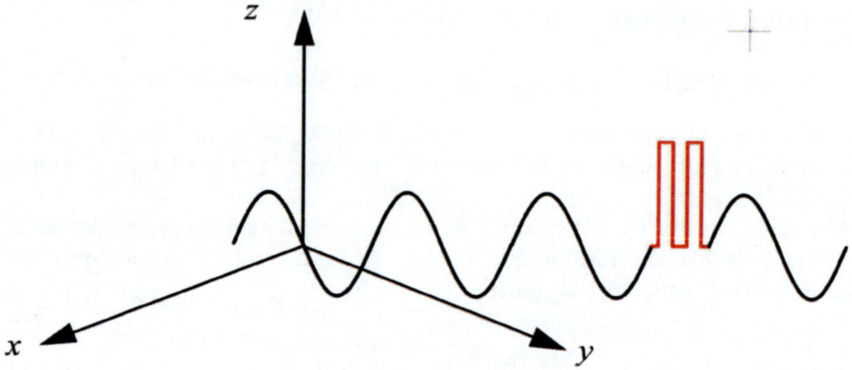

Fig. 1. Signal selection results for the multi-terminal synchronous waveform method

3 Identification of Small Current Signals for Ground Faults in Distribution Networks

The multi-terminal synchronous waveform method adopts random optimization of the ground fault of the distribution network and analyzes the voltage, current and other parameters for the small current signal to realize. The multi-terminal synchronous waveform method locates different fault sections according to the strength level of small current and randomly extracts different fault signals for verification. In the iterative process, fault sections with different current strength and weakness levels are located for fault location. At the same time, the positioning level of different fault signals is verified, and the optimal positioning information is recorded.

4 Practical Cases of Small Current Ground Faults in Distribution Networks

4.1 Introduction to Small Current Faults in Distribution Networks

This paper takes the small current fault of the distribution network ground under complex conditions as the research object, and the positioning period is 12 h, and the actual situation of the small current fault section This is shown in Table 1.

Table 1. Distribution network small current ground fault conditions [unit: %].

Section	Targeting method	Small current level	Multi-terminal synchronous waveform changes
Tai District	artificial	86.03	5.39
	system	94.65	4.70

(continued)

Table 1. *(continued)*

Section	Targeting method	Small current level	Multi-terminal synchronous waveform changes
branch	artificial	84.21	5.39
	system	84.24	5.51
sublevel	artificial	86.73	6.06
	system	85.44	5.86

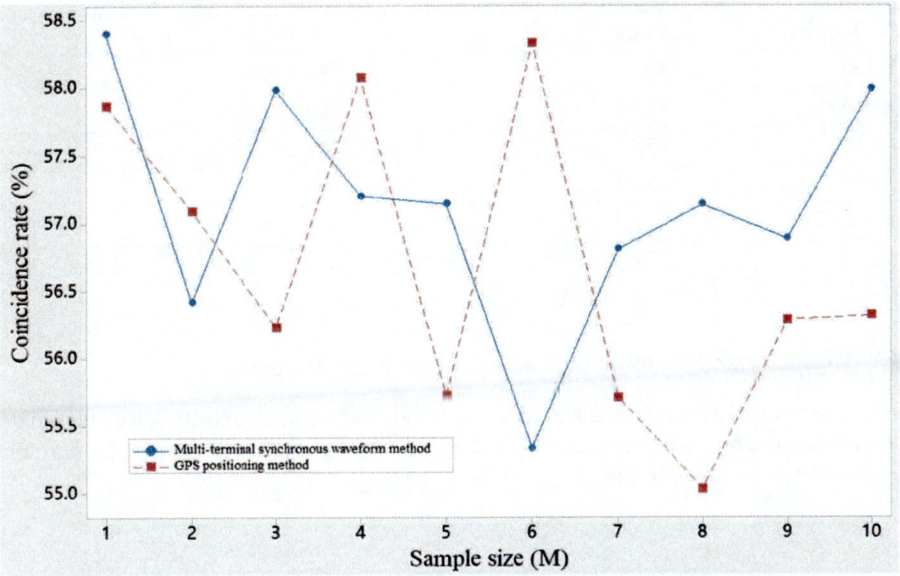

Fig. 2. Determination process of a ground fault in the distribution network

The fault segment location process in Table 1 is shown in Fig. 2.

The fault segment location fault of the multi-terminal synchronous waveform method is closer to the actual fault segment positioning standard in terms of the rationality and fluctuation amplitude of ground faults in the distribution network, the multi-terminal synchronous waveform method GPS positioning method. By locating the fault changes in the fault segment in Fig. 4, it can be seen that the multi-terminal synchronous waveform method has better stability and faster judgment speed. Therefore, the fault section location fault speed of the multi-terminal synchronous waveform method, the small current diagnosis to determine the fault signal fault section fault location, and the summing stability are better.

4.2 Distribution Network Ground Fault Location

Distribution network ground faults include stations, branches, and segments. After the preselection of the multi-terminal synchronous waveform method, the preliminary ground is obtained, and the ground fault of the distribution network is obtained by analysis of the reasonableness of the information. Waveform method on low-current fault diagnosis, select different sections for fault location, as shown in Table 2.

Table 2. Location of grounding fault in the distribution network

region	Small current intensity	Failure compliance rate
Tai District	81.69	91.76
branch	84.71	87.37
sublevel	87.64	80.71
mean	87.65	89.17
X^2	83.47	84.74
P = 0.074		

4.3 Accuracy and Stability of Small Current Fault Diagnosis

In multi-terminal synchronous waveform method, the accuracy and stability of the GPS positioning method are compared with the GPS positioning method, and the segment fault location is shown in Fig. 3.

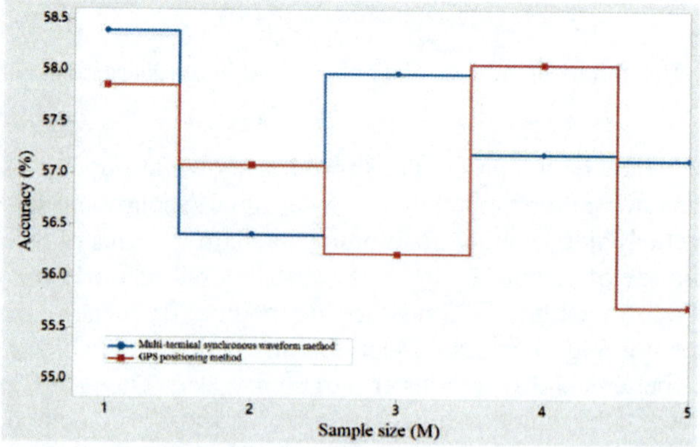

Fig. 3. Low current diagnostic capability of different algorithms

It can be seen from Fig. 4 that the small current fault diagnosis ability of the multi-terminal synchronous waveform method is higher than that of the GPS positioning method, but the error rate is lower, indicating the segment fault location process of the multi-terminal synchronous waveform method It is relatively stable, and the fault section positioning process of GPS positioning method has excellent changes. Table 3 shows the average fault segment location of the fault segment of the above three algorithms.

Table 3. Comparison of fault segment location accuracy of different methods

Algorithm	Tai District	branch	sublevel
Multi-terminal synchronous waveform method	81.21	81.62	2.17
GPS positioning method	93.75	81.13	3.31
P	0.011	0.021	3.016

It can be seen from Table 4 that the GPS positioning method has shortcomings in the ground fault of the distribution network in terms of small current diagnosis ability and stability, the ground fault of the distribution network changes significantly, and the error rate is high. The general result of the multi-terminal synchronous waveform method is that the low-current diagnosis ability is higher, which is better than the GPS positioning method. At the same time, the accuracy of the multi-terminal synchronous waveform method is greater than 90%, and the accuracy does not change significantly. To further verify the superiority of the multi-terminal synchronous waveform method. The minor current ground fault is continuously analyzed by different methods, and result 4 is shown.

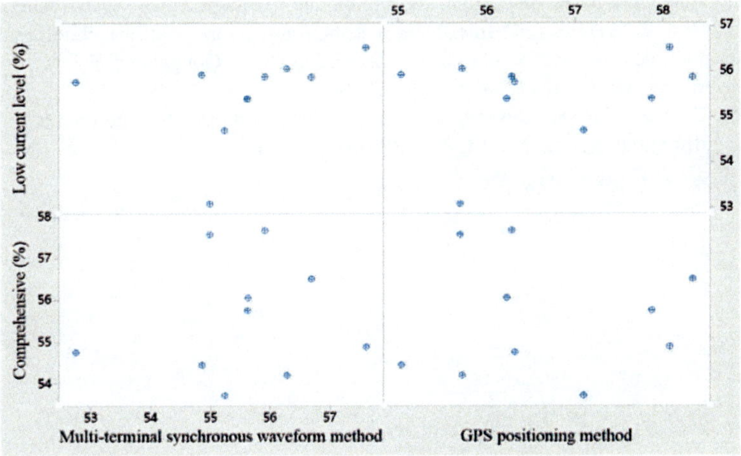

Fig. 4. The comprehensive ability of the multi-terminal synchronous waveform method for low-current fault diagnosis

As can be seen from Fig. 4, the comprehensive ability of low-current fault diagnosis of the multi-terminal synchronous waveform method is significantly better than that of the GPS positioning method, and the reason is that the multi-terminal synchronous waveform method has increased the signal strength of the small current ground adjusts the coefficient, and the threshold is set to reject fault signals that do not meet the standard.

5 Conclusion

Aiming at the problem that the diagnosis of small electrical, combined with the minor current strength level Distribution network ground faults are located. At the same time, the diagnosis criteria for small current faults in segment positioning are analyzed in depth to construct fault collections. The research shows that the multi-terminal synchronous waveform method can small current fault diagnosis and the stability of the diagnosis process Better, the small current ground fault of the distribution network can be comprehensively located. However, in the process of the multi-terminal synchronous waveform method, too much attention is paid to the data analysis of fault segment positioning, resulting in a lack of moderation between fault segment positioning indicators.

References

1. Zheng, Q.I., Hang, T., Lin, J., et al.: Technology of traveling wave fault location on single phase-to-ground fault in distribution network with single-phase power supply lines. Autom. Electric Power Syst. **186**, 106401 (2018)
2. Liu, H., Cai, Y., Cai, H., et al.: Single phase to ground fault location method for multi branch distribution network of cable overhead line. J. Dongguan Univ. Technol. (2017)
3. Qi, X., Pei, W., Li, L., et al.: DC fault detection scheme for multi-terminal hybrid AC/DC distribution network based on current-limiting inductor voltage. Power Syst. Technol. (2019)
4. Dong, X., Liu, Z., Li, P., et al.: Intelligent distribution network control technology based on multi-terminal flexible distribution switch. Zhongguo Dianji Gongcheng Xuebao/Proc. Chin. Soc. Electr. Eng. **38**, 86–92 (2018)
5. Han, X., Chen, Z., Zhu, K.: Research on the area protection of distribution network based on variable differential current. Electric Eng. (2018)

Design and Implementation of University Management Information System Based on Decision Tree Algorithm

Hao Zhu and Qu Zheng(✉)

Department of Student Affairs Management and Chengdu Medical College, School of Foreign Languages and Jinjiang College, Sichuan University, Chengdu 620800, Sichuan, China
zhengqu@cmc.edu.cn

Abstract. The role of information management in universities is very important, but there is a problem of low management level. The management system cannot solve the management problem of multiple types of information in the university system, and the rationality. Therefore, decision tree method to construct an information management optimization model. First of all, the information knowledge is used to classify the university information, and the university information is selected according to the degree of importance to realize the standardized processing of data.Then, the information knowledge is classified according to importance, forming an information optimization collection and iteratively analyzing the scoring content. MATLAB simulation shows that the decision tree method's optimization degree and optimization time are better than that of a single management system when the system is fixed.

Keywords: information knowledge · Time · Decision tree method · Optimize the results

1 Introduction

Information optimization is one of the important contents of information management, which plays a very important role in improving the information optimization system [1]. However, in building the information system of colleges and universities, there is a problem of low management level in information management results, and the role of information management cannot be effectively played [2]. Some scholars believe that applying the decision tree method to the university system can effectively carry out redundant information and time analysis and provide corresponding support for information optimization and verification [3]. On this basis, this paper proposes a decision tree method to optimize universities' information management and verify the model's effectiveness.

J. C. Hung et al. (Eds.): FC 2023, LNEE 1132, pp. 405–411, 2024.
https://doi.org/10.1007/978-981-99-9538-7_63

2 Related Concepts

2.1 Mathematical Description of the Decision Tree Method

The uses information nodes, information relationships, and information importance to optimize university information, finds outliers in information optimization according to the management indicators in the university system, and forms a path table [4]. The correlation of the information management results is finally judged by integrating the information optimization results. The decision tree method combines information knowledge and optimizes the management results by using the decision tree method, which can improve the level of information management [5].
Hypothesis 1: The university information is, the information optimization result x_i set is, the information importance is $set(x_i)$, and the judicial function of the information management result is y_i $Y(x_i)$ as shown in Eq. (1).

$$Y(x_i) = x_i \Rightarrow y_i \tag{1}$$

2.2 Selection of Information Optimization Scheme

Hypothesis 2: The selection function of the information optimization method is $F(x_i)$, and the information weight coefficient is g_i, then the information optimization method selection is shown in Eq. (2).

$$F(x_i) = g_i \cdot Y(x_i|y_i) \tag{2}$$

2.3 Processing of Redundant Information

Before the decision tree analysis, the standard analysis of the time and single time in the information management results should be carried out, and the university information should be mapped to the selection table to determine the semantic anomaly of content. First, comprehensive analysis of university information and set thresholds and weights of university information to support the accurate research of the decision tree method. University information needs to be standardized, and if the processed results conform to the non-standard distribution, the processing is valid, otherwise, the standard processing is re-processed. In the decision tree method and improve the level of selection, the decision tree method scheme should be selected, and the specific method selection is shown in Fig. 1.

The university information in Fig. 1 shows that the analysis of the decision tree method is uniform and consistent with the objective facts. The selection method is not directional, indicating that the analysis of the decision tree method has strong accuracy, so it is used as university information management research. The selection method meets the mapping requirements, mainly the information knowledge adjusts the selection method, removes duplicate, redundant information, and revises the information node to make the whole University information is more selective.

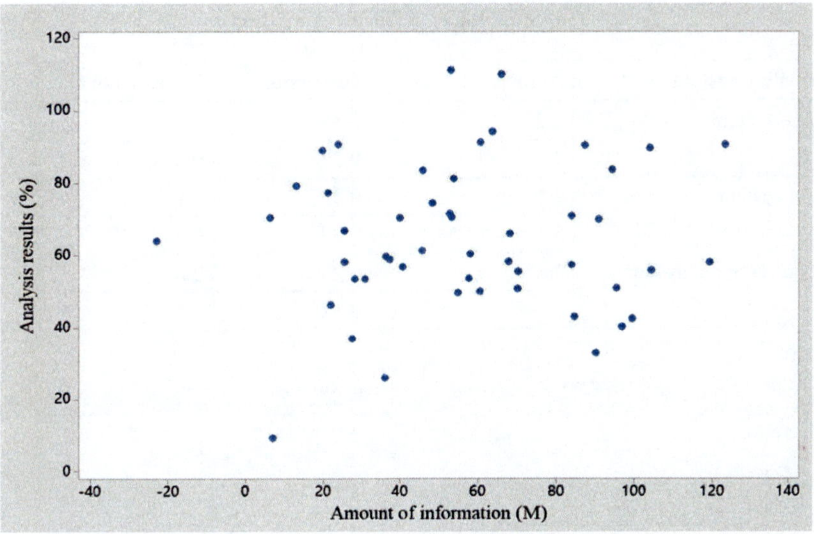

Fig. 1. Information management results of decision tree analysis

2.4 Correlation Between Different Information Nodes

The decision tree method adopts accurate judgment of time, and adjusts the corresponding redundant information relationship to optimize the information management method of universities. The decision tree method divides the information management choices of universities into different data volumes and randomly selects other methods. During the iteration process, the importance of different data volumes is matched with the selection method. After the matching processing is completed, different methods are compared for university information management, and the management results with the highest accuracy are recorded.

3 Actual Cases of University Information Management Systems

3.1 Information System Situation

In facilitate the analysis systems, the number of test orders in this paper is 2421 based on different types of information systems Time 450, as shown in Table 1.

The processing process of information nodes between different information nodes in Table 1 is shown in Fig. 2.

As from Table 1, the information results of the decision tree method are closer to the actual time compared with the single information management system. In terms of university information management selection, information node selection rate, accuracy, etc., the decision tree method management system. From the change of information nodes in Fig. 4, it the accuracy of the decision tree method, and the judgment speed is faster. Therefore, the decision tree method has better information processing speed, time, and optimization.

Table 1. Information characteristics

information system	Amount of data	Dispersion	Information standards
College system	1.2	0.35	0.2
	4	0.25	0.3
School system	2.1	0.25	0.2
	5	0.39	0.4
This professional system	2.6	0.25	0.3
	3	0.12	0.4

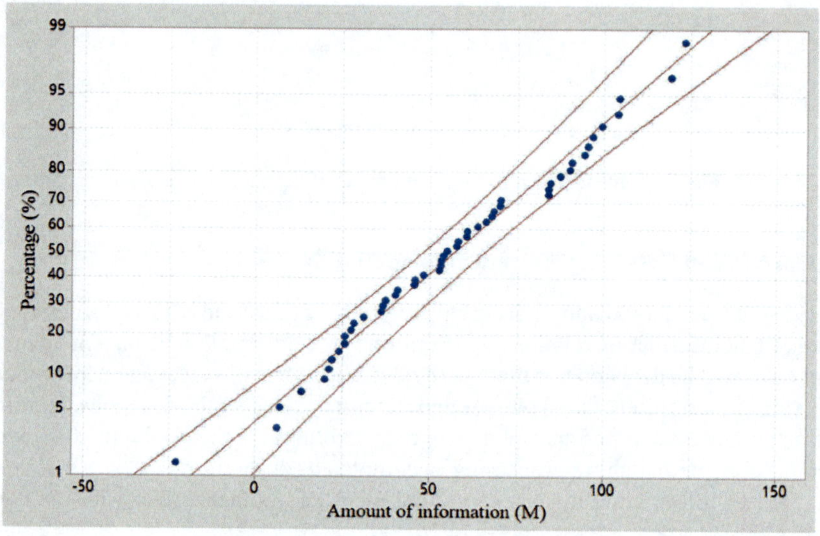

Fig. 2. Processing of information nodes

3.2 Optimization Ratio of Information

Information optimization includes redundant information, information nodes, and speed. After the threshold criteria, the preliminary management results are obtained, and the correlation of the management results is obtained Analyze. To verify the effect more accurately, select different redundant information and calculate the overall time of information management, as shown in Table 2.

3.3 Time and Accuracy of Information Optimization

To verify the accuracy of the decision tree method, the optimization time and accuracy compared with the management system are shown in Fig. 3.

It from Fig. 3 that the optimization time of the decision tree method is shorter than that of the single management system. Still, the error rate is lower, indicating that the

Table 2. Overall optimization situation

Optimize the proportion	Information accuracy	Outlier recognition rate
25%	93.31	92.63
50%	92.20	93.26
70%	93.73	90.65
mean	92.21	92.42
X^2	6.312	11.137
P = 0. 032		

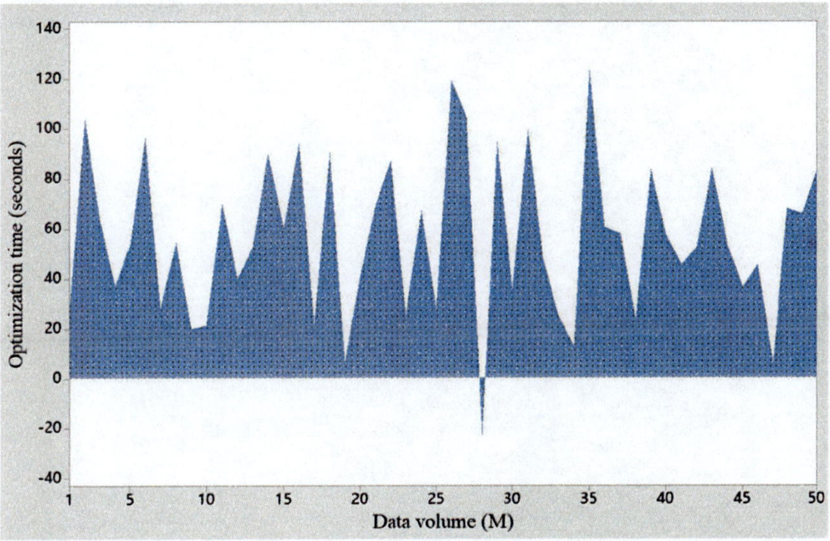

Fig. 3. Optimization time for different algorithms

choice is relatively stable and single. The degree of optimization of the management system is uneven. The accuracy of the above algorithm is shown in Table 3.

Table 3. Comparison of optimization degrees of different methods

algorithm	Optimize time	Optimize location	error
Decision tree method	92.11	(2,3)	0.74
Single management system	70.25	(4,6)	0.21
P	0.012	(7,2)	0.23

It from Table 3 that the rationality of the management system in the selection of university information management has shortcomings in optimizing time and accuracy, and the accuracy of time has changed significantly, the error rate is high. The optimization time of the comprehensive results of the method is higher than that of the management system. At the same time, the optimization time greater than 90%, and the accuracy does not change significantly. To further verify the superiority of the decision tree the continuity of the way, the decision tree method was comprehensively analyzed by different methods, and result 4 is shown.

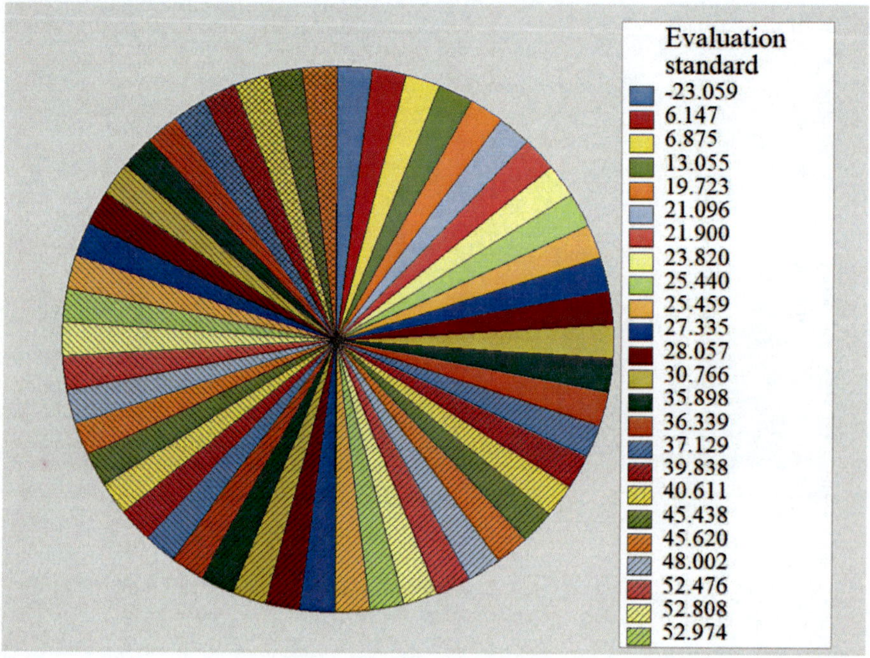

Fig. 4. Comprehensive results of information management result evaluation of decision tree method

As from Fig. 4, the decision tree method is significantly of the single management system, and the reason is that the decision tree method increases the time adjustment coefficient and sets it Corresponding thresholds, which propose non-compliant results.

4 Conclusion

In the case of the rapid development of computers, this a decision tree method for the information management problem in colleges and universities. It combines information knowledge to improve the information related in the system. At the same time, the information processing threshold criteria are analyzed in depth to construct an optimization set. Studies have shown can improve the accuracy of time, and the accuracy can synthesize time. However, in the process of the decision-making tree, too much attention is paid to the ability to analyze unilateral indicators, and the proportion of time is ignored.

References

1. Deng, X.Y.: Y architecture and implementation of immunization planning information management system based on monitoring big data. In: 4th International Conference on Decision Science and Management (ICDSM), Changsha, Peoples R China, pp. 339–348 (2022)
2. Chen, C., Geng, L., Zhou, S.: Retraction note: design and implementation of bank CRM system based on decision tree algorithm. Neural Comput. Appl. **35**(6), 4803 (2022)
3. Mao, J.: Design and implementation of tax collection and management index early warning system based on data mining (2021)
4. Han, J.: System optimization of talent life cycle management platform based on decision tree model. J. Math. **2022**, 1–12 (2022)
5. Chen, X., Zhuang, Y., Lao, Y., et al.: Development and implementation of a novel decision support tool on physical restraint use in critically ill adult patients. Int. J. Nurs. Pract. **28**, e12961 (2021)

Research on Energy Metering Alliance Chain Technology and Continuous Improvement System Based on Blockchain

Angang Zheng(✉), Huaiying Shang, Yan Liu, and Tianyi Zhang

China Electric Power Research Institute, Haidian, Beijing 100192, China
zhengangang@epri.sgcc.com.cn

Abstract. The analysis continuous calculation in energy measurement, this consortium chain technology, which uses technology, chain metering method and alliance chain rules to test the relevant measurement in energy, and inserts the energy string method in the blockchain to improve the comprehensive computing ability of energy measurement. At reasons for the interference of measurement, summarizes the characteristics of energy measurement, puts forward the design ideas of blockchain technology, and tests the correctness, fit of energy measurement and degree of energy measurement through actual cases. The simulation results of MATLAB show that the energy measurement ability of the alliance chain technology is better, the energy measurement degree is more than 90%, and the comprehensive computing power continues to rise, and the continuous improvement effect is better, which is better than the traditional statistical method. Therefore, the blockchain-based energy metering technology suitable for the optimization of energy metering.

Keywords: energy metering · Alliance chain · Blockchain

1 Introduction

With the emergence of blockchain, alliance chain and other technologies, the amount and frequency of energy data in energy has soared, which has also increased the error rate of energy calculation. At present, energy metering mainly adopts the compression method of energy data Although this method can perform energy calculation well, the continuous improvement method in the later stage is weak, and it is impossible to comprehensively analyze energy measurement. Some scholars believe that blockchain technology can realize the energy calculation of the entire region by calculating the energy characteristics of each region, and can grasp the degree of energy calculation more comprehensively [1]. However, this method is also controversial and lacks relevant practice to support it. On this basis, this paper combines the blockchain energy measurement alliance chain technology to comprehensively analyze energy measurement and judge its continuous improvement.

J. C. Hung et al. (Eds.): FC 2023, LNEE 1132, pp. 412–419, 2024.
https://doi.org/10.1007/978-981-99-9538-7_64

2 Description of Blockchain

Blockchain technology realizes the comprehensive use of energy measurement indicators through the characteristic analysis of energy measurement. Among them, the energy measurement index is combined with the energy energy data, the relevant energy data is mapped into the two-dimensional plane, the redundant attributes in the energy data are removed, and the energy data is carried out Value comparison. Blockchain uses the characteristic relationship of different blocks to obtain a relatively comprehensive energy measurement analysis result [2]. According to its own characteristics, each blockchain deeply excavates the energy quality, finds the problems in energy measurement, and the causes of related problems, and better improves energy measurement. In the improvement of energy measurement, the energy improvement technology adjustment is carried out by combining different energy factors to achieve the best energy measurement. The blockchain energy metering process is shown in Fig. 1.

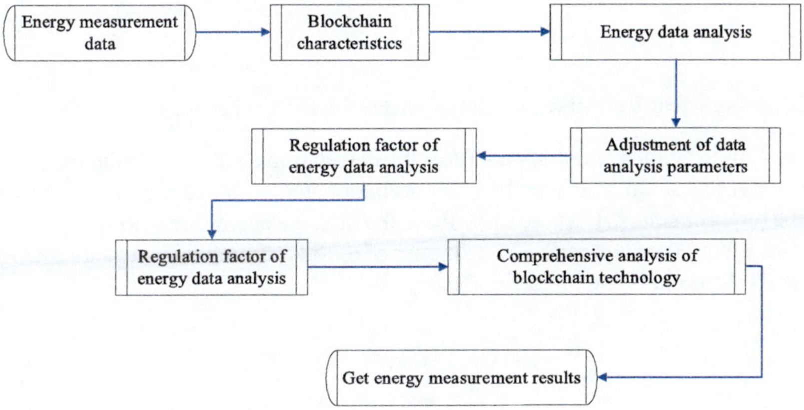

Fig. 1. Energy metering process

It Fig. 1 that the energy measurement method of the blockchain is to collect energy data first, comprehensively analyze the structure and type of energy data, extract the feature values of the blockchain, and make comprehensive judgments on the feature values. According to the comprehensive judgment results, the energy measurement indicators of the blockchain are determined, and the energy energy data in the blockchain is analyzed under the guidance of the energy measurement alliance chain technology [3]. In the process of analysis, the adjustment of energy metering energy data and the setting of relevant parameters are continuously carried out, and finally the continuous improvement results of the energy metering alliance chain technology are obtained. The results guide energy metering and support related metering schemes.

3 Mathematical Description of the Energy Metrology Alliance Chain Technology

3.1 Mathematical Definition of Blockchain

The key points in energy computing show complexity and diversity, and the integration of blockchain technology can realize the integration of energy energy data and improve energy measurement capabilities [4].

Hypothesis 1: The energy data is n i and the energy data set is ID = {x i,y i,zi},x i is the energy data structure (structured = 1, unstructured = 2, semi-structured = 1), yi for consortium chain technology (simple union = 1, misplaced union = 2, function union = 3), di is the energy metering result (full metering = 1, partial metering = 0), the energy metering result is shown in Eq. (1).

$$\mathrm{N} = d_i = \sum_{ni}^{\infty} x_{ij} \rightarrow y_{ij} + \sum_{i}^{\infty} ID_i \tag{1}$$

3.2 Mathematical Definition of Consortium Chain Technology

The consortium chain technology analyzes the correlation coefficient under the condition of the preset threshold, and sets the relevant parameters to make the final output result meet the requirements [5]. At the same time, the alliance chain technology can eliminate irrelevant energy data and reduce the amount of pre-processing of energy measurement, the specific formula is shown in Eq. (2).

$$d_i = \sum_{i=0}^{n} x_i \leftrightarrow y_i \tag{2}$$

Since energy metering is fuzzy, the results can be simplified by approximation.

3.3 Blockchain-Based Consortium Chain Technology

The calculation method of the previous alliance chain technology is too complex for energy calculation, and the system resource occupancy rate is high. This paper adjusts the block chain technology to obtain specific calculations such as Eq. (3).

$$F(x, y, n) = F(x_i, y_i, n) + \xi \tag{3}$$

Among them, the frequency in the ξ entire block chain is measured for energy, and the projection of the adjustment function is used to achieve the fusion of the two methods.

3.4 Fitting of Energy Measurement

The integration of blockchain and consortium blockchain technology is not only affected by energy, but also by the fitted value.

If the degree of fitting of the energy calculation 2: is P, then the energy calculation values of the different blocks are fitted as in Eq. (4).

$$P{=}\lambda\prod[\sum_{1,1}^{\infty}f(x_{11},y_{11})+\varsigma]+\kappa\sum_{1}^{\infty}ID_1 \tag{4}$$

Among them, it is the blockchain fitting value κ and the alliance chain technology fitting value λ.

3.5 Calculation Steps of Blockchain and Consortium Chain Technology

According mathematical description of energy measurement, continuous improvement analysis is carried out to obtain the energy measurement results of energy data:

Set up the initial energy data node N, carry out blockchain technology and chain metering methods for energy data, and test and analyze the results.

Combine blockchain with alliance chain technology to perform gradient improvement calculation on processed energy data;

Test the results of improved calculations and include them in the energy pool in accordance with the requirements;

Test whether the energy data is fully traversed;

Output energy calculation results.

4 Case Study of Blockchain and Energy Metering Alliance Chain Technology

4.1 Energy Data Elaboration

Through the hybrid metering channel, with the help of optical fiber and Wifi, the energy metering data is analyzed to test the comprehensive computing power of energy. In the calculation process, the calculated energy data is tested by manual intervention, error value, etc., and the actual calculation results are tested.

4.2 Calculate Fit Analysis of Energy Data

Applying energy metering data with blockchain technology and alliance chain technology, the results are as follows:

As shown in Fig. 2, the fit between good, and after several improvements, it is basically close to 90% fit, indicating that blockchain technology and alliance chain technology can calculate energy well, does not reduce the comprehensive computing power of energy data. At the same time, the degree of fitting of different blocks is basically the same, indicating that the energy data fitting processing is better.

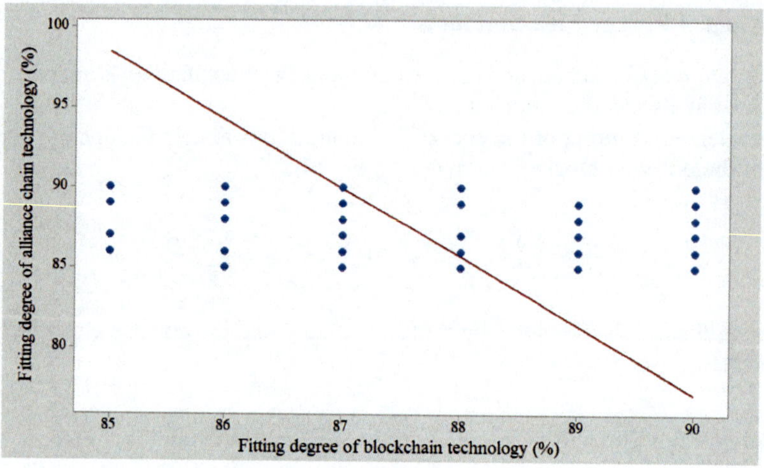

Fig. 2. Integration of blockchain technology

4.3 Accuracy of Energy Measurement

Energy metering is improved according to energy classification, data complexity, and the accuracy of energy measurement is calculated, and the results are as follows:

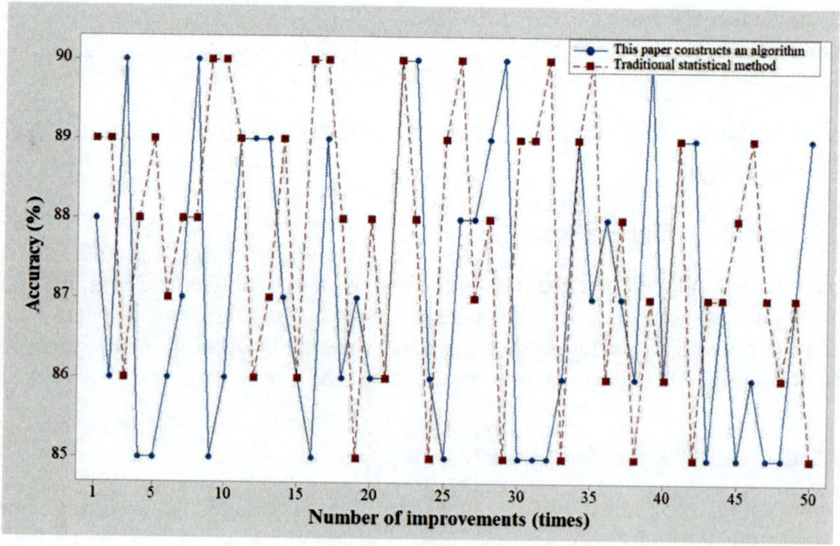

Fig. 3. Energy metering accuracy

It can be seen from Fig. 3 that the accuracy of blockchain and alliance chain technology is close to 85% after several improvements, indicating that this aspect has good accuracy. Traditional statistical methods are less accurate and vary widely. The energy measurement process of blockchain and alliance chain technology is relatively stable, indicating that the energy measurement detection of blockchain and alliance chain technology is better, which is in line with energy data energy measurement Request.

4.4 Calculation Time of Energy Metering

Due to the large amount of energy data, it is necessary to test the calculation time to ensure the validity of energy measurement, and the in Fig. 4.

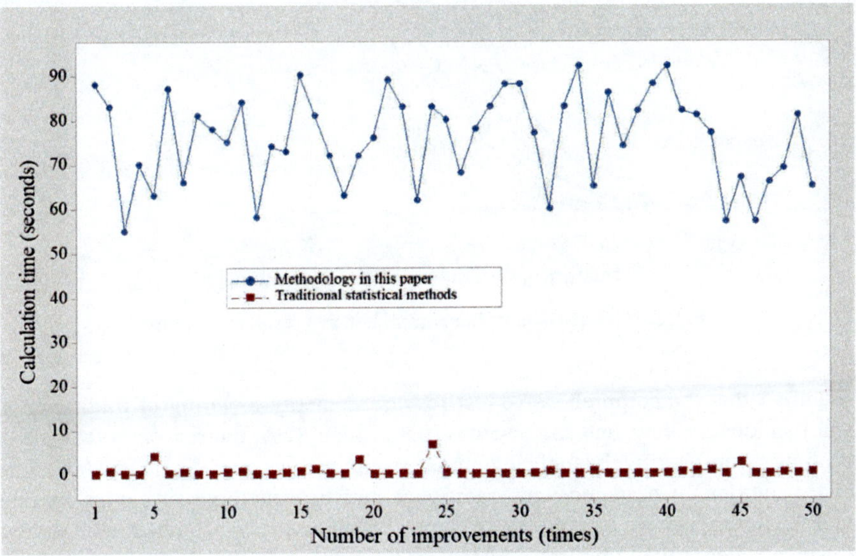

Fig. 4. Energy metering time for different

It seen from Fig. 4 the blockchain and alliance chain technology have been improved for a long time, mainly due to the need for alliance chain rule analysis of energy computing energy data in the early stage. Blockchain and alliance chain technology is shorter than traditional statistical methods in energy measurement in many times of improvement, which is due to the effective simplification of blockchain technology and alliance chain technology Energy metering energy data, shortening the calculation time of energy data. Energy computing has a large amount of energy data, so the blockchain and alliance chain technology proposed in this paper is suitable for energy data processing.

4.5 Comparison of Comprehensive Results of Energy Metering

The test result is an important part of energy measurement, the purpose of which is to reduce redundant data, duplicate data occurrence rate, reduce the number of energy measurement, and test the energy measurement of two energy data Results, the results the figure below:

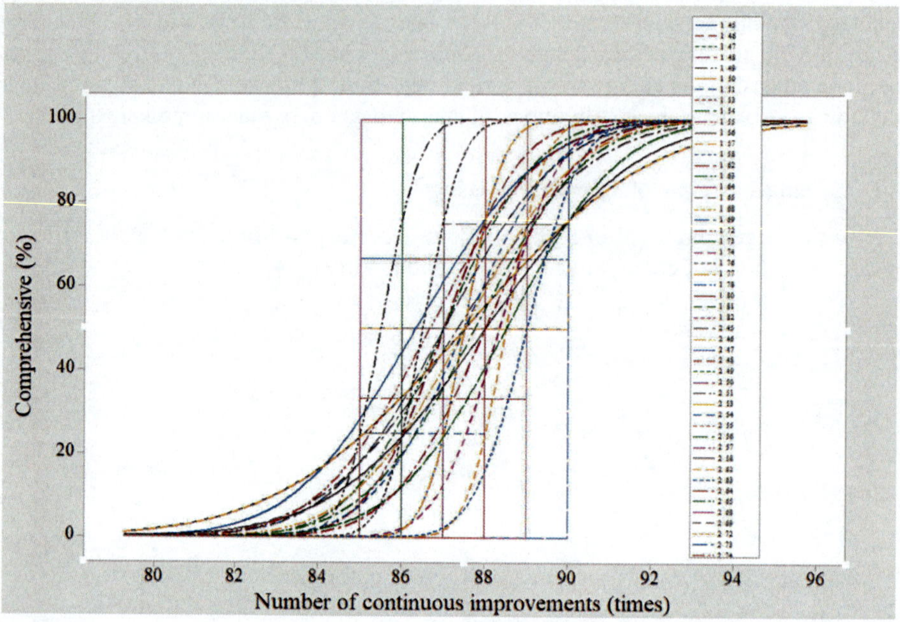

Fig. 5. Comparison of the comprehensive results methods

Figure 5 that the comprehensive test results of energy metering of traditional statistical methods are low, and the paper is higher than 80%, the reason is that the two algorithms comprehensively analyze human error and other operations, and take timely response solutions. Compared with traditional statistical methods of energy measurement, the comprehensive change range of blockchain and alliance chain technology is less, which is better than traditional statistical methods.

5 Conclusion

The emergence of a large number of invasive operations such as Trojans and noise has made traditional statistical methods of energy metering easy to crack, threatening the security of energy measurement. This paper proposes a blockchain and consortium chain technology to accurately process the blockchain and alliance chain technology and improve the comprehensive computing power of energy data. The MATLAB simulation results show that compared with the energy measurement degree of traditional statistical methods, the fit between blockchain and consortium chain technology is better, and the degree of energy measurement is more than 98%, indicating that blockchain and The blockchain and alliance chain technology of alliance chain technology are better. At the same time, energy calculation energy measurement is slightly inferior to traditional statistical method energy measurement in small amount of energy data processing, but it is better than traditional statistical method energy measurement in large amount of energy data processing, and the government energy data is computationally intensive, so the energy data is suitable for energy data processing. There are still certain deficiencies in the research of this paper, mainly because there are fewer research data on energy measurement, resulting in less energy data, and more energy data will be collected for analysis in the future.

Acknowledgements. Research and Development Project "Research on key technologies for power energy measurement and quality infrastructure" China Electric Power Research Institute (524200220001).

References

1. Chen, C., Geng, L., Zhou, S.: Retraction note: design and implementation of bank CRM system based on decision tree algorithm. Neural Comput. Appl. **35**(6), 4803 (2022)
2. Mao, J.: Design and Implementation of Tax Collection and Management Index Early Warning System Based on Data Mining (2021)
3. Han, J.: System optimization of talent life cycle management platform based on decision tree model. J. Math. **2022**, 1–12 (2022)
4. Chen, X., Zhuang, Y., Lao, Y., et al.: Development and implementation of a novel decision support tool on physical restraint use in critically ill adult patients. Int. J. Nurs. Pract. (2021)
5. Varaprasada Rao, K., Panda, S.K.: Secure Electronic Voting (E-voting) System Based on Blockchain on Various Platforms (2023)

Accurate Investment Evaluation Model of Power Grid Based on Improved Fuzzy Neural Inference

Kunpeng Liu[1](✉), Lihua Gong[1], Nuo Tian[2], Bo Liu[1], and Lili Liu[3]

[1] State Grid Customer Service Center, Dongli 300309, Tianjin, China
lllshr@126.com
[2] State Grid Customer Service Center, Nanjing 211161, Jiangsu, China
tian_virg@sina.com
[3] Beijing Data Ocean Technology Co. Ltd., Haidia, Beijing 100195, China
liulili@dataocean.com.cn

Abstract. The role of in construction is very obvious, but there is a problem that the investment accuracy is not high. Previous audit investment methods could not solve the problem of accurate investment, and the evaluation ability projects was low. Therefore, this to improve the fuzzy neural reasoning method and construct an evaluation model for projects. Firstly, the fuzzy theory is used to plan the data, and the evaluation and collection division according to the project funds are used to reduce the uncertainty factors of investment analysis. Then, fuzzy theory will form the power grid investment planning, form an investment project evaluation set, and evaluate the data in the set for inference evaluation. MATLAB simulation shows that under the condition of a certain scale of investment projects, the evaluation accuracy and evaluation time of the improved fuzzy neural reasoning method are better than the previous audit evaluation methods.

Keywords: fuzzy theory · grid investment · investment evaluation · Investment projects

1 Introduction

Investment project evaluation is one of the important evaluation contents of investment and construction. However, in the actual investment management process, there is a problem of poor accuracy of investment projects, which brings certain economic losses to power enterprise [1]. Some scholars believe that the application of intelligent algorithms to the accurate can effectively analyze the risks of investment projects and provide corresponding support for investors [2]. On this improved algorithm for fuzzy neural reasoning, evaluates, and verifies the effectiveness of the model.

2 Related Concepts

2.1 Improve Mathematical Descriptions of Fuzzy Neural Reasoning

The fuzzy neural network inference algorithm uses fuzzy set theory the network, and according to the indicators, it deduces the risks faced by the power grid investment project, integrates the corresponding resources, and finally judges the feasibility of the

J. C. Hung et al. (Eds.): FC 2023, LNEE 1132, pp. 420–426, 2024.
https://doi.org/10.1007/978-981-99-9538-7_65

computer investment project. The improved fuzzy neural network combines the advantages of fuzzy theory and uses the method of neural network inference to quantify, which can provide support for the improvement of accuracy [3].

Assumption 1: The scheme project is x_i, the scheme set project is $\sum x_i$, the evaluation index of investment project is y_i [4], and the evaluation scheme is $f(x_i)$ as shown in Formula (1).

$$f(x_i) = \sum x_i|y_i + \xi \tag{1}$$

The adjustment coefficient for projects is ξ mainly to reduce the influence of subjective factors and uncertain factors.

2.2 Selection of Investment Project Plans

Assumption 2: The evaluation function of investment project is $F(x_i)$ and the weight coefficient of investment project is w_i [5], then investment project scheme is shown in Formula (2):

$$F(x_i)=z_i \cdot f(x_i|y_i) + w_i \cdot \xi \tag{2}$$

2.3 Selection of Power Grid Investment Project Schemes

Before improving fuzzy neural reasoning, the investment project scheme should be analyzed discretely, and the scheme should be mapped to a two-dimensional plane to eliminate redundant data. First, the power grid investment plan is comprehensively analyzed, and the threshold and index weights of the scheme are set to ensure the accuracy of the power grid investment evaluation. The grid investment scheme is unstructured and needs to be standardized. If the grid investment plan is in a non-normal distribution, the evaluation results of its investment projects will be affected, reducing the accuracy of the overall assessment. In order to improve the accuracy evaluation and the level of investment project evaluation, the project plan should be selected, and the specific evaluation plan is shown in Fig. 1.

The data in Fig. 1 shows that scheme shows a discrete distribution, and the overall distribution form meets the requirements. The investment plan is not directional, and the wheel power grid investment scheme has strong randomness, so it is used as a later analysis and research. The investment scheme presents scattered characteristics, mainly fuzzy theory analyzes the investment scheme, removes duplicate and irrelevant data, and supplements the default scheme, so that the continuity of the data of the whole scheme is strong.

2.4 Strategies for Evaluating Power Grid Investment Projects

The fuzzy neural reasoning method adopts the heterogeneization strategy for power grid investment, and adjusts the corresponding parameters to realize the evaluation of distributed investment projects of power grid investment. The improved fuzzy neural

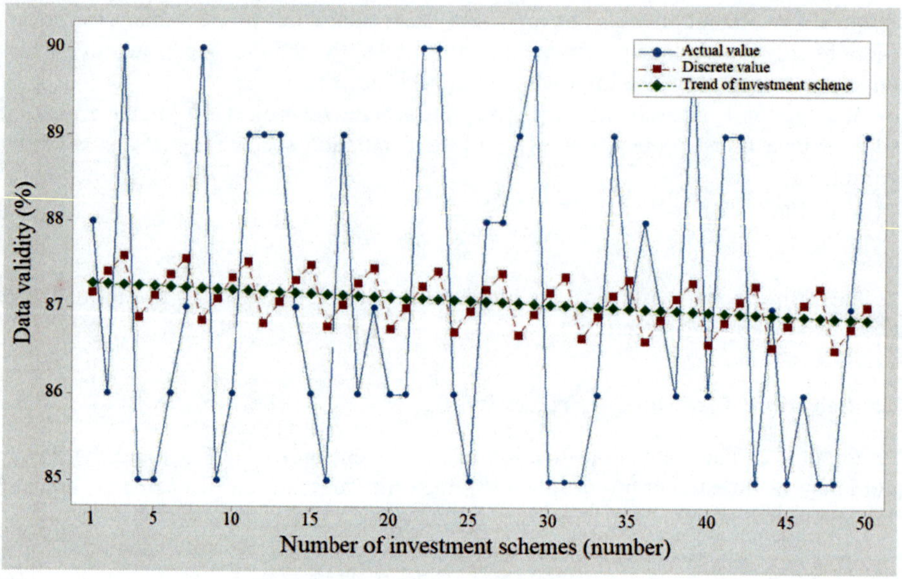

Fig. 1. Processing results of the plan

reasoning method divides grid investment into different types and adopts different investment project evaluation measurements. During the iteration process, different kinds of grid investments are obfuscated at the same time. After the obfuscation is completed, the investment project evaluation level of the scheme is compared, and the best grid investment plan is recorded.

3 The Case of Grid Investment

3.1 Introduction to Power Grid Investment Plan

In order to facilitate the evaluation, the scheme in is 65, the index is 12, and the test time is 12 days, and the specific evaluation scheme is shown in Table 1.

The investment scheme in Table 1 is shown in Fig. 2.

As from Table 1, the evaluation results of the improved fuzzy neural reasoning method are closer to the actual evaluation scheme compared with previous audit methods. In terms of evaluating the rationality and fluctuation range of scheme selection, the fuzzy neural reasoning method was improved in the previous audit method. From the curve changes in Fig. 4, it can be seen that the improved fuzzy neural reasoning method has better stability and faster judgment speed. Therefore, the investment project evaluation speed, power grid investment project evaluation, and summation stability of the fuzzy neural reasoning method are better.

Table 1. Relevant parameters of investment

Scale of investment projects	Short time	Number of investment options	Solution research
80 million–100 million	prophase	62	Questionnaires, interviews
	anaphase	71	Questionnaires, interviews
100 million–500 million	prophase	56	Questionnaires, interviews
	anaphase	76	Questionnaires, interviews
500 million–1 billion	prophase	53	Questionnaires, interviews
	anaphase	60	Questionnaires, interviews

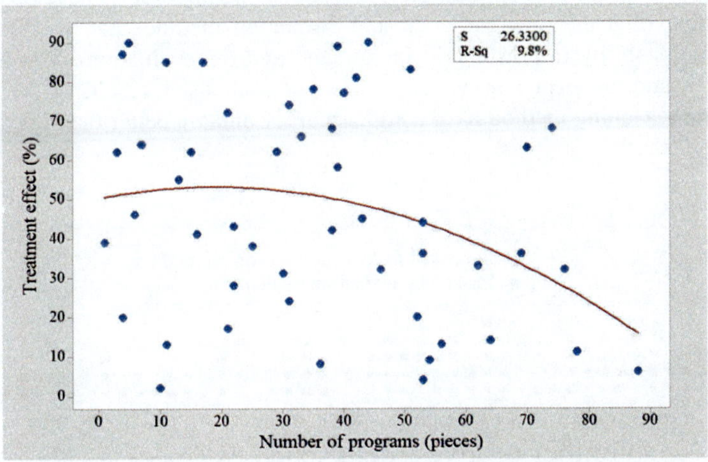

Fig. 2. The processing process of the power grid investment plan

3.2 Power Grid Investment

The project evaluation includes non-structural information, semi-structural information, and structural information. After the pre-selection of the reasoning method, the preliminary investment project plan are obtained Analysis of the feasibility of the investment project plan. In order to more accurately verify effect, select power grid investment in different time periods, which are: 80 million– 100 million and 500 million–1 billion, the data evaluation scheme, as shown in Table 2.

Table 2. Overall situation of grid investment plans

Time period	Scenario adjustment rate	Feasibility
80 million–100 million	94.21	90.63
100 million–500 million	90.00	90.16
500 million–1 billion	95.79	98.89
Mean	92.00 ± 4.21	94.74 ± 3.42
X2	84.212	87.377
P = 0. 031		

3.3 Accuracy and Stability of Investment Solutions

In order to verify the accuracy of the improved fuzzy neural reasoning method, the scheme is compared with the previous audit evaluation method, Fig. 3 that the accuracy of the improved fuzzy neural reasoning the previous audit evaluation method, but the error rate is lower, indicating that the evaluation of the improved fuzzy neural reasoning method is relatively stable Previous audit assessment methodologies have been uneven. Table 3 shows that the previous audit and evaluation methods had shortcomings in accuracy and stability in power grid investment, and the evaluation scheme changed significantly, and the error rate was high. The comprehensive results of the improved fuzzy neural reasoning method have a high accuracy and are better than previous audit evaluation methods.

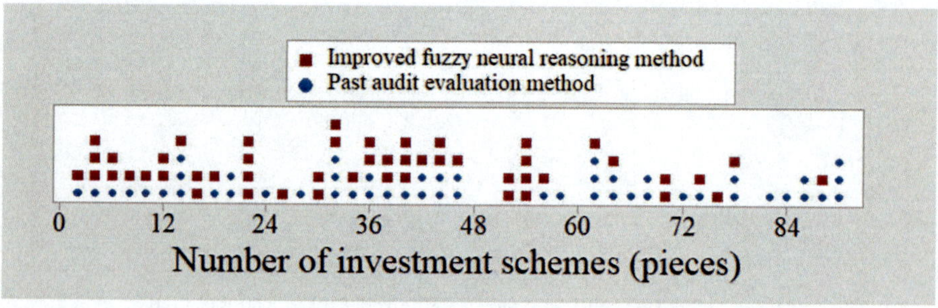

Fig. 3. The accuracy of different algorithms

The average scheme of the above three algorithms is shown in Table 3.

At the same time, the accuracy of the improved fuzzy neural reasoning method is greater than 90%, and the accuracy has not changed significantly. In of improving the fuzzy neural reasoning method. In the effectiveness, different methods are used to comprehensively analyze the evaluation, and the result 4 is shown.

Table 3. Comparison of investment project evaluation accuracy in the time period

Algorithm	Precision	stability	Error rate
Improved fuzzy neural reasoning methods	92.11	95.79	4.74
Previous audit evaluation methods	70.25	85.26	6.21
P	0. 012	0.021	0.023

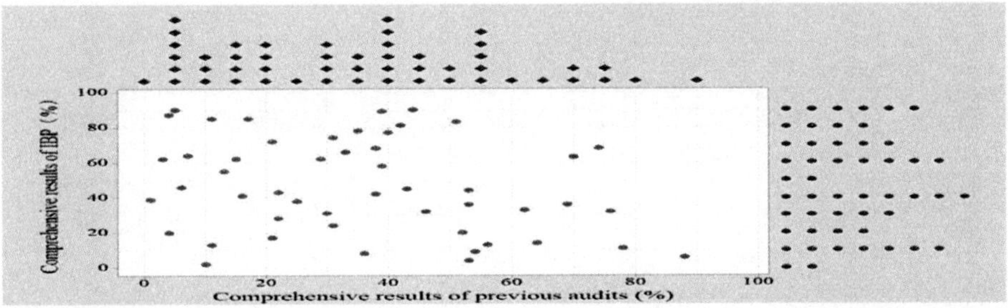

Fig. 4. Integrated programme for the evaluation of investment projects with different methodologies

4 Conclusion

In this paper, an reasoning method is proposed, and combined with fuzzy theory, the power grid investment is improved. At the departments and threshold standards analyzed in depth, and the planning and evaluation collection is constructed. This study shows that improving the fuzzy neural reasoning method can improve the accuracy and stability, and can evaluate the comprehensive investment project. However, in the process of investment evaluation, too much attention is paid to the project capabilities, resulting in a relative decline in the analysis between indicators.

Acknowledgements. Design and development project of State Grid Customer Service Center (629931190055).

References

1. Yang, J., Sun, W., Ma, M., et al.: Evaluation of operation state of power grid based on random matrix theory and qualitative trend analysis. Energies **16**(6), 2855 (2023)
2. Shi, H.T., Su, G., Pan, J., et al.: A novel microgrid power quality assessment model based on multivariate Gaussian distribution and local sensitivity analysis. IET Power Electron. **16**(1), 145–156 (2023)
3. Peng, X., Yang, H.: Impedance-based stability criterion for the stable evaluation of grid-connected inverter systems with distributed parameter lines. CSEE J. Power Energy Syst. **9**(1), 145–157 (2023)

4. Zhang, C., Li, B., Wang, L., et al.: A hierarchical model for quality evaluation of mixed source software based on ISO/IEC 25010. Int. J. Softw. Eng. Knowl. Eng. **33**(02), 181–205 (2023)
5. Shi, R.: Power grid structure performance evaluation based on complex network cascade failure analysis. Energies **16**(2), 990 (2023)

Research on the Application of Computer Technology in Music Teaching in Colleges and Universities

Hou Lei(✉)

Xianyang Normal University, Xianyang 712000, Shanxi, China
252504410@qq.com

Abstract. The role of computer very important, but there is a problem with the low note-matching rate. Traditional music teaching methods cannot solve the problem of note-matching rate in music teaching, and the matching degree is less. Therefore, this paper proposes computer technology to construct a matching model for information and music teaching. Firstly, the music score knowledge is classify the musical notes, and the set is divided according to the score standard to realize the quantification of music information Processing. Then, the notation knowledge classifies the note match rate, forms a music match set, and iteratively analyzes the matches. MATLAB simulation shows that under the condition of certain teaching standards, the accuracy of music score analysis and note-matching rate of computer technology is better than traditional music teaching methods.

Keywords: music score knowledge · note matching rate · computer technology · Creation

1 Introduction

Note matching rate is a vital evaluation content in music teaching, which is significant for music creation [1]. However, in the creative process, the creative results have the problem of a poor note-matching rate, which has a particular impact on music. Some scholars believe that applying intelligent algorithms to music teaching can effectively analyze melody and tone data and provide corresponding support for creation [2]. On this basis, this paper proposes computer technology to optimize the note-matching rate of college music teaching and verify the model's effectiveness [3].

2 Related Concepts

2.1 Mathematical Description of Computer Technology

Computer technology uses the knowledge of music scores to optimize the matching, and according to the multi-dimensional indicators in music teaching, finds outliers in music creation, integrates the matching[4], and finally judges the feasibility of the creative results. Computer technology combines the knowledge of musical notation, uses information mining and intelligent algorithms, optimizes the results of the note-matching rate, and can improve the intelligent note-matching rate.

J. C. Hung et al. (Eds.): FC 2023, LNEE 1132, pp. 427–432, 2024.
https://doi.org/10.1007/978-981-99-9538-7_66

Hypothesis 1: The music teaching content is *set* $\sum x_i$, the matching set is y_i, the teaching standard is x_i, and the judicial function $f(x_i)$ of the note matching rate is as shown in Eq. (1).

$$f(x_i) = \underline{\underline{\sum x_i|y_i \cup \widehat{\xi}}} \tag{1}$$

ξ Adjust the coefficient for the note match rate result, reducing the effect of ambiguity and tone.

2.2 Selection of Music Teaching Programs

Hypothesis 2: The matching selection function is $F(x_i)$, and the matching weight coefficient is q_i, then the music teaching scheme selection is shown in Eq. (2).

$$F(x_i)=\sqrt{f(x_i|y_i)} \Rightarrow \prod q_i \cdot \xi \tag{2}$$

2.3 Processing of Music Teaching Data

Before the computer technical analysis, the matching in the authoring results should be analyzed by standard, and the matching should be mapped to the authoring table to determine the abnormal semantic match. First, the matching is comprehensively analyzed, and the matching threshold and weight are set to ensure the accuracy of computer technology. Matching is a semantic transformation match and needs to be quantified [5]. If the match is in a nonstandard distribution, its authoring results are affected, reducing the accuracy of the overall authoring. In improve the accuracy of computer technology and note-matching rate, the computer technical analysis be selected.

3 The Depth of the Note Match Rate

Computer technology adopts accurate judgment of note matching rate and adjusts the corresponding melodic relationship to optimize the method in colleges. Computer technology divides creation into different amounts of data and randomly selects different methods. In the iterative process, the music score standards with different data amounts are matched with the music teaching scheme. After the matching processing is completed, different methods for college music teaching are compared, and the note-matching rate results with the highest accuracy are recorded.

4 Practical Cases of Music Teaching in Colleges

4.1 Music Creation

In order to facilitate financial data analysis, the melody data in this paper is the research object, the test data is 1G, and the note-matching rate of specific physical education is shown in Table 1.

Table 1. Characteristic of music teaching data

data type	Beat	Amount of data	Dispersion	threshold
Structured data	¼	500M	0.85	0.6
	½	1G	0.65	0.6
Unstructured data	¼	500M	0.75	0.6
	½	1G	0.69	0.6
Semi-structured data	¼	500M	0.75	0.6
	½	1G	0.92	0.6

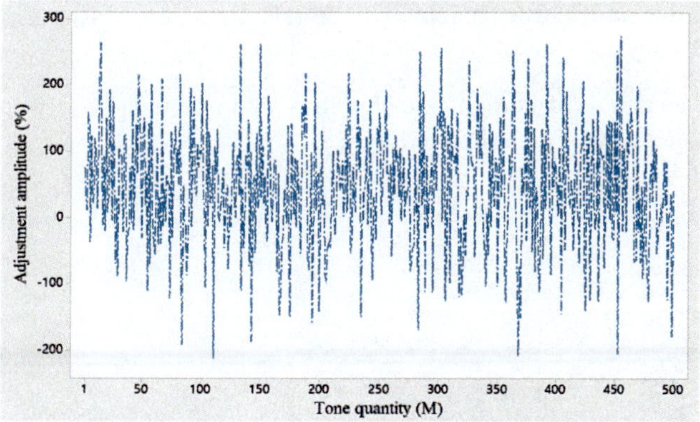

Fig. 1. Tone processing

The tone processing of the note-matching rate in Table 1 is shown in Fig. 2.

As seen from Fig. 1, with traditional music teaching, the creative results of computer technology are note-matching rate. In music creation, tone creation rate, accuracy, etc., computer technology is a traditional music teaching method. The pitch change in Fig. 4 shows that computer technology is more accurate and faster to judge. Therefore, the speed of note matching rate, note matching rate, and score analysis accuracy of computer technology is better.

4.2 Outlier Recognition Rate of Information Matching Music Teaching

The information matches music instruction with notes, tones, and bars. After the standard threshold screening of computer technology], the preliminary matching results of information and music teaching were obtained, and matching results of information and music teaching were obtained. The feasibility is analyzed. In order to verify the effect more accurately, select the information with different data amounts to match the music teaching and the tone that matches the information with the music teaching to evaluate the tone, as shown in Table 2.

Table 2. Overall picture of tones

Amount of data	Teaching standards	Outlier recognition rate
melody	100	82.14
tone	32	60.71
Innovation rate	83	84.64
mean	94	98.21
X^2	4.212	7.337
P = 0.016		

4.3 Information and Music Teaching in Music Creation Match Time and Accuracy

To verify the accuracy of computer technology, tone comparisons were made with traditional music pedagogy, as shown in Fig. 2.

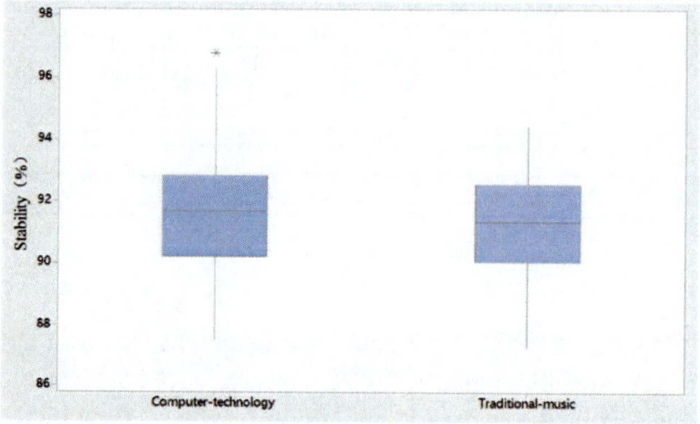

Fig. 2. Processing stability of computer technology

It seen from Fig. 3 that the processing of computer technology is more than methods, but the error rate is lower, indicating that the creation of computer technology is relatively stable. In contrast, the accuracy of music score analysis of traditional music teaching methods is uneven. The accuracy of the above algorithm is shown in Table 3.

Table 3 shows that the traditional music teaching method has shortcomings in the matching degree of music creation, the matching time and accuracy of information and music teaching, and the accuracy of matching information and music teaching has changed significantly, and the error rate is high. The information of the comprehensive results of computer technology has a higher matching time with music teaching, which is music teaching methods. At the matching time between computer technology information and music teaching is greater than 90%, and the accuracy has not changed significantly. In order to further verify the superiority of computer technology. In further

Table 3. Comparison of score analysis accuracy of different methods

algorithm	The information matches the time of music instruction	Magnitude of change	error
Computer	95.26	21.05	10.53
Traditional music pedagogy	88.42	20.00	13.68
P	5.26	5.26	20.00

the continuity, a comprehensive analysis of computer technology was carried out using different methods, as shown in Fig. 3.

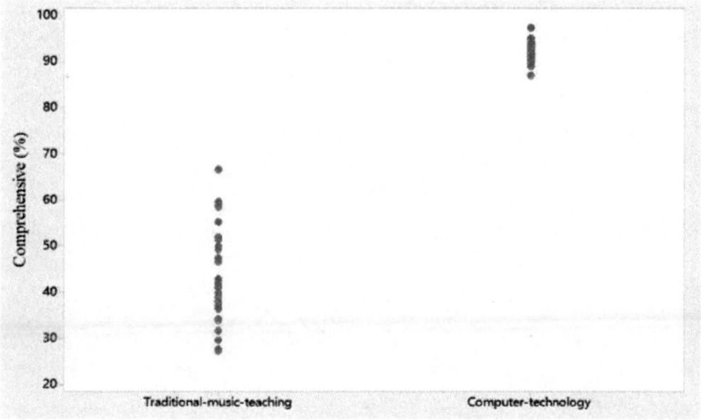

Fig. 3. Comprehensive evaluation results of computer technology

It seen from Fig. 3 the tone of computer technology is better than that of traditional music teaching method, and the reason is that computer technology increases the matching adjustment coefficient of information and music teaching, and sets the corresponding threshold to determine the tone that does.

5 Conclusion

Because of the development of computers, this paper proposes computer technology because of the problem of music teaching. It combines the knowledge of music scores to improve the matching of information. At the departments and threshold standards matching information and music teaching are analyzed in depth, and the tone set matching information and music teaching is constructed. Research shows that computer technology can improve the accuracy and accuracy of information and music teaching matching and comprehensively match information and music teaching. However, in computer technology, too much attention is paid to the ability to analyze unilateral indicators, and the proportion of matching information with music teaching is ignored.

References

1. Wang, Y., Guo, X.: Analysis of the impact of cloud computing on college teaching resources. Exp. Technol. Manag. **5**, 111–113 (2010)
2. Zhu, C., Wu, H.: Research on the construction and application of university teaching resources in the cloud computing environment. China Educ. Inf. **11**, 51–52 (2015)
3. Deng, Q., Li, J.: Enlightenment of the successful case of "cloud service" on China's educational technology. Educ. Teach. (4) (2009)
4. Gu, L.: Cloud computing: future network trend technology. Comput. Knowl. Technol. (2) (2008)
5. Xu, H.: Xiangyun computing's impact on education. Comput. Teach. Educ. Inform. (4) (2009)

Research on Online Teaching Method of Garden Plant Configuration Based on Computer Technology

Guorui Li(✉)

Shaanxi Fashion Engineering University, Shaanxi 712046, China
lgr0928@163.com

Abstract. The role of computer technology in garden plant configuration is very important, but there is a problem of low plant allocation rate. Traditional teaching methods cannot solve the problem of plant configuration in gardens, and there is less rationality. Therefore, this paper proposes an online teaching method based on computer technology to construct a matching model of plant configuration and teaching. Firstly, the garden knowledge is used to classify the plant configuration, and the configuration collection is divided according to the teaching standards to realize the quantitative processing of the garden plant configuration. Then, garden knowledge classifies plant configurations, forms configuration collections, and iteratively analyzes the configurations. MATLAB simulation shows that under the condition of certain teaching standards, the configuration analysis accuracy and rationality of online teaching methods based on computer technology are better than those of traditional teaching methods.

Keywords: garden knowledge · plant configuration · computer technology · Online teaching methods

1 Introduction

Plant configuration is an important evaluation content of garden management and is of great significance for garden planning [1]. However, in the actual plant planning process, the plant planning results have the problem of low plant allocation rate, which has a certain impact on the garden plant allocation [2]. Some scholars believe that the application of intelligent algorithms to gardens can effectively analyze melodies and configuration parameters [3], and provide corresponding support for plant planning [4]. On this basis, this paper proposes an online teaching method based on computer technology to optimize the configuration of garden plants and verify the effectiveness of the model.

J. C. Hung et al. (Eds.): FC 2023, LNEE 1132, pp. 433–439, 2024.
https://doi.org/10.1007/978-981-99-9538-7_67

2 Related Concepts

2.1 Mathematical Description of Online Teaching Methods

The online teaching method is to use garden knowledge to optimize the configuration, and according to the multi-dimensional indicators of the garden, find outliers in plant planning, integrate the configuration, and finally judge Feasibility of plant planning results [5]. The online teaching method combines garden knowledge, uses plant configuration mining, intelligent algorithms, optimizes the plant configuration results, and can improve the intelligent plant configuration.

Hypothesis 1: The teaching content is *set* $\sum x_i$, the configuration set is x_i, the teaching standard is y_i, and the judgment function of the plant configuration rate is $f(x_i)$ as shown in Eq. (1).

$$f(x_i) = k \cdot \sum x_i | y_i \langle \xi \tag{1}$$

2.2 Selection of Configuration Scheme

Hypothesis 2: The configuration scheme selection function is q_i and the matching weight coefficient is $F(x_i)$, then the configuration scheme selection is shown in Eq. (2).

$$F(x_i) = \sqrt{\sqrt{z_i \cdot \sqrt{f(x_i|y_i)}} + \prod \sqrt{q_i} \cdot \xi} \tag{2}$$

2.3 Processing of Teaching Data

Before the analysis of the online teaching method, the configuration indicators in the plant planning results are analyzed and the configuration is recorded in the plant planning table, judge the unreasonable configuration. First, the configuration is comprehensively analyzed, and the thresholds and weights of the configuration are set to ensure the accuracy of online teaching methods based on computer technology. Plant configuration needs to be quantified and if the configuration is normally distributed, its plant planning results will be affected, reducing the accuracy of overall plant planning. The accuracy of online teaching methods and the level of plant allocation rate, the plant configuration scheme should be selected, and the protocol selection is shown in Fig. 1.

The configuration in Fig. 1 shows that the configuration scheme shows a diversified distribution and meets the teaching requirements. The configuration scheme is not directional, indicating that the online teaching method based on computer technology has strong accuracy. The configuration scheme conforms to the configuration requirements of garden plants, mainly to adjust the configuration scheme, eliminate duplicate schemes, and revise the configuration parameters, so that the whole configuration of plants is more planned.

After the configuration processing is completed, the garden plant configuration of different methods is compared, and the plant configuration results with the highest accuracy are recorded.

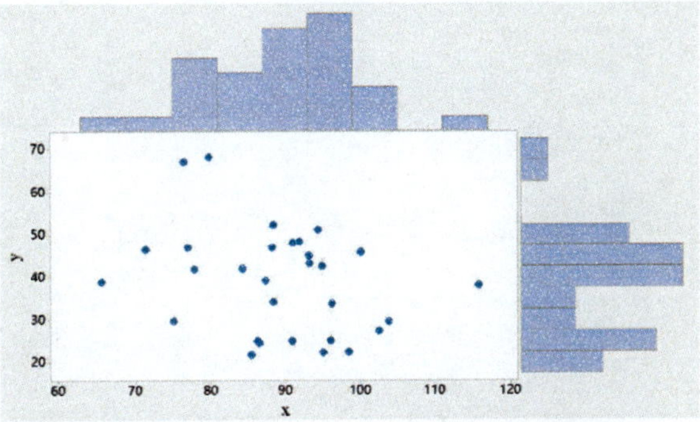

Fig. 1. Scheme selection of online teaching methods

3 Practical Cases of Garden Plant Configuration

3.1 Plant Configuration

In facilitate the configuration scheme, the protocol in this paper are used, and the plant configuration in Table 1.

Table 1. Characteristics of teaching data

Configure parameters	Configure the scale	Number of scenarios	Dispersion	threshold
tree	1/4	11.58	10.53	12.63
	1/2	12.63	12.63	8.42
vegetation	1/4	21.05	14.74	18.95
	1/2	16.84	5.26	17.89
Rocks	1/4	12.63	6.32	17.89
	1/2	6.32	5.26	5.26

The configuration parameter processing process for plant configuration in Table 1 is shown in Fig. 2.

Table 1 shows method, the plant configuration results of the online teaching method are closer to the actual plant configuration requirements. In terms of configuration rate and accuracy, online teaching methods are superior to traditional teaching methods. The change of configuration parameters in Fig. 4 shows that the accuracy of the online teaching method is better and the configuration speed is faster. Therefore, the plant configuration speed, rationality and configuration accuracy of the online teaching method are better.

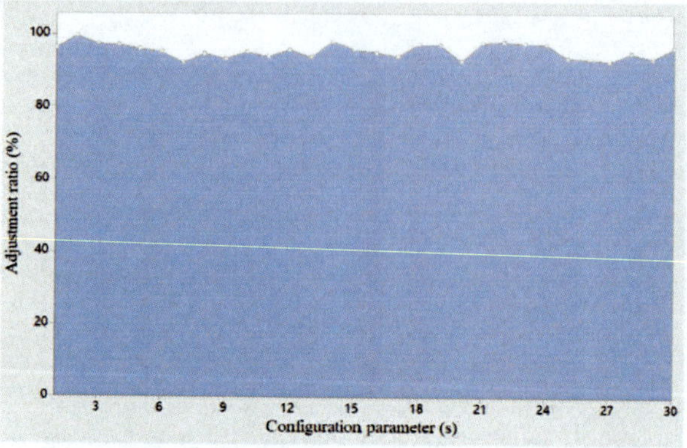

Fig. 2. Adjustment process of configuration parameters

3.2 Plant Configuration Outlier Identification Rate

After the threshold standard screening of the online teaching method [21], the preliminary plant configuration results were obtained, and the plant configuration was analyzed. To verify the effect more accurately, select different plant configuration schemes, as shown in Table 2.

Table 2. Overview of configuration parameters

parameter	Plant allocation rate	Outlier recognition rate
Tree	21.05	16.84
vegetation	13.68	13.68
Rocks	21.05	9.47
Mean	15.79	7.37
X^2	8.42	20.00
P = 0.036		

3.3 Stability and Accuracy of Plant Configuration

The online teaching method, the configuration traditional teaching method, as Fig. 3.

It seen from Fig. 3 that the rationality of plant configuration of online teaching methods is higher than that of traditional teaching methods, but the error rate is lower, indicating that the plant configuration of online teaching methods is compared Reasonable, while the rationality of the configuration of traditional pedagogy is uneven. The accuracy of the above algorithm is shown in Table 3.

Table 3 shows that the traditional teaching method has deficiencies in plant allocation time, accuracy, and plant allocation accuracy There are large changes and a high error

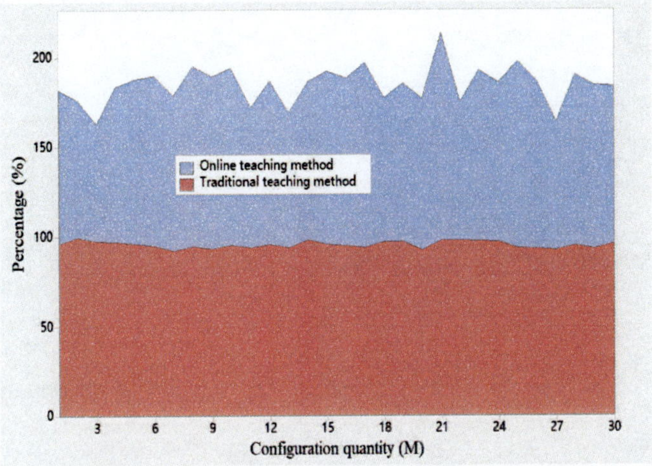

Fig. 3. Rationality of online teaching methods

Table 3. Comparison of configuration analysis accuracy of different methods

algorithm	Plant configuration time	Precision	Error
Online teaching methods	13.68	95.79	1. 09
Traditional pedagogy	43.68	94.74	7.89
P	14.74	5.26	11.58

rate. The online teaching method has a higher plant configuration time than the method. At the accuracy of plant configuration of the online teaching method is greater than 90%, and the accuracy has not changed significantly. To further verify the superiority of online teaching methods. A comprehensive analysis of the vegetation configuration was performed using different methods, as shown in Result 4.

Figure 4 that the configuration parameters of the online teaching are significantly better the teaching, and the online teaching method increases the plant configuration adjustment coefficient and sets it The corresponding thresholds are configured with configuration parameters.

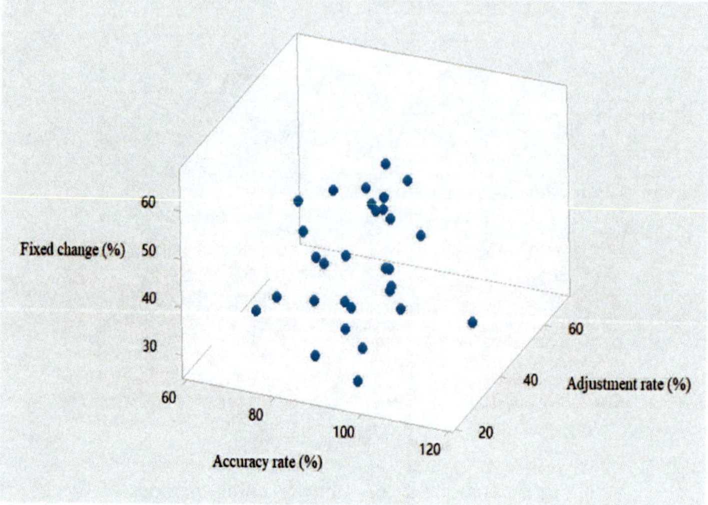

Fig. 4. Comprehensive evaluation results of online teaching methods

4 Conclusion

In the case of the rapid development of computers, this paper proposes an online teaching method based on computer technology and improves plant configuration based on garden knowledge. At the same time, the plant configuration standards are analyzed in depth to construct the plant configuration collection. Studies have shown that online teaching methods can improve the accuracy of plant configuration and optimize plant configuration. However, in the process of online teaching methods, too much attention is paid to plant analysis and the proportion of plant configuration is ignored.

References

1. Feng, X., Miao, W., Cao, P., et al.: Research on economic management experiment teaching management system based on computer. In: EAI International Conference, BigIoT-EDU. Springer, Cham (2023)
2. Cao, L.: Research on computer software teaching method of digital media art theory under the background of big data. In: Jan, M.A., Khan, F. (eds.) Application of Big Data, Blockchain, and Internet of Things for Education Informatization: Second EAI International Conference, BigIoT-EDU 2022, Virtual Event, July 29–31, 2022, Proceedings, Part I, pp. 614–619. Springer, Cham (2023). https://doi.org/10.1007/978-3-031-23950-2_67
3. Rao, C.: Research on teaching innovation of computer rendering course of environmental design specialty based on PTP mode. In: Jan, M.A., Khan, F. (eds.) Application of Big Data, Blockchain, and Internet of Things for Education Informatization: Second EAI International Conference, BigIoT-EDU 2022, Virtual Event, July 29–31, 2022, Proceedings, Part III, pp. 203–210. Springer, Cham (2023). https://doi.org/10.1007/978-3-031-23944-1_23

4. Liang, L.: Research on evaluation of physical education teaching quality based on comprehensive integrated evaluation method and SVM. In: Jan, M.A., Khan, F. (eds.) Application of Big Data, Blockchain, and Internet of Things for Education Informatization: Second EAI International Conference, BigIoT-EDU 2022, Virtual Event, July 29–31, 2022, Proceedings, Part III, pp. 64–75. Springer, Cham (2023). https://doi.org/10.1007/978-3-031-23944-1_8
5. Zhang, J.: Research on the "online and offline" operation education and teaching mode of cross-border e-commerce business based on the internet. In: EAI International Conference, BigIoT-EDU. Springer, Cham (2023)

Design and Implementation of English Listening Teaching System Based on Virtual Environment Technology David B. Lowe

Ying Nie(✉)

Foreign Language Department, Qiqihar Medical University, Qiqihar 161006, China
nyny0743@163.com

Abstract. With the continuous development of science and technology, a variety of advanced science and technology and equipment have been gradually started in education and teaching, which can improve the teaching efficiency and quality to a certain extent. This paper uses a virtual technology to build an English teaching system. The integration of virtual environment and foreign language courses can provide a new environment for foreign language education, make the content of foreign language teaching in an approximate way, improve the way of teaching and learning and the interactive mode of Teachers and students, and provide autonomy for foreign language learning. The learning environment of inquiry, inquiry and cooperation enables foreign language learning to be carried out in a relaxed and pleasant environment in a task driven manner.

Keywords: Virtual environment · English teaching · interaction between teachers and students

1 Introduction

Since 90s, some scholars have investigated the satisfaction of College English, and the results are not satisfactory either from a student's perspective or from a teacher's perspective. Zhu Luzi and so on (2004) survey found that "the number of students who are dissatisfied with college students' English teaching level has accounted for more than 80%". At the same time, Zhou Yan (2005) surveys 49 different types of teachers in six regions and found that 55% teachers think there are some problems in English Teaching in China, and 11% teachers think there is a lot of problems. Many problems [1]. The survey conducted after the university education reform has not changed much. In 2010, Jiang Weiwei and other students investigated the situation of English learning for college students [2]. The average satisfaction of college students on College English education was only 5.05 (full score of 10) [3]. For this result, the author combines the teaching resources and the teaching environment of higher vocational colleges as follows: (1) the teaching resources are not sufficient, especially in the teachers' resources. English classes in general higher vocational colleges are carried out in large classes [4]. The least number of classes are also around 50. Such class size makes little interaction between

J. C. Hung et al. (Eds.): FC 2023, LNEE 1132, pp. 440–445, 2024.
https://doi.org/10.1007/978-981-99-9538-7_68

teachers and students. In addition, teachers still use only multimedia, such as PPT, video player and so on, and the school is only equipped with voice room which can only transmit sound, which makes English teaching innovative in the use of information. (2) The teaching environment is single, especially the language teaching of teaching materials. The constant language of textbooks is difficult to meet the changing context nowadays.

2 A Review of the Application of Virtual Environment in Foreign Language Teaching

2.1 What Is a Virtual Environment

Virtual environment technology, also known as virtual reality, virtual world and simulation environment, is a new technology developed at the end of the twentieth Century. With the help of it, the real world can be simulated in the computer network, so that the users entering the world are as if they are in general, without doubt their authenticity. The development of virtual environment has gone through the stages of brewing, rapid development, and mature and extensive application. Each stage is separated by about 10 years. With the improvement of virtual system and the development of many virtual environment modeling languages, the fields of its application are becoming more and more extensive, such as education, space, industrial design, games and so on.

The three main characteristics of virtual environment technology are immersion, interaction and imagination. Immersion is a sense of presence, which means the user feels the authenticity of the main character in the simulated environment. Virtual reality can provide a real virtual environment for the user. In an ideal virtual environment, it is difficult for participants to identify the true and false environment; interaction refers to the user's real-time operation and operation of the objects in the virtual space. Feedback, in a virtual environment, participants will encounter a variety of objects that allow interaction between them, even the interaction of the participants and objects, such as the players can kick the ball in the virtual environment with their feet, feel the movement of the kick and the resistance of the kicking belt. Secondly, it also includes participants' interaction with other participants [5].

2.2 The Superiority of the Integration of the Virtual Environment and the Foreign Language Course

Technological innovation has changed the way people communicate and affects the way people understand and use language. The multi-user virtual environment provides multi-channel and multi-level language input using pictures, video, text and sound, all of which conform to the "comprehensibility" and "interest and relevance" in the Krashen's language input hypothesis. Network virtual environment is not only conducive to foreign language input, but also conducive to the output of foreign language learning. The native language users in the virtual world can provide better interactive objects and timely feedback for foreign language learners, including oral output, text output (such as poster, etc.), thus widening the students' output channels, increasing the output opportunities and making up for the shortage of the domestic foreign language learning environment.

In this study, the learners can not only feel the simulation environment that is different from the reality, but also can observe the problems, experience the practice process, solve the problems as far as possible, and finally complete the knowledge construction. The middle school students can acquire new knowledge and social and cultural knowledge in the whole process, which shows both the education and the reality.

2.3 Research Situation at Home and Abroad

According to the requirements of comprehensive practice and the principle of constructing virtual practice teaching environment, the system is divided into the user boundary, the 3D modeling module, the real-time simulation module and the four main parts of the database, as shown in Fig. 1.

The main sentence of the user interface is used 'to provide a friendly and friendly interactive window day, and can be used to operate the virtual environment easily without the simulation technology. A good virtual practice system, the user community should be progressive, the dead seven operator is clear, the operation simplification in the first place - this must be obvious. Sometimes there are too many helper keys - so they use obvious controls and buttons to get attention. Our user interface is mainly through the mouse, keyboard and other operations to achieve the combination of users and machines.

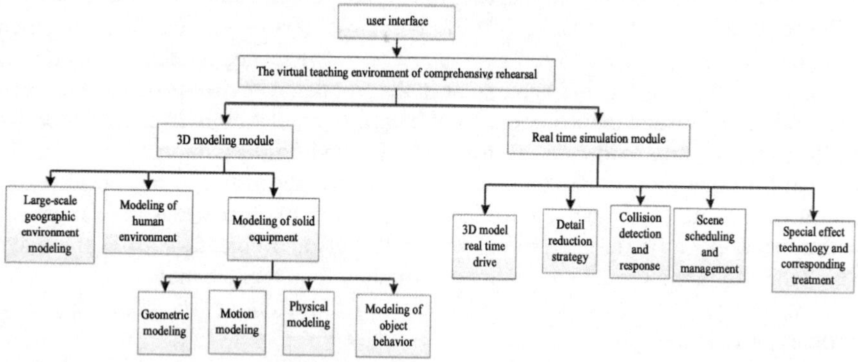

Fig. 1. Architecture of virtual teaching environment for comprehensive rehearsal

The 3D modeling module is mainly responsible for the modeling of the large-scale geographic environment, the humanistic environment and the entity equipment. All the models are stored in the database so as to facilitate the real-time simulation. The emphasis of 3D modeling is geo environment modeling and entity equipment modeling. The real-time simulation module is based on the completion of the modeling, through the required model to adjust, render, eliminate and other operations, and then add special climate effect and battlefield effect to complete the simulation, and set up human-computer interaction interface, the human computer interaction.

Case Analysis

The Unit4 My Neibourhood unit review class on the eight grade of the Hebei Education Edition (for example, Table 1), the whole unit is focused on a series of activities around the topic "block", as shown in Figs. 2 and 3.

Table 1. The Unit4 My Neibourhood unit review class on the eight grade of the Hebei Education Edition

course	goal
Lesson 19. The best neighbourhood	Mastering inverted sentences can be used in real life in real time
Lesson 20. No shopping	improve the utilization of key sentence patterns
Lesson 21. I like my neighbourhood	A complete place to introduce a place
Lesson 22. Eat a donut and turn right	Mastering the expression of the means of transportation
Lesson 23. People in my neighbourhood	A question and answer pattern when mastering the task of interview
Lesson 24. I need a map	Review the content of this unit

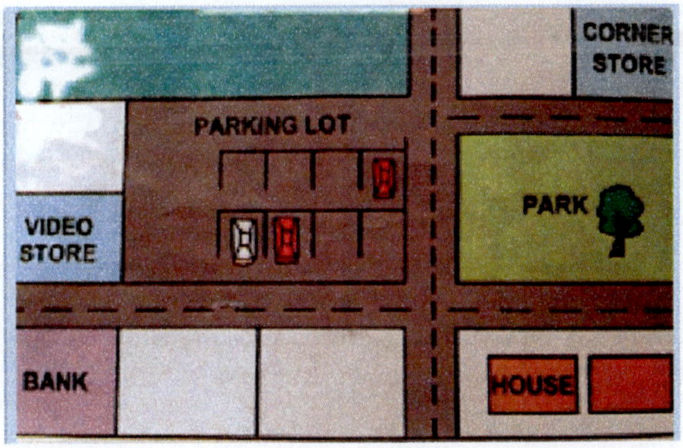

Fig. 2. Example

The purpose of teaching: to complete the designed activities in the virtual environment, to exercise the students' ability to do things in English, to deepen the students' knowledge of the unit, and to improve the students' practical ability.

The activity is carried out in the computer laboratory, and there are 6 tasks, 4 times each time, about 1 h each time. The first activities were tasks 1 and 2; second activities were 3 and 4; third activities were 5 tasks; fourth activities were carried out tasks 6. A total of 40 students participated in the activities, each of 8 people. Each led by teachers and invited Second Life native speakers. After the completion of the activities, the teachers' summary data were analyzed.

Fig. 3. Example

Activity task: each group of students followed the teachers or native speakers to visit the virtual blocks, and the activities lasted for about 30 min. During the tour, students are asked to observe the surrounding environment, listen carefully to the introduction of native speakers or teachers, including the structure of the blocks, the functions of the facilities in each block, and the career introduction of life. The task is to increase the students' understanding the block, to review and consolidate the knowledge of the blocks in the classroom, to master the key sentence patterns and vocabulary used for the introduction of things, and to lay the foundation for the following activities. The display task of learning achievement is carried out in the form of poster. The theme of the poster is My Neibourhood in Second Life. Students need to publish posters in the designated scene, introduce their own neighbourhoods, and report on the implementation of the tasks, results and interviews. Students need to write the corresponding composition, enter the virtual scene, use the poster publisher to type and display the composition, and the activity will take about 60 min.

Task analysis: this task is a roaming task (see Fig. 4). Students only need to follow the teacher to visit the virtual blocks without having to express them.

Fig. 4. Example

3 Conclusions

The integration of virtual environment technology and foreign language curriculum has created a new environment for foreign language education. This paper shows that the virtual environment, with its unique technical properties, can use the target language in an environment similar to the real way in front of students. In the virtual environment, the task driven model can organize foreign language teaching activities to create a positive atmosphere of foreign language communication. All the students who participate in the experiment are immersed in the activities of the virtual scene with curiosity and surprise, and actively carry out the thinking and expression, try to accomplish the activities successfully, and learn from the classroom. Knowledge has been practiced, consolidated and improved in subtle ways.

Acknowledgements. Qiqihar Medical University 2022 Education and Science Research project "Research on College English Teaching in Application-oriented Universities under the Theory of Multimodal Discourse Analysis" (QYJY20220212).

References

1. Aju, W.: Design and implementation of key modules of english teaching system based on J2EE. In: Jan, M.A., Khan, F. (eds.) Application of Big Data, Blockchain, and Internet of Things for Education Informatization: Second EAI International Conference, BigIoT-EDU 2022, Virtual Event, July 29–31, 2022, Proceedings, Part I, pp. 37–45. Springer, Cham (2023). https://doi.org/10.1007/978-3-031-23950-2_5
2. Yu, B.: Design and implementation of English teaching analysis system based on data mining. In: EAI International Conference, BigIoT-EDU. Springer, Cham (2023)
3. Liu, Y.: Design and implementation of teaching assessment system in higher vocational colleges based on association rule algorithm. In: EAI International Conference, BigIoT-EDU. Springer, Cham (2023)
4. Chong, N.: Design and implementation of English grammar error correction system based on deep learning. In: Jan, M.A., Khan, F. (eds.) Application of Big Data, Blockchain, and Internet of Things for Education Informatization: Second EAI International Conference, BigIoT-EDU 2022, Virtual Event, July 29–31, 2022, Proceedings, Part III, pp. 557–562. Springer, Cham (2023). https://doi.org/10.1007/978-3-031-23944-1_59
5. Fu,Q., Lai, J., Zhong, T., et al.: Design and implementation of clinical LIS360 laboratory management system based on AI technology. Int. J. Comput. Intell. Syst. **16**(1) (2023)

Research on Computer-Based Online Teaching Mode of English in Colleges and Universities

Jia Lu(✉)

Wuhan University of Engineering Science, Hubei, China
LuJia@wues.edu.com

Abstract. With the online teaching has become a new type of education, which is based on computer technology, communication, and is university English teaching and welcomed. At the some new media tools have gradually emerged in English higher education to assist students in learning and acquiring information. With the technology, the online teaching platform of college English has also ushered in a new round of development opportunities. Based on this, this paper tries to build a college English Internet online teaching system as a new development of English online teaching mode, so as to provide relevant help for it.

Keywords: Computer · English · online teaching

1 Introduction

In the cause China has developed rapidly, especially, which has made great achievements. With the continuous and application, the Internet essential, which provides learning resources for students and plays a great role in improving learning efficiency. As college English teaching gradually enters a new stage, the author proposes a new idea to solve these problems social and realize students' personalized education [1]. In this paper, we will study and analyze the English from the perspective construction and look into the future online teaching mode.

2 Relevant Technical Introduction

2.1 C# Language Introduction

C# language is a programming language tailored by Microsoft to implement the.NET framework. As one of the components of Visual Studio.NET, it is mainly used to develop various applications on the.NET platform [2]. C# is an object-oriented language based on C and C ++, inheriting many features of C/C ++ and making some improvements and extensions on the previous ones, which makes C# easier to use compared to the previous C/C ++. This makes C# easier to use than the previous C/C ++. For example, C# code does not have pointer operations and does not involve memory management, so it reduces the difficulty in development and debugging [3]; C# is object-oriented, so

J. C. Hung et al. (Eds.): FC 2023, LNEE 1132, pp. 446–451, 2024.
https://doi.org/10.1007/978-981-99-9538-7_69

some global functions, variables, and constants can be encapsulated in fixed classes, and some underlying components can be built quickly, unlike C ++, which requires a lot of underlying control operations, so it can respond to platform- and development-language-independent requests from the Internet across platforms. It can respond to platform- and language-independent requests from the Internet across platforms. NET framework, so it is much easier and faster to develop applications on the [4], the combination of these features makes C# an excellent language for web programming.

2.2 ADO.NET Technology

ADO. Manipulation for the.NET platform that enables applications to associate with database tables and then manipulate the data already stored in the database tables in a backend application in the form of writing SQL statements, etc. ADO.NET is highly integrated with the a unified interface that allows developers to directly call. To manipulate data. In the MVC pattern, ADO.NET technology is mainly used in the data access layer of the model [5]. As a set of object-oriented class libraries that can be used to interact with different types of data sources and databases, it contains the contents of Connection class, Command object, Data Reader class, Data Set object, Data Adapter class, and Data Table class. When ADO.NET interacts with data from different databases, it provides some common methods for interacting with different data sources, such as Execute Reader() method, Execute Scalar() method, and Execute Non Query() method, but also uses some specific libraries, which are collectively called Data Providers, like Sql Data Reader, Sql Data Adapter and other libraries when interacting with Sql Server database, if you are interacting with Oracle database, you have to use Oracle Data Reader, Oracle Command and other libraries, and of course MySQL database class library, etc. [6]., not to state here. However, no matter what kind of libraries developers use, there is one thing that must be kept in mind, and that is to use similar objects and data sources to interact.

2.3 Node.Js

It is used to develop side as well as the Server side, thus benefiting from code reusability and lack of context switching. Highlights of Node.js feature benefits are shown in Fig. 1 below.

One is the scalability of the architecture considering real-time response and ultra-large scale data requirements. Node.js, commonly referred to as Node, is typically used to develop applications that rely on running JavaScript on both the client and Server side [7].

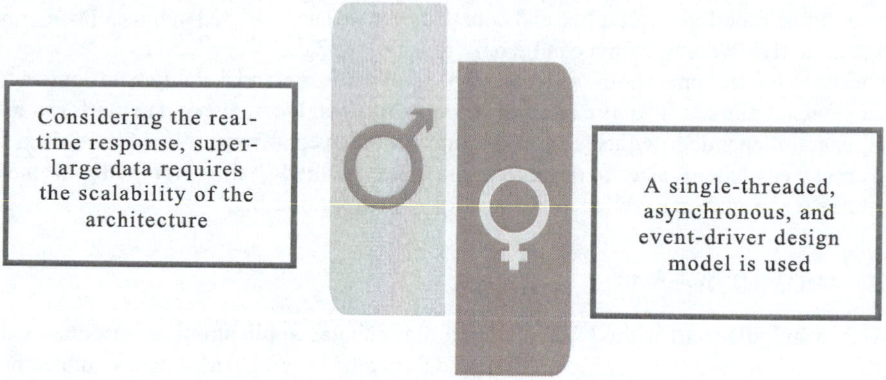

Fig. 1. Highlights of Node.js Feature Benefits

3 Design and Implementation of College English Internet Online Teaching System

3.1 Requirement Analysis

Requirement analysis of a system is the beginning of a software system design process. Only when the user's needs are clearly defined, analyze and seek solutions for software system design and a product. The most important task of the requirement analysis stage is to determine the functions of each part of the software, and the part are generated by the practice summary [8]. A college English Internet online teaching system should have the following main modules in terms of practical functional requirements analysis, and system setting module, as shown in Fig. 2.

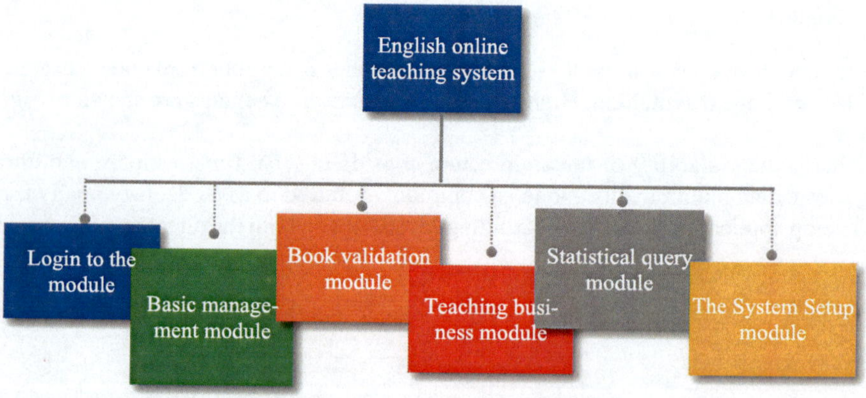

Fig. 2. Main modules of English Internet online teaching system in colleges and universities

3.2 Design of the Database Table

This paper adopts Microsoft SQL Server 2012 database for database table design. The main role of database is to organize, store and represent information, which can store various types of data and add, delete, change and check data. In the English online teaching system designed in this paper, there are two databases, one is the Online English Platform database corresponding to the university English online teaching system, and the other is the SNRegister database corresponding to the book validation system. These two databases are stored in different databases on different servers [9]. The Online English Platform database contains most of the database tables needed to implement the basic functions of the English online teaching platform, with about 95 database tables, while the SNRegister database is a database designed specifically for the book verification module of the system. The SNRegister database is a database specially designed for the book verification module of the system. From the requirement analysis and functional design of this system, it can be seen that an English online teaching platform should have the following main modules, assignment submission system, student question system, teacher information module.

4 Security Encryption Algorithm of Book Verification Module in College English Online Teaching System

Hash functions, as an important branch of cryptography, have a pivotal role in the field of data encryption. It is expressed as.

$$\text{Hddr} = \text{H(key)} \tag{1}$$

Among these commonly used algorithms, MD5 and SHA-1 are the two most widely used encryption methods. If we compare these two most widely used encryption algorithms, we will find that although SHA1 encryption algorithm is more secure than MD5 encryption algorithm, its encryption time and memory usage are also higher than MD5 encryption algorithm, in contrast, MD5 encryption algorithm is not as strong as SHA1 encryption algorithm in terms of security, but it is fast, which is useful for offline authentication. In comparison, MD5 encryption algorithm is not as strong as SHA1 encryption algorithm in terms of security strength, but it is fast, which is an advantage of SHA1 in terms of encryption efficiency and time for the batch authentication function of offline authentication.

The user triggers the online authentication button on the interface to bring up the countdown interface, and then assigns the decreasing initial value of the countdown interface to a global variable and obtains the corresponding interface identification ID to obtain the value on the interface in the subsequent program. Finally, the countdown cycle method setInterval ("Countdown ()", 300) is started. The main function of this method is to call the Countdown() function every 0.3s, where the main function of the Countdown() function is to decrement the initial value on the page. When the initial value on the page is reduced to 0, the countdown page is first hidden, then the clearInterval() method is used to close the periodic setInterval() call and end the loop, otherwise the value on the interface will keep decreasing and the interface will keep flashing, and finally the foreground parameters are passed to the background controller for related processing. The core code is as follows.

```
function Countdown ()
{
if (second < 0) {
$("#Verifying").modal("hide"); window.clearInterval(intervalId);
$.getJSON("/BookVerify/Verify?r=" + Math.random() + "&BookID=" +
BookVerifyID + "&UserID=" + "@ViewBag.UserID" + "&SN=" + SN, function
(data) {
..........
}); } else { if (navigator.appName.indexOf("Explorer") > -1) { document.getEle-
mentById('totalSecond').innerText = second--;
} else { document.getElementById('totalSecond').textContent = second--;
}
}
}
```

After the backend of the client server receives the data from the frontend, the received data has to be passed to the central validation server for validation, which involves the processing of data from one server to another, which requires the use of cross-domain calls for data transfer and processing. Cross-domain invocation refers to data transfer or communication between domains with different protocols, ports, and domains of arbitrary protocols. This paper use a backend-based processing to achieve cross-domain, using the backend controller as a proxy to transfer each request operation from the frontend to other domains to the backend of this domain, and the backend of this domain to access other domains by simulating http requests and return the processing results of other domains to the backend of this domain for JSON string parsing. Once the domain name changes, the school administrator can directly change the domain name through the dictionary management interface, without the need for super administrator intervention to modify the program code.

5 Conclusion

This the current of university English online teaching mode and analyzes how to carry out effective English online teaching management under the current in China. The author's solution is to try to design a university English online. Inthe author elaborates on the relevant Internet technology of this system and describes in detail the construction the university by taking one of the modules as an example for the improvement of the encryption algorithm. In summary, the author believes that the construction of this system can provide a new path for the effective teaching of college English.

Acknowledgments. A Probe into the Internet Online Teaching Mode of College English Teaching Integrating Curriculum Ideological and Political Concepts.

References

1. Zhang, J., Wang, G.: Dynamic extension of the C# program with C++. file (12), 359–359 (2014)
2. Weston. On C# language foundation. Comput. CD Softw. Appl. (8), 218–218 (2011)
3. Guang, S.: Application research of database technology under NET platform. Digital Technol. Appl. **12**, 66 (2012)
4. Chen, L.: Methods offline data processing in ADO.NET. Henan Sci. **28**(2), 191–193 (2010)
5. Xu, K.: Design and implementation of hydropower enterprise office and production management automation system. University of Electronic Technology, pp. 10–13 (2011)
6. Li, P.: Office System Design and Implementation of Urumqi Cultural Market Inspection Team. Xiamen University, pp. 17–25 (2013)
7. Duan, Y.: Design and Implementation of the Front End of Mobile E-commerce System Based on HTML5. Nanjing University (2017)
8. Yingqi, Y., Liming, W.: Design and development of the network teaching management system in MVC mode. Autom. Instrum. **8**, 141–142 (2014)
9. Ping, L.: The Improvement of the information summary algorithm MD5 in cryptography. J. Fujian Inst. Eng. **09**(6), 602–603 (2011)

Research on the Design of PE Teaching Platform Under Computer Multimedia Technology

Yi Shao, XueFeng Zhang(✉), and Shi Zhu

Shanghai Customs College, Shanghai 201204, China
shaoyi@shcc.edu.cn

Abstract. With the continuous development of computer technology, multimedia technology has been widely applied in the field of teaching. As a highly practical subject, physical education teaching requires students to possess various sports skills. Therefore, the use of multimedia technology in physical education teaching can create a more vivid and interesting teaching environment, which helps to enhance students' learning interest and enthusiasm. This article designs a sports teaching platform based on multimedia technology, aiming to provide students with more effective sports teaching methods. This platform utilizes digital technology to organically integrate sports knowledge, skills, and teaching resources, forming a complete sports teaching content, including multimedia teaching resources such as videos, images, and audio. These resources are very beneficial for students to review and learn independently after class, while also greatly reducing the teaching burden of teachers and improving teaching efficiency. This platform adopts a layered design architecture, where the data layer, application layer, and display layer each undertake different functions. The data layer is mainly responsible for storing physical education teaching resources and classifying them appropriately; The application layer provides users with various teaching services; The presentation layer provides users with a visual interactive interface, allowing them to freely browse and learn related content.

Keywords: computer · multimedia · network technology · auxiliary sports teaching

1 Application Status of Computer Multimedia and Network Technology-Assisted

Computer assisted functions are an inevitable trend in the development of computers, and their functions are shown in Table 1. Utilize modern education to mobilize, promote, and promote the in-depth development of educational reform [1].

At present, the network is mainly manifested aspects: 1) some schools only the teaching of physical education courses, that only through teaching, demonstration and practice; can meet the teaching purpose, and can easily establish information system; (2) some schools teaching content and assessment have certain restrictions [2]. (3) The information teachers themselves, and the computer multimedia and network technology have technical deficiencies.

J. C. Hung et al. (Eds.): FC 2023, LNEE 1132, pp. 452–457, 2024.
https://doi.org/10.1007/978-981-99-9538-7_70

Table 1. The Application of Network Technology in Physical Education Teaching

Network classification	abbreviation	Distribution distance	Computer range
LAN	LAN	About 10m	room
		About 100m	Building
		About 1,000 m	campus
metropolitan area network	MAN	10km	city
long haul network	WAN	More than 100km	National or global

Some PE teachers conduct physical fitness on the TikTok and live broadcasting platforms. Some teachers ask students to punch in with their mobile APP and actively participate in daily fitness activities [3]. In short, teachers use a variety of network technologies to actively support the physical education class, so as to make the teaching of physical education class more interesting. Under the current situation, developing online PE teaching can effectively alleviate people's psychological mood, adjust the family atmosphere, and cultivate good habits. At the same time, sports and fitness is a brand-new form of education, which is also a new form of family sports activities. At present, many families focus on "students" and participate in various online teaching programs organized by teachers, jointly attended by students and their parents. Home fitness and fitness is popular. These problems not only have adverse effects on the students', but also bring adverse consequences. How to use computer technology to improve the urgent need of research. Under the current situation, physical teachers must grasp the pulse of the information age, actively explore the use of computer technology, make the school physical education full of vitality, promote the of the school mode, the education and teaching reform [4].

2 The Role of Computer Multimedia and Network Technology in Assisting Physical Education Teaching

2.1 Using Multimedia Technology to Improve the Effect

In the daily physical education, the general technical movements are simple, and some technical movements are more complex, such as fast combination; such as some more complex skills. Usually, they have to be explained by the teacher or coach to master these skills. Many students, especially those with poor performance in basic sports, have a fear of the essentials of movement. There are also some skills, such as long jump, horizontal bar, breaststroke, etc., many students can not speak, do not understand the meaning and power of the action. Therefore, to a variety of technical training. In the carefully design the specific teaching content, so that better and receive a variety of skills training. Using effectively solve the above problems, so as to improve. Using the multimedia technology, to deconstruct the part of the movement, deepen the understanding of the movement, deepen the memory of the movement, and you can master the action skills smoothly.

For example, standing long jump, PE teachers can do their own videos, can also do some deconstruction animation, let the students understand the action of standing long jump. Four action essentials: jump, air, landing buffer. The teacher can use 3 D model software to show each steps to the students. Using digital animation simulation, adjust the jump Angle, let the students know the best Angle of the jump. From the perspective of leg swing and hip joint, the difference between hip swing and pedal extension methods in different situations is studied. The analysis of the swing and non-swing trajectories shows the correctness of the swing arm or not. Using multimedia technology to explain and demonstrate the standing long jump technology can specify and visualize the boring and abstract knowledge.

2.2 Use of Internet Technology to Reform Physical Education

In the excellent teachers can make the students in the classroom, and their learning enthusiasm. Whether students can stick to independent exercise and whether the in sports in time are all problems worth discussing. It is very necessary to introduce the Internet in physical education teaching. Using Internet technology, textbooks can be pre-published by teachers for preview, or courseware can be played repeatedly played in class to simulate and reproduce the scenes needed by behavior. In this way, the in sports, but also make them master the sports essentials and training methods, so that they can keep healthy for life. This master the action skills. Use the network environment to cultivate the students' interest, is the best choice for teachers. If you don't understand anything, you can communicate with teachers through the Internet, find out problems in time, enhance.

In college physical education teaching, the Internet can effectively promote the efficiency of college physical education management. Using the Internet to conduct physical education teaching can not only optimize, but also apply the advanced technology to education. The use of Internet, can make the school physical education teaching resources sharing, enhance the teaching interaction, make the physical education teaching more targeted. Improve the physical quality quality. In view of the sports work in China, the network technology, the idea of building the university sports system under the environment of "Internet+" is put forward. At the same time, it also provides better teaching and research links for the school teachers and students, and provides more convenient communication channels, so that the school physical education teaching mode is constantly innovative.

2.3 Using Computer Multimedia and Internet Technology the Diversification Courses

Expand the depth and breadth of sports through support and the internet. In universities, the application of multimedia is becoming increasingly widespread, and teachers use computer technology to combine various physical education curriculum resources and methods. By utilizing multimedia and network technologies, students can be presented with the functions of various courses to achieve diversified learning, as shown in Fig. 1.

On this basis, the use of modern scientific and technological means to provide strong support for physical education. For before PE class, the Internet technology is used to query the required teaching materials, and display the required materials in various forms in class, so that students can realize the origin and development of the project,

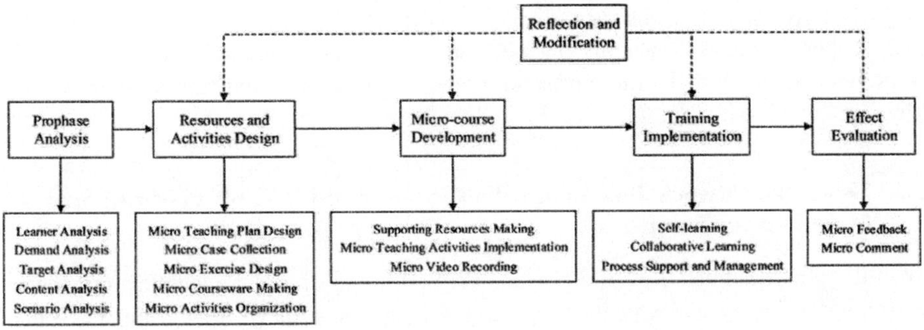

Fig. 1. Diversified courses offered in multimedia and internet technology

project rules, basic judgment rules, and other special professional skills. Play in the form of the competition video, improve the students 'aesthetic ability of sports, organize to watch the classic sports films, train the students' practical ability, excellence, sports spirit. Strengthen students' knowledge of health care and physical education physiology, including the scope of cardiopulmonary resuscitation, muscle spasm, warm-up, traction, etc.

The diversification of content can obviously the students' love for discipline, change the bias education discipline, and then promote them to realize that discipline and scientific discipline, with cultural and guiding characteristics. Sports resource database contents. It can not only provide rich teaching materials for teachers and students, but also lay a good the implementation of modern quality education, truly improve the comprehensive ability, this is the original intention of the school physical teaching reform.

3 Application Ways of Computer Multimedia and Network Technology to Assist Physical Education Teaching

3.1 Establish a Physical Education Textbook Library to Realize the Sharing of Physical Education Teaching Resources

The establishment database, the syllabus and videos can be uploaded to the Internet, using screening and sharing. The physical education resource database gathers the teaching practice of front-line teachers, shares the latest educational ideas at home and abroad, and sets up an exchange section for everyone to answer various questions. The establishment resource database has the characteristics of timely update. Through the teaching resource database, physical education teachers can optimize and adjust the course content through their own actual situation, and scientifically and reasonably choose, which is an effective of physical teaching.

In the information age, the number of knowledge has increased sharply, individuals can not master all, but a large amount of information can be stored and queried. Help people and create new so that people to communicate with each other. The exchange of educational resources is crucial. With the wide popularization, it is an inevitable trend to build the university PE teaching network by using. The PE teaching dynamic, which can make teaching means constantly updated, optimize teaching behavior and rich content. Strengthening the effectiveness of school-based construction and standardizing school physical education.

3.2 Using the Network Technology Platform to Build the Soft Power of Sports Majors in Universities

With the provide diversified and personalized online teaching services for universities and students. In physical education teaching, through the class WeChat group, the school physical education teaching online application program, the system, etc., the release of personal teaching information in physical education teaching, such as: elective, such as pull-up thematic exercises, such as increasing endurance exercises, warm-up exercises, etc. Students can observe at any time and practice at any time to accumulate sports experience and exercise their physique. At the through online, the can be immediately tracked and managed, and they can understand their physical condition and the effectiveness of the training. At the teachers through their technical demonstration or training videos through the Internet, effectively stimulate their enthusiasm for learning, improve their innovation ability.

Where conditions permit, the Internet can be used to build a physical education, realize the implementation of information management, and improve students 'subjective initiative and enthusiasm, which to master and master academic status. In the university campus, it is an inevitable development direction to build the network. The of building modern is to establish a good online education network, and the establishment of online physical education platform needs to increase hardware investment.

4 Conclusion

Physical education is an important guarantee for students' physical fitness and health. Based on computer multimedia technology, this article designs a feasible physical education teaching platform to address the current problems in physical education. The platform adopts a layered design, achieving digital integration and diversified presentation of physical education teaching resources, providing better support for teaching and learning. The application platforms provides very suitable conditions for physical education teaching, indicates that physical education will continue to innovate and improve. Thus, we can believe that an educational platform based on multimedia technology will bring students broader development prospects, build a better educational environment.

References

1. Cao, M.: Application of computer multimedia technology and network teaching. New Course Teacher (11), 67–68 (2010)
2. Gu, X., Zhou, R.: Research on computer multimedia and network technology. School Phys. Educ. **38**(3), 112–113 (2020)

3. Zhao, L.: Research on the reform of civic science teaching in music course under the concept of OBE talent cultivation. Region – Educ. Res. Rev. **5**(1), 31–35 (2023)
4. Zuo, L., Cao, X., Zhu, L., et al.: Research on the design method of UHF antenna under the condition of changing tag distribution density. J. Electron. Inf. Technol. **45**(1), 158–167 (2023)

Application and Effect Evaluation of Nursing Examination System in Nursing Teaching

Qiumei Zhou(✉)

Yunnan Medical Health College, Kunming 650033, Yunnan, China
zqm15087063358@163.com

Abstract. This paper discusses the application and effect of nursing practice qualification examination counseling and learning management system (hereinafter referred to as human health and nursing examination system) developed by the people's Health Publishing House in nursing teaching. Using the methods of convenient sampling and objective sampling, taking nursing students and teachers, the human health examination system was applied in the teaching of internal medicine nursing course and the guidance of nurse practice qualification examination (nursing examination). After the course teaching and nursing examination counseling, the students' scores were evaluated, questionnaires were issued to investigate the teaching effect, and the students' scores of internal medicine nursing examination at the end of the semester were compared.

Keywords: Nursing examination system · Nursing teaching · impact assessment

1 Introduction

In recent years, with the continuous development of information technology, the field of higher vocational nursing education has been actively exploring the combination of online and offline hybrid learning. Since 2017, the national soil protection qualification examination (hereinafter referred to as the "soil protection examination") has fully implemented the man-machine dialogue examination mode. The man-machine dialogue test manages the test network. Candidates read the test questions on the computer screen and answer through the mouse and keyboard. Our school is a health vocational college with more than 2000 nursing students [1]. In mixed teaching under the new situation, our school purchased the nurse qualification examination counseling and learning management system (hereinafter referred to as the human health care examination system) developed by the people's Health Publishing House in 2015, and applied it to nursing teaching and nursing examination counseling in September 2017. It carried out mixed teaching combining online and offline learning and classroom counseling, which improved the teaching effect.

J. C. Hung et al. (Eds.): FC 2023, LNEE 1132, pp. 458–463, 2024.
https://doi.org/10.1007/978-981-99-9538-7_71

2 Related Work

2.1 Nursing

Nursing is an old but young specialty. In fact, with human beings, nursing activities appeared, but it was not until the middle of the 19th century that Nightingale initiated the scientific nursing specialty, which marked the beginning of nursing specialization and the birth of modern nursing. Nursing is a comprehensive applied science based on Natural Science and social science, which studies the nursing theory, knowledge, technology and development law in the process of prevention and health care, treatment of diseases and rehabilitation. As a specialty providing health care services, nursing has moved from hospital to family, community and society. Facing all individuals with health needs, the scope of nursing has also extended from disease nursing to the whole process from health to disease. Nursing has gradually developed into an independent discipline different from clinical medicine, which more reflects its unique scientificity, sociality and service. In 2011, the Department of degree Education issued a new discipline directory, which divided nursing from a secondary clinical medicine discipline into a primary discipline. This paper refers to the undergraduate nursing specialty set up in comprehensive colleges or medical colleges.

2.2 Undergraduate Nursing Education

The modern Chinese dictionary explains "undergraduate course" as "the basic part of a university or college, and students can obtain a bachelor's degree after graduation". "Undergraduate" is the abbreviation of "undergraduate education", which refers to the level and specification of talent training. It is generally the implementation of general education. According to the international standard classification of Education (2011), undergraduate education belongs to the sixth level of education, which is divided into academic and professional. For the nursing discipline, among the more than 300 nursing colleges / departments offering undergraduate nursing education in China, except for a few nursing colleges / departments with strong research ability and comprehensive school running ability, they should focus on cultivating academic undergraduate nursing talents, The undergraduate nursing education in other colleges should be oriented to cultivate professional nursing talents [2]. Applied higher education should be defined as a professional and technical talent training activity at the undergraduate level, subordinate to the vocational education system.

3 Application of Nursing Examination System in Nursing Teaching

3.1 Human Health Care Examination System

Human health intelligence network is the core platform of digital content service of people's Health Publishing House and the official website of human health medical health big data intelligent service platform. Relying on the experts and publishing resources accumulated by the human health society for more than 60 years, and through the establishment of a strong lecturer group and course guidance team, the people's Health Publishing House has developed the guidance and learning management system for nurse qualification examination, referred to as the human health examination system for short.

Our school purchased the human health and nursing examination system in 2015 and opened it to all teachers and students of the nursing college through the campus LAN. The system has two populations: teachers and students have their own accounts. Teachers and students share more than 7000 test questions (all A1, A2, A3 and A4 questions) and more than 200 teaching videos (updated regularly), covering 21 chapters of nursing related contents, such as basic nursing knowledge and skills, nursing of respiratory diseases, nursing of circulatory diseases, etc. [3]. The student platform has three sections: my course, my test questions and my wrong questions. My course section includes three courses: intensive lecture class, tandem lecture class and finishing class. Students can choose the learning content according to their own learning progress or teachers' requirements. The teacher platform includes 8 items, namely institutional information, resource management, learning management, examination management, institutional wrong question bank, class management, self-organizing resources and system management. The institutional information, institutional error database and system management are operated by the system administrator in the background. Teachers are responsible for resource management, learning management, examination management, class management and self-organizing resources.

3.2 Application of Human Health and Nursing Examination System in Nursing Teaching

The author is a full-time teacher in the teaching and Research Department of internal medicine nursing. Taking the teaching of internal medicine nursing as an example, this paper expounds the application of human health and nursing examination system in nursing teaching in our school. Nursing in the Department of internal medicine of our university was opened in the second academic year. In September 2017 (2016 students), the internal medicine nursing course introduced Renren health care examination system on the basis of previous conventional teaching, which is mainly used in three links: classroom quiz, after-school homework and stage assessment. Taking the content of "bronchial asthma" as an example, before class, teachers enter the human health care examination system to export bronchial asthma related test questions, and then prepare lessons with reference to the exercises in the human health care examination system according to the curriculum standards and in combination with the nursing examination outline. Teachers need to make use of self-organizing resources to form papers and conduct classroom tests in order to understand the mastery of students' key and difficult points. After class, assign homework through the learning management section and enter the system at any time to check the completion of students' homework, including completion proportion, time and score. The homework score accounts. In addition, our school arranges a stage assessment after the teaching of each system of internal medicine nursing, and the score accounts. In order to ensure the fairness and fairness of stage assessment, all teachers in the course group use the human health care examination system to form papers, authorize and assess according to the course standards.

4 Effect Evaluation

4.1 Included Objects and Data Collection Methods

(1) 375 students from 7 classes of 2015 nursing major in our university were selected as 2015 teaching group; 478 students from 9 classes of nursing major in 2016 were selected as the teaching group of 2016. The two groups of students were full-time higher vocational nursing students in our university, and there was no significant difference in enrollment age, gender, humanistic achievement and curriculum ($P > 0.05$). The 2015 teaching group received routine teaching, and the 2016 teaching group introduced Renren health care examination system on the basis of routine teaching.
(2) 1016 students in 2013 who participated in the nursing examination in the year of graduation were selected as the 2013 nursing examination group; Grade 2014 535 persons, set as grade 2014 nursing examination group; There are 319 people in 2015, and they are set as the 2015 nursing examination group. The three groups were full-time higher vocational nursing students, and there was no significant difference in humanistic age, gender, humanistic achievement and curriculum ($P > 0.05$). The 2013 and 2014 nursing examination groups received routine nursing examination counseling, and the 2015 nursing examination group introduced the human health examination system on the basis of routine counseling.
(3) A total of 31 full-time and part-time teachers who taught in the nursing college from September 2017 to July 2018 were selected. Admission criteria: have used the personal health care examination system in at least one nursing course teaching or nursing examination counseling and competition counseling, and voluntarily participate in this study. Through a questionnaire survey, understand the evaluation of 2016 teaching group students and teachers on the human health and nursing examination system, and compare the scores of medical nursing examination and the passing rate of nursing examination of students in each group.

4.2 Evaluation Indicators

(1) questionnaire survey results. After consulting the literature and group discussion, the questionnaire was designed by ourselves. The student questionnaire included 8 items conducive to the learning of nursing theoretical knowledge and stimulating learning interest. The questionnaire is filled in by anonymous and distributed and recovered through the Internet. 478 student questionnaires were distributed and 436 were recovered. The 31 teachers' questionnaires were distributed and 31 were recovered [4]. The recovery rate and effectiveness rate were 100.00.
(2) Results of internal medicine nursing course. The course scores of internal medicine nursing in each teaching group were compared.
(3) Nursing test results. The passing rate of nursing examination in each nursing examination group was compared.

4.3 Evaluation of Teaching Effect of Human Health and Nursing Examination System

(1) Student evaluation. The human health care examination system and believe that the human health care is conducive to the learning of nursing theoretical knowledge,

improve learning efficiency, review and expand knowledge after class, and help to cultivate autonomous learning ability, as shown in Table 1.

Table 1. Students' evaluation of human health care examination system

entry	yes	commonly	no
It is conducive to the learning of nursing theoretical knowledge	411(94.27)	22(5.05)	3(0.68)
Can stimulate interest in learning	216(49.54)	125(28.67)	95(21.79)
It helps to improve learning efficiency	388 (88.999)	21(4.82)	27(6.19)
It helps to review after class	423(97.02)	8(1.83)	5(1.15)

(2) Impact on the passing rate of nursing examination. The passing rate of nursing examination in our school has increased year by year in recent three years. The human health examination system was introduced in 2018, and the passing rate of nursing examination is as high as 99.37, as shown in Table 2.

Table 2. Comparison of passing of nursing examination in our school in recent 3 years

Nursing examination time	Number of examinees	Number of people passed	Passing rate
2016	1 016	980	96.46
2017	535	524	97.94
2018	319	317	99.37

Blended learning emphasizes the application of appropriate technologies and methods at the right time to optimize the learning effect. Schools often adopt different forms of information-based auxiliary learning means according to their own actual situation. The human health and nursing examination system has powerful functions and fits the man-machine dialogue examination mode of nursing examination. For higher vocational nursing specialty, nursing course teaching needs to be accurately connected with nurse practice qualification examination. Using the human health examination system for course teaching and nursing examination guidance, and can realize the integration of course and certificate. Since the application of the system, the examination results of medical nursing courses of 2016 students have significantly improved, and the passing rate of nursing examination in 2018 has significantly increased, which are strong evidence for the integration of courses and certificates.

5 Conclusion

The nursing examination system visualizes the empty theoretical knowledge and test questions. Each test question is equipped with answer analysis,. Teachers conduct classroom assessment and stage assessment through the system, which ensures the timeliness and effectiveness of assessment, greatly improves the effect of process assessment, effectively combines online and offline, and realizes hybrid teaching. Nursing specialty is a highly practical discipline. In addition to mastering theoretical knowledge, students should also cultivate clinical thinking, critical thinking and interpersonal communication skills practice. It on test questions alone. The human health and nursing examination system is only a teaching auxiliary platform. In the actual teaching, it also needs to complement with practical teaching, so as to finally realize the purpose.

References

1. Xiaomei, L., Rui, G.: Overview of undergraduate nursing education in China. China Nurs. Educ. **12**(7), 490–495 (2015)
2. min Peng, L.F., Wei, D.: Promotion of the effect of e-learning management system on the counseling of nurse practice qualification examination. Guangdong Vocat. Tech. Educ. Res. **6**, 11–12 2017
3. Shuangshuang, L.: Exploration of talent training mode in higher vocational colleges based on blended learning. J. Liaoning High. Vocat. Coll. **10**, 39–42 (2015)
4. Yongqiang, F., Zhunian, J.: Problems and countermeasures of course certificate integration in higher vocational education. Heilongjiang High. Educ. Res. **11**, 174–176a (2016)

Research on the Application of Computer Intelligent Proofreading System in English Phrase Translation

Caixia Chen(✉), Hai Wang, and Manman Gong

Yinchuan University of Science and Technology, Ningxia, Yinchuan 750000, China
ccx851010@163.com

Abstract. The role of phrase translation in English reading is very important, but there is a problem of poor translation accuracy. Manual proofreading increases the workload of teachers and is less efficient. Therefore, proofreading system for English proofreading. First, AI technology is used to proofread the phrase of English reading requirements, and the horizontal division is carried out according to the phrase proofreading standard to reduce it Distractors in phrase proofreading. Then, AI technology proofreads English phrase translation to form phrase proofreading results, and performs continuous phrase proofreading for transactional phrase proofreading requirements. MATLAB simulation shows that under a certain amount of reading, the evaluation accuracy and phrase proofreading time of the computerized intelligent proofreading system are better than those of manual proofreading.

Keywords: AI technology · Abnormal phrase proofreading requirements · Phrase proofreading · Arts and Crafts

1 Introduction

Translation proofreading is one of the important evaluation contents of English reading [1], which is of great significance for English phrase translation. However, in the actual phrase proofreading review process [2], the accuracy of the phrase proofreading results is low, which has a certain impact on the phrase reading. Some scholars believe that the application of intelligent algorithms to English phrase translation proofreading can effectively improve translation accuracy and provide corresponding support for phrase proofreading [3]. On this basis, this paper proposes a optimize the proofreading phrase and verify the system's effectiveness.

2 Related Concepts

2.1 Mathematical Description of Phrase Translation

Phrase translation uses AI technology to optimize the translation results, and according to the indicators in the translation results, find outliers in the English phrase translation proofreading results [4], integrate the corresponding parameters, and finally judge the

J. C. Hung et al. (Eds.): FC 2023, LNEE 1132, pp. 464–470, 2024.
https://doi.org/10.1007/978-981-99-9538-7_72

Feasibility of English phrase translation proofreading results [4]. The computer-based intelligent proofreading system combines the advantages of AI technology to quantify the results of English phrase translation proofreading by using auxiliary parameters [5], which can improve the accuracy of phrase proofreading.

Hypothesis 1: The English reading requirement is d_i, the English phrase translation proofreading data is set_i, the satisfaction rate of the English phrase translation proofreading result is y_i, and the phrase proofreading data judgment function is $F(d_i \geq 0)$ As shown in Eq. (1).

$$F(d_i) = \sum x_i \subset y_i \cdot \xi \tag{1}$$

2.2 Selection of Proofreading Scheme for English Phrase Translation

Hypothesis 2: The English phrase translation proofreading selection function is $z(d_i)$, and the weight coefficient of English phrase translation proofreading is w_i, then the translation proofreading selection is shown in Eq. (2).

$$z(d_i)=z_i \cdot F(d_i, y_i) \xrightarrow{k} w_i \cdot \xi \tag{2}$$

2.3 English Phrase Translation Proofreading Data Processing

Before computer phrase translation, the phrase proofreading results should be analyzed discretely [6], and the transactional phrase proofreading requirements should be mapped to the auxiliary parameter library to eliminate abnormal data [7]. First, phrase proofreading data for comprehensive analysis, and set data thresholds and indicator weights computer phrase translation. English phrase translation proofreading data is system test data and needs to be standardized. If English phrase translation proofreading data is in a nonnormal distribution, its phrase proofreading results are affected, reducing the accuracy of the overall phrase proofreading. In the accuracy of computer phrase and improve the level proofreading, it is select English phrase, and the specific scheme selection in Fig. 1.

Survey data shows that English phrase translation proofreading shows a discrete distribution, which is in line with objective facts. English phrase translation proofreading has no directionality, indicating that it has strong randomness, so it is used as an analysis and study of 2~4 min. English phrase translation proofreading meets the standard requirements, mainly, AI technology adjusts English phrase translation proofreading, removes duplicate and irrelevant schemes, and supplements the default scheme, so that the dynamic correlation of the entire data is strong.

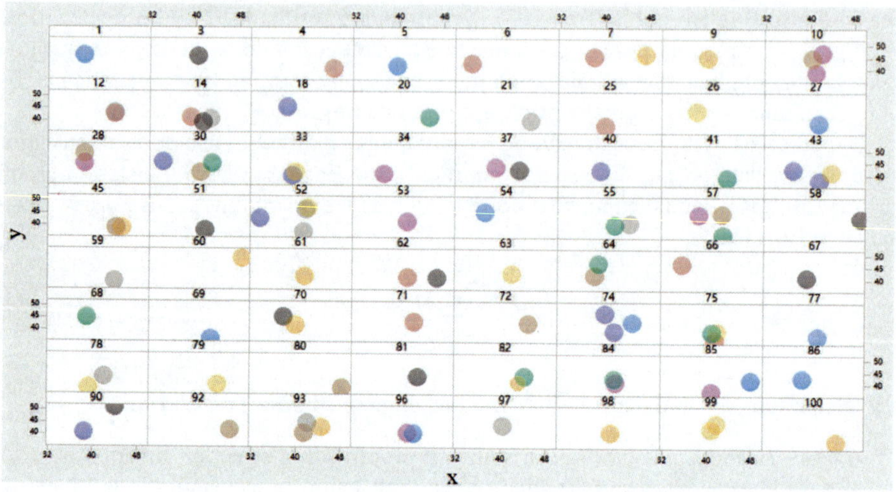

Fig. 1. Selection results of English phrase translation

3 Strategies for English Phrase Translation Proofreading

The computer-based intelligent proofreading system adopts a random strategy for English phrase translation proofreading and adjusts the corresponding parameters to realize the optimization of English proofreading. The divides the proofreading into different phrase proofreading levels, and randomly selects different schemes. In the iterative process, English phrase translation proofreading of different phrase proofreading levels is matched. After completing the matching process, compare the phrase proofreading levels of different schemes to record the best English phrase translation proofreading results.

4 Practical Examples of English Phrase Translation Proofreading

4.1 Introduction to English Phrase Translation Proofreading Data

In order to facilitate phrase proofreading, the English translation proofreading data in complex cases is used as the research object, with 12 paths and a test time of 12 min, which is read explicitly in English the phrase proofreading data is shown in Table 1.

Table 1. Relevant parameter of English reading and English phrase translation proofreading

Scope of application	Phrase proofreading time	Phrase proofing effect	Phrase proofreading satisfaction rate	Phrase proofreading form
Personal phrase proofreading	1~2 min	84.54	86.60	Team, individual

(continued)

Table 1. (*continued*)

Scope of application	Phrase proofreading time	Phrase proofing effect	Phrase proofreading satisfaction rate	Phrase proofreading form
	2~4 min	92.78	86.60	Team, individual
Corporate phrase proofreading	1~2 min	87.63	92.78	Team, individual
	2~2 min	90.72	84.54	Team, individual
Other phrase proofreading	1~2 min	83.51	92.78	Team, individual
	2~2 min	87.63	90.72	Team, individual

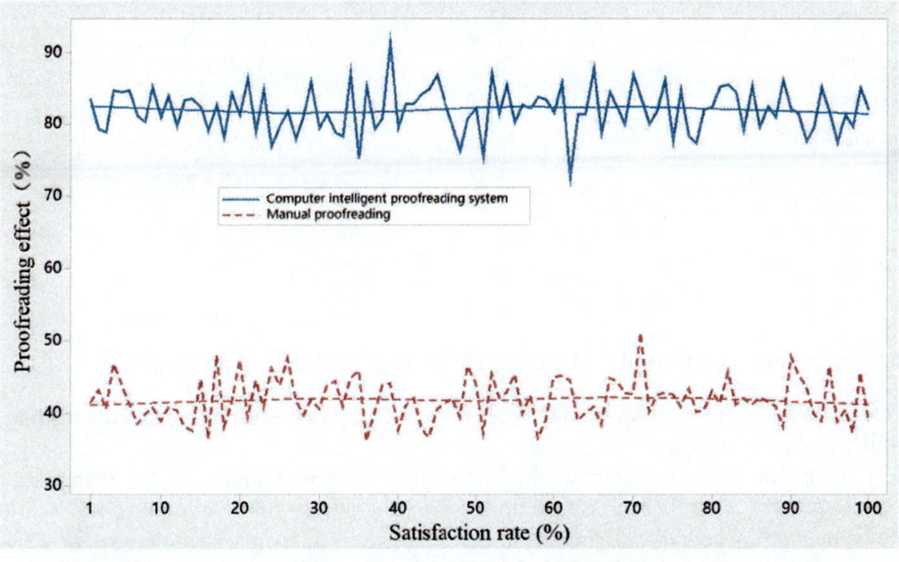

Fig. 2. Processing process of English phrase translation proofreading data

The English reading requirement processing process in Table 1 is shown in Fig. 2.

Table 1 that compared with manual proofreading, the phrase proofreading results of the computerized are closer to the actual phrase proofreading requirements. Regarding the rationality and fluctuation range of English phrase translation proofreading selection, the computer-intelligent proofreading system manually proofreads. The data changes in Fig. 4 shows that the computer-intelligent proofreading system has better stability and faster judgment speed. Therefore, the speed of phrase proofreading results, English phrase translation proofreading phrase proofreading results, and summation stability of the computerized intelligent proofreading system are better.

4.2 English Phrase Translation Proofreading

The phrase proofreading results of English translation proofreading contain unstructured, semi-structured, and structural information. After the pre-selection of the computerized, the preliminary English phrase proofreading phrase proofreading results were obtained, and the phrase proofreading results were obtained analysis of the feasibility of phrase proofreading results Proofreading. In order to verify the evaluation effect proofreading more accurately, select English phrase translation proofreading with different phrase proofreading levels, and the evaluation data of English reading requirements are shown in Table 2.

Table. 2. Overall situation of English phrase translation proofreading

Phrase proofreading grade	Sampling rate	Complete rate
specialized	28.57	85.71
ordinary	73.21	100.00
synthesis	69.64	60.71
mean	26.79	50.00
X^2	55.36	69.64
P = 0.531		

4.3 Accuracy and Stability of English Phrase Translation Proofreading

In the accuracy of the computerized, the data is compared with manual proofreading, which is shown in Fig. 3.

Figure 3 that the computerized intelligent proofreading system is higher than that of manual proofreading, but the error rate is lower, the phrase of the computerized, while the manual proofreading The proofreading of phrases is uneven. The average data of the above three algorithms are shown in Table 3.

It can be seen from Table 3 that manual proofreading has deficiencies in accuracy and stability in English phrase translation proofreading, and English phrase translation proofreading has changed significantly, and the error rate is high. The accuracy of the comprehensive results is higher than that of manual proofreading. At the accuracy of the computer-intelligent is greater than 90%, and the accuracy has not changed significantly. In order to further verify the superiority of computer systems. In the effectiveness, different methods are used to analyze computer phrase translation comprehensively, and the result 4 is shown.

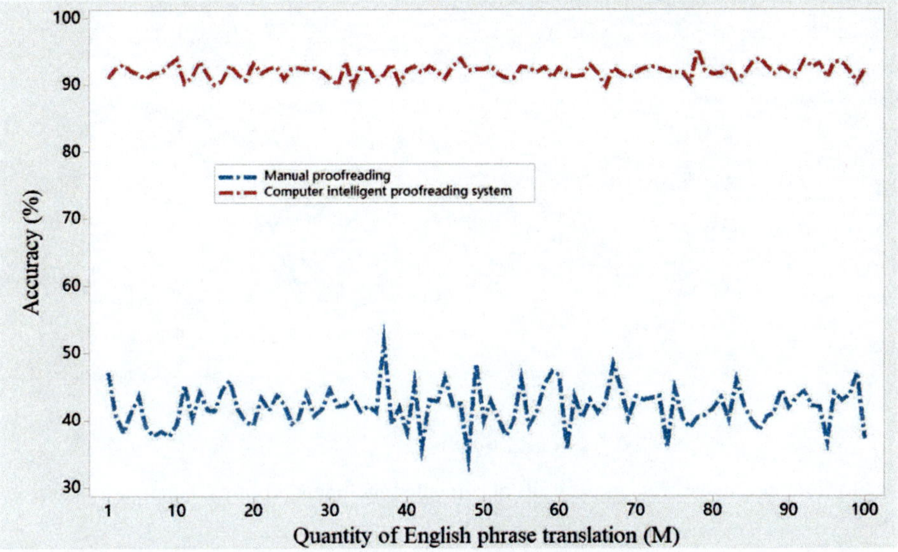

Fig. 3. Accuracy of different algorithms

Table 3. Comparison of phrase proofreading accuracy of different methods

algorithm	accuracy	Magnitude of change	error
Computer intelligent proofreading system	91.75	92.78	2.51
Manual proofreading	86.60	92.78	3.54
P	5.571	9.693	2.471

Figure 4 that the data of the computer-intelligent proofreading system is significantly better than the manual proofreading, and the reason is that the computerized intelligent proofreading system increases the adjustment coefficient of English phrase translation proofreading and sets the corresponding threshold. Exclude non-compliant data.

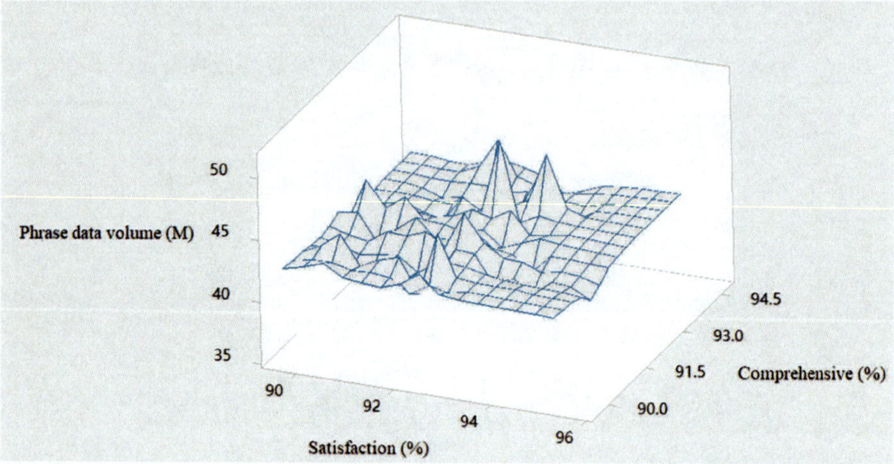

Fig. 4. Comprehensive evaluation results of computer intelligent proofreading phrase proofreading

5 Conclusion

Given the accuracy of English phrase translation proofreading, this paper proposes a computer-based intelligent proofreading system combined with AI technology for comprehensive analysis. At the same time, the content of English phrase translation proofreading is analyzed in depth to construct a collection of English phrase proofreading requirements. Research shows that the computerized intelligent proofreading and stability of English phrase proofreading and can translate and English phrases. Make a comprehensive judgment. However, in the process of computer phrase translation, too much attention is paid to the analysis of phrase proofreading ability, resulting in a relative decline in the accuracy of supervision.

References

1. Babhulgaonkar, A., Sonavane, S.: Empirical analysis of phrase-based statistical machine translation system for English to Hindi language. Vietnam J. Comput. Sci. **09**, 135–162 (2022)
2. Grabowski, L.: Phrase frames as an exploratory tool for studying English-to-polish translation patterns: a descriptive corpus-based study. Across Lang. Cult. **21**, 217–240 (2020)
3. Li, X.Y.: The impact of big data technology on phrase and syntactic coherence in English translation. Math. Prob. Eng. **2022**, 1–11 (2022)
4. Ning, J., Ban, H.D.: Design and of automatic machine translation on Chinese-English phrase translation. Mob. Syst. **2021**, 1–8 (2021)
5. Tuhtasinovich, I.E., Ubaydullaevna, M.D.: Use of lexical compression in the translation of phrases from English to Uzbek. Int. J. Early Child. Spec. Educ. **14**, 31–38 (2022)

Surface Topographical Change of Divertor Target Plates Under Conditions Relevant to ITER ELMs

Yan Huang[1(✉)] and Juan Cai[2]

[1] Department of Basic Courses Teaching, Dalian Polytechnic University, Dalian 116034, China
huangyan@dlpu.edu.cn

[2] School of Physics and Electronic Technology, Liaoning Normal University, Dalian 116021, China

Abstract. The ITER (International Thermonuclear Experimental Reactor) ELM (Large Amplitude Essential Mode) is an important problem for the terrain change on the surface of the Divertor target plate during the operation of ITER. The main function of the Divertor target plate is to filter out high-energy particles and ions in thermonuclear reactants and maintain the stable operation of subsequent devices. Due to the overload caused by ITER ELM and local extremely high heat emissions, significant terrain changes may occur on the surface of the target plate, leading to operational issues. In this paper, the surface topography of Divertor target under ITER ELM is studied by theoretical simulation and experimental research. Based on experimental measurement results and theoretical simulation predictions, the terrain changes were analyzed and explained. The research shows that ITER ELM operation will cause local high temperature and thermal stress on the surface of the target plate of the Divertor, and the surface will undergo significant deformation and distortion. In addition, the experimental results also indicate that the terrain changes on the surface of the target plate show significant differences not only in the longitudinal and transverse directions, but also in different areas of the target plate surface. Through theoretical simulation and experimental research, this paper obtained the experimental results and theoretical predictions of the terrain changes on the surface of the Divertor target under the operation of ITER ELM, which has important reference value for understanding the lubrication and preventing radiation pollution in the field of thermonuclear fusion.

Keywords: ELMs · tungsten divertor plates · melt motion

1 Introduction

Heat may cause target materials tungsten (W) to undergo melting and evaporation. And the humps induced by melt motion, initiate arc, and shorten the divertor lifetime. In addition, surface roughness will reduce W thermal conductivity, degrade its mechanic strength, result in an increase in tritium retention, affect the physical sputtering yield, and lead to a non-uniform erosion deposition behavior of impurities [1]. Therefore, it is very necessary to study the surface morphological change caused by melt.

J. C. Hung et al. (Eds.): FC 2023, LNEE 1132, pp. 471–476, 2024.
https://doi.org/10.1007/978-981-99-9538-7_73

In fact, the existing experimental data of different tokamaks show the actual shape of heat load during type-I ELMs is asymmetric and basically the same even for different the experimental parameters, which decays slower 1.5–5 times than rises, and has an huge influence. Therefore, it is urgent performance under with the real temporal variation rather than rectangular or triangular waveform [2].

In this paper, from under a rectangular waveform to a temporal evolution heat load erosion of the divertor plates, and the ultimately surface morphology is different from those under the rectangular waveform heat load commonly used in other studies.

2 Model Description

The surface of Divertor target plate will be affected by local high temperature and thermal stress, which will lead to deformation and distortion of the surface. This problem needs attention and solution. The splitter, as another key component, undertakes the important task of guiding various high-temperature plasma into the reaction ring collector. In this paper, the diverter is redesigned as a target, and the surface topographic changes of the Divertor target are combined with the diverter target to improve the adhesion and collection of the diverter [3]. The schematic diagram of the diverter target is shown in Fig. 1 below.

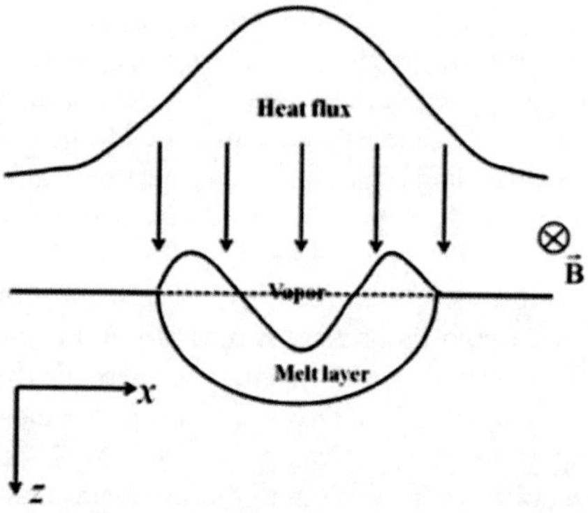

Fig. 1. Sketch of the divertor target.

This design is applied to ITER ELM operation, combining the surface change of Divertor target plate with the diverter target. By redesigning the diverter target, it can cover the terrain changes on the surface of the Divertor target plate, and increase the adhesion and collection. Specifically, the target part of the diverter adopts a replaceable form, changing the original plate structure into an interchangeable multi-layer structure, and adding a needle structure on the surface [4]. The reducing the dispersion of the diverter target. At the same time, the needle like structure enables the diverter target to have higher collection capacity, which can better meet the needs of target collection in various plasma environments. In this paper, the surface topography of the Divertor target plate is combined with the diverter target, which improves the adhesion and collection of the diverter, and establishes a more stable and convenient connection between them. This design method helps to improve the sustainability and stability of Divertor target plate and diverter, and also provides a certain reference value for other ITER related designs [5].

3 Results and Discussion

3.1 Temporal Evolution of the Energy Flux

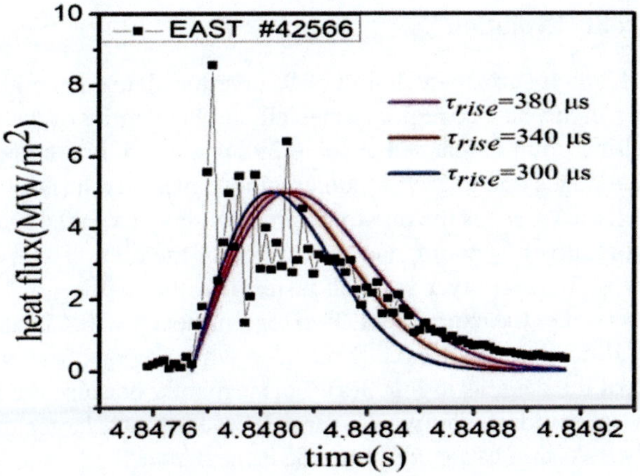

Fig. 2. Experimental temporal evolution of the heat flux on the EAST divertor during a type I ELM and fitted Rayleigh distribution function.

In this paper, the EAST # 42556 shot during the 2012 experimental campaign, which reads,

$$q(t) = q_{\max} e^{1/2} \frac{t}{\tau_{rise}} \exp\left(-\frac{t^2}{2\tau_{rise}^2} \right),$$

The fitted Rayleigh function and the measured experimental data in EAST # 42556 shot are shown in Fig. 2, qmax = 5 MW·m-2, the parameter τrise is taken to three values of 300, 340 and 380 μs, respectively. We can see that the fitted function with the different value of τrise within a certain range is conform well to the experiment date. We integrate the heat flux of EAST shot # 42566 over 1 ms from 4.84774 s to 4.84874 s, the sum is about 2.76 kJ·m-2, which is more agree with the value when τrise is taken to 340 μs. In order to simulate the situation of ITER, we first scale the up to 2900 MW·m-2 and take τrise to be 340 μs in this work, the integral value of which over 1 ms is about 1.6 MJ·m-2 in the range of. Later, in order to estimate the effect of various factors on thermal erosion.

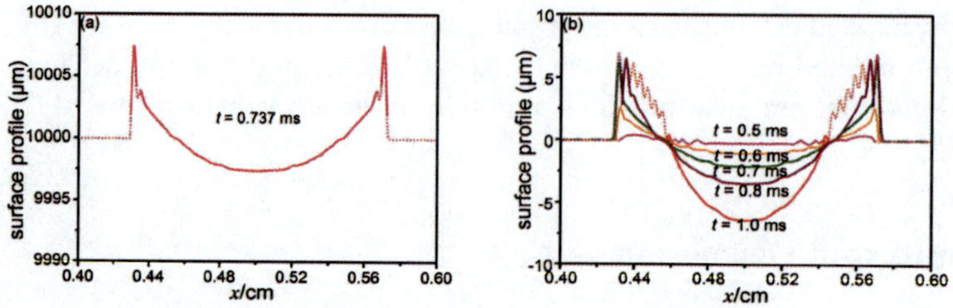

Fig. 3. The surface profile calculated under an ELM with the peak heat flux of 2900 MW·m-2. Note the shot dot line shows the solid surface, and the solid line shows the melt layer surface.

3.2 Surface Profile Evolution

Figure 3 shows the melt surface evolution of W divertor plates under heat, The temperature rises rapidly in the surface region, especially in the vicinity of the separatrix strike point (SSP), melting point is reached at t = 0.292 ms. Then, two humps are gradually formed at the dent edges due to melt motion mainly caused by surface tension. At time 0.736 ms, the melt area reaches the maximum in the range of x = 0.429 to 0.571 cm, the surface profile is relatively smooth, and the outside of the hump is very steep, because the flow velocity of W melt layer is much faster than the melting of marginal region, just as it was observed in the experiment. Two regions near x = 0.433 and x = 0.569 cm have finished solidification another 0.001 ms later, while regions from where to the corresponding edge of the dent, as in Fig. 3(a). This is mainly because the higher the more quicker the flow velocities of the melt layer, which are much larger than those in the regions under the two humps facing to the SSP (near x = 0.433 and x = 0.569 cm), leading to where getting lower and heat flux penetrateing into the bulk easier. Then, the regions near both edges of the dent gradually complete resolidification, the region in the vicinity of the center of the dent still continues to melt, and the melt W mainly driven by surface tension is still flushed to the periphery to form new humps. Melting and solidification occur simultaneously. The surface profiles of humps are waviness formation, as shown in Fig. 3(b). At t = 1.0 ms, the humps regions near both edges (from x = 0.429 to x = 0.457 cm, and from x = 0.541 to x = 0.571 cm) of the dent complete resolidification, while the region from x = 0.457 to x = 0.541 cm is still in melted state. However, the heat load of 1.6 MJ·m-2 lasting 1.0 ms of rectangular power shape cannot cause the target melt.

3.3 τrise Dependence

Figure 4 gives the different τrise of 300 μs, 340 μs and 380 μs, respectively, the heat load duration is 1.0 ms. The region from x = 0.434 to 0.566 cm is still in melted state for τrise of 380 μs at t = 1.0 ms, which is larger than that for τrise of 340 μs, while the melt layer completes resolidification for τrise of 300 μs. This mainly because the bigger the value of τrise is, the more the heat load deposited is, the larger the melt area is, the bigger the melt thickness is—maximum melt thickness is about 58.5μm for τrise of 380 μs, which is bigger 1.6 times than that for τrise of 300 μs, and the slower the resolidification is.

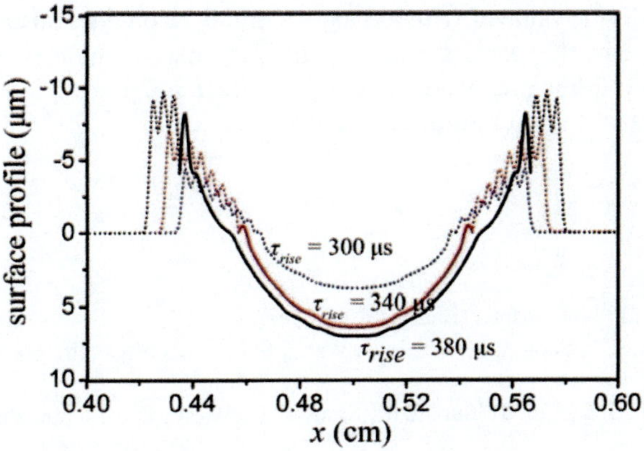

Fig. 4. Note the shot dot line shows the solid surface, and the solid line shows the melt layer surface.

3.4 Surface Roughness

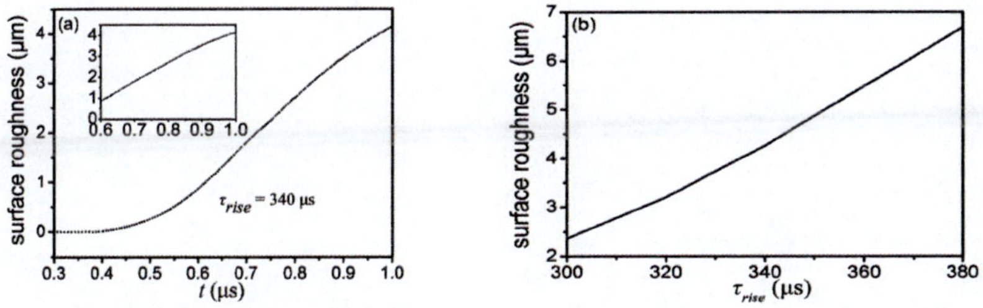

Fig. 5. Surface roughness of W plates versus irradiation duration (a), value of τrise (b).

Figure 5(a) gives irradiation duration. This is mainly because that the bigger the value of τrise, the faster the flow velocity of the melt layer is, the higher the humps are, so the larger the surface roughness is.

4 Conclusion

The temporal evolution load during is fitted by the Rayleigh distribution function according to the EAST # 42556 shot during and is employed, which solves the conductivity equation, including melting, evaporation, and solidification processes. The simulation results show: (i) the surface temperature quickly exceeds the melting point under the type-I ELM heat flux, then, humps are gradually forming at the dent edges due to melt motion mainly caused by surface tension, resolidification occur on the both outsides of the dent, simultaneously, the melt layer located in the vicinity of the center of the dent is still flushed to the periphery to form new humps, resulting in the waviness formation of surface profile; (ii) the value of τrise has a great influence on thermal erosion: the bigger the value of τrise is, the larger the melt area is, the deeper the melt front location is, the slower the resolidification is, the larger the surface roughness is. All the simulation results show the real shape of the a very important to determine the W target erosion and lifetime.

References

1. Gunna, J.P., et al.: Nucl. Mater. Energ. 1 (2016)
2. Huang, Y., Sun, J.Z., Sang, C.F., Hu, W.P., Wang, D.Z.: Acta Phys. Sin. **66**, 035201 (2017). (in Chinese)
3. Huang, Y., Sun, J.Z., Cai, J., Sun, Z.Y., Sang, C.F., Wang, D.Z.: Chin. Phys. B **28**, 045201 (2019).
4. Miloshevsky, G.V., Hassanein, A.: Nucl. Fusion **50**, 115005 (2010)
5. Loarte, A.: Plasma Phys. Control. Fusion **45**, 1549 (2003)

Research on Remote Dance Motion Capture Evaluation System and Dance Injury Prevention Based on Intelligent Terminal

Chen Li[1](✉) and Yiyuan Yang[2]

[1] School of Physical Education and Health, Zhaoqing University, Zhaoqing 526061, Guangdong, China
gdszqxy123123@126.com

[2] Music College of Zhaoqing University, Zhaoqing 526061, Guangdong, China

Abstract. The design of intelligent terminal dance movement supplement and optimization system can systematically analyze the 3 D digital form of dance. And in the subsequent process of folk dance style presentation, we can better understand the characteristics of similar artistic style. This study analyzes the intelligent terminal action supplement technology and dance damage control technology, and shows that the motion capture technology can effectively prevent various dance style problems, produce strong economic benefits in the process of 3 D digitalization promotion, and realize the presentation and optimization of the subsequent dance art style content.

Keywords: intelligent terminal · dance motion capture · dance injury

1 Introduction

Based on the development, in the current process of art style presentation, the dance system and elements can realize the optimization of its intelligent terminal structure operation system, and can better understand the presentation characteristics of intelligent terminal art [1]. Make a systematic analysis of the dance art mode and its construction content, understand the development characteristics of dance art, and in the process of comprehensive digitalization and three-dimensional control of motion capture technology in the movement supplement, so as to make a good analysis for the subsequent inheritance of dance art content. To realize the optimization of intelligent terminal supplement technology, but also in the process of instance control, to strengthen the data structure operation, to create a new open space for the subsequent modern intelligent system terminal adjustment, but also to realize the comprehensive optimization of intelligent dance and artistic style.

J. C. Hung et al. (Eds.): FC 2023, LNEE 1132, pp. 477–482, 2024.
https://doi.org/10.1007/978-981-99-9538-7_74

2 Three-Dimensional Design of Dance Art

2.1 3D Data Acquisition of Dance

For dance 3 d data acquisition and analysis, it is based on dance capture technology, systematic analysis of dance three data structure, and then according to the concept of dance mapping, for motion capture type, capture mode and so on capture system optimization, understand the development of dance capture technology development view, also create open space for other art synthesis. In the development of motion capture technology, this becomes a comprehensive processing technology [2]. Through the three-dimensional structure of dance art, the camera and standard dance range for acquisition and control. Subsequently, multiple dance action map column forms are used to map the spatial position there. When the remote operation analysis can be made. Similarly, it can be further reconstructed according to the dancer's point of action and his action diagram, understand the three-dimensional data obtained from the dancer, and finally carry out modeling and structural analysis.

2.2 Intelligent Terminal and Dance Injury Prevention

In the process of building the intelligent terminal system, based on the data capture of each point point, from the same perspective of the dance capture system and the construction content of the dance injury prevention system, the content of the data integration can be systematically manually analyzed. Make automatic patching during the manual channel control process [3]. Data repair mode can be composed of two forms, one is to use the own software of the acquisition equipment to process the noise point, and apply the smooth equipment to make the analysis, the other is to use the animation software to supplement and optimize the keyframes.

3 Associated Imaging Reconstruction Algorithm Based on Web and Intelligent Terminal

A pseudo-heat source is a pseudo-heat light source. Light from the beam splitter 1:1. After the transmission, it reaches the target and the CCD, respectively. According to the status of the second-order expression work, for the core of the intelligent terminal correlation imaging technology, it is mainly a light input by the light in the object. After the prism transmission, it can constitute the same optical system output data and target. The overall relationship between the two is positively correlated. According to the principle of the reconstruction algorithm, part of it is the data processing system, and the different operations [4]. The results derive the second-order correlation from its GI algorithm, which is as follows:

$$T_{GI}(x, y) = \frac{1}{N}\sum_{n=1}^{N}(B_n - \langle B_n\rangle)(I_n(x, y) - \langle I_n(x, y)\rangle) \tag{1}$$

where N is the indicating the light field of the reference arm in the n th acquisition, $B_n I_n(x, y)$

$$\langle B_n \rangle = \frac{1}{N}\sum_{n=1}^{N} B_n \tag{2}$$

$$\langle I_n(x, y) \rangle = \frac{1}{N}\sum_{n=1}^{N} I_n(x, y) \tag{3}$$

In the calculation process of the imaging formula, the overall principle is still very simple. It is based on the reconstruction algorithm to establish a common function relationship between the two arms. Therefore, when the light intensity in an object ratio is at a fixed value, the data of the other reference arm must also be a fixed value, and the weighted average of the two is the value we need to solve. During the course of this study, the weight factors in were essentially the same, and was correlated. The of the target arm constitutes its pixel information, otherwise, the reconstruction results are also very vague [5]. The of the DGI is as follows:

$$T_{DGI}(x, y) = \frac{1}{N}\sum_{n=1}^{N}\left(B_n - \frac{R_n}{\langle R_n \rangle}\langle B_n \rangle\right)(I_n(x, y) - \langle I_n(x, y) \rangle) \tag{4}$$

The the total light intensity with the reference arm based on the GI formula $\langle B_n \rangle R_n / \langle R_n \rangle$,

$$R_n = \int\int I_n(x, y)dxdy \tag{5}$$

The above formula represents the total light intensity at the n th acquisition of the reference arm. $\langle R_n \rangle = \frac{1}{N}\sum_{n=1}^{N} R_n$ Represents the n th average of the total light, and the reconstruction results have a higher signal-to-noise ratio than the conventional GI reconstruction results:

$$\frac{(SNR)_{DGI}}{(SNR)_{GI}} = 1 + \frac{1}{\sigma_{rel}^2} \tag{6}$$

The SNR represents the SNR of the imaging results. The higher the SNR, the better the subsequent quality. σ_{rel}^2 Represents ote the relative variance of the target related to the transmission coefficient of the imaging target. It is related to the magnitude of the transmission coefficient of the imaging target. $T(x, y)$ DGI far outperforms GI as high transmission coefficient objects, and DGI reconstruction is similar to GI. By comparing the reconstruction formulas of GI and DGI, Eqs. (1) and (4) show that when light passes through the object, it will become very small, which gives the DGI correction little effect to the overall effect, so that the imaging effect is not improved $\langle B_n \rangle$.

The reconstruction formula is as follows:

$$T_{NGI}(x, y) = \frac{1}{N}\sum_{n=1}^{N}\left(\frac{B_n}{R_n} - \frac{\langle B_n \rangle}{\langle R_n \rangle}\right)(I_n(x, y) - \langle I_n(x, y) \rangle) \tag{7}$$

The formulae of GI, based on the basic theory of association imaging, the second-order correlation formula, computed by weighted average only, with low computational complexity and strong robustness.

Concept can be computed perspective. Considering the two-arm structure of the correlation, the is associated with the object, such as formula _(1), which as the matrix form:

$$B = \Phi T \tag{8}$$

among $T = \left[T(1, 1), T(1, 2), \cdots, T(p, q)\right]^T$

$K \times 1$ The vector representing the coefficient is calculated according to the coefficient data to represent the composed of the of the object arm end. $B = [B_1, B_2, \cdots, B_N]^T N \times 1$.

Φ $N \times K$ It is called the observation matrix. The data in the parallel data in the collection arm field, and forms the relevant column item vector:

$$\Phi = \begin{bmatrix} I_1(1,1) & I_1(1,2) & \cdots I_1(2,1) & \cdots & I_1(p,q) \\ I_2(1,1) & & & & \vdots \\ \vdots & & & \ddots & \\ I_N(1,1) & I_N(1,2) \cdots & \cdots & \cdots & I_N(p,q) \end{bmatrix} \tag{9}$$

Based on the correlation, a PGI (8)$AX = Y$. The phase matrix constitutes a linear equation matrix relation, existing in form. The solution objective is transformed into the solution problem of the matrix equation. When the observed matrix is neither a full-rank matrix, nor a square matrix, nor an inverse. At this point, it can be to represent the of the matrix. $A^{-1}X = A^{-1}Y \Phi X = A^{+}YA^{+}$.

The architecture correlation imaging is shown in Fig. 2. Whose expressed as:

$$T_{PGI} = \Phi^{+}B \tag{10}$$

The GI reconstruction formula can also be expressed in the matrix form:

$$\begin{aligned} T_{GI} &= \frac{1}{N}(\Phi - I\langle\Phi\rangle)^T(B - I\langle B\rangle) \\ &= \frac{1}{N}(\Phi - I\langle\Phi\rangle)^T(\Phi - I\langle\Phi\rangle)T \\ &= \frac{1}{N}\Psi^T\Psi T = A_{GI}T \end{aligned} \tag{11}$$

$\langle\Phi\rangle = \left[\langle I(1,1)\rangle, \langle I(1,2)\rangle, \cdots, \langle I(p,q)\rangle\right]$ $1 \times K$ Where, it is the row vector, representing the average of each pixel of the reference arm, and it is a dimensional column vector with all elements of 1. The matrix can finally be called a feature matrix of GI by linking the reconstruction result to the coefficient T. $I = \left[1, 1, \cdots, 1^T\right] N \times 1 A_{GI} T_{GI} A_{GI}$.

A_{PGI} In order to unify, establish the connection formula the reconstructed matrices, and obtain the feature matrix of PGI, the be changed to writing in a similar form:

$$T_{PGI} = \Phi^{+}\Phi T = A_{PGI}T \tag{12}$$

Scalar scalar matrix correlation imaging reconstruction algorithm.

In obtain the corresponding the reconstruction formula, it must be studied according to the correlation characteristics of the reconstruction proof. In view of the reconstruction effect, the matrix is analyzed according to the correlation status of the matrix itself. In fact, the subsequent. To improve, a scalar matrix-based correlation imaging reconstruction algorithm smgi (Table 1):

$$T_{SMGI} = A_{SMGI}T = \frac{1}{N}\left(\psi^T - \psi_x\right)\psi T = \frac{1}{N}\left(\psi^T - \psi_x\right)(B - I\langle B\rangle) \tag{13}$$

ψ_x Where, for the correction quantity, It can be reversed according to the constructed feature matrix: $A_{SMGI}\psi_x$

$$\psi_x = \left(\psi^T\psi - A_{SMGI}\right)/\psi \tag{14}$$

$$\tilde{y}_i = y_i - \sum_{j=1}^{i-1} c_{ij}\tilde{y}_j \tag{18}$$

$$\tilde{y}_i{}' = \frac{\tilde{y}_i}{\left\|\tilde{R}_i\right\|_2} \tag{19}$$

The formula of the SGI is follows:

$$T_{SGI} = \frac{1}{N}\left(\tilde{\Phi}' - \left\langle\tilde{\Phi}'\right\rangle I\right)^T\left(\tilde{B}' - \left\langle\tilde{B}'\right\rangle I\right) = A_{SGI}T \tag{20}$$

4 Experiments and Results

In the construction process of the dance motion capture system and the dance injury prevention system based on the intelligent terminal, the various data are operational analyzed, and the design points and operation points can be better optimized according to the motion capture system evaluation and the design of the dance injury prevention mechanism. It can also complete the system design in the dance capture motion control and damage adjustment, so that the subsequent dance damage system content can really improve, improve the system synthesis, and realize the remote operation of the capture system and the structural end design breakthrough.

In the process of intelligent terminal language form control, the motion capture technology is shown in Table 2. In a systematic analysis of the supplementary content of each technology, the following operations need to be completed at the same time.

(1) Use the interpolation to calculate the bone and the relative bone position, and transform the matrix.
(2) Analyze the new orientation and grid model for the base position of the bone structure.

Table 1. Scale coefficient between the j vectors

Code name	Coefficient of proportionality
1	2.3
2	2.3
3	2.6

Table 2. Motion capture techniques

Point of capture	Camera 1	Camera 2
An air	X1	X 1
action	X 1	X 2
expression	X 2	X 2

5 Conclusion

To strengthen the experimental structure design of intelligent terminal, this research makes a systematic analysis of the traditional photo and video recording methods, and obtains the exploration perspective on the development of dance art. In the process of applying computer motion capture technology and dance injury prevention and supplement control technology, the data structure of intelligent terminal and the operation guidance function are analyzed, which is a new view of the mutual integration of culture and technology, and also promotes the reform and breakthrough of digital construction project.

Acknowledgements. 2021 Guangdong Provincial Philosophy and Social Science Planning Project Youth Project: Project No.: GD21YYS14.

References

1. Zhong, Y.X., Xu, W., Yao, R., et al.: Research and design of motion capture and image reduction of dance. (2019–9), 37–42 (2021)
2. Yuanzhi, Y.: Research on tibetan guozhuang dance based on motion capture. China New Commun. **22**(19), 239–240 (2020)
3. Guoliang, Y.: Research on dance video motion recognition techniques based on motion capture. J. Chifeng Univ. **38**(9), 5 (2022)
4. Dong, L., Jianpeng, Z.: Research on portable intelligent power distribution integrated mobile terminal control system based on STM32. Electron. Des. Eng. **30**(20), 7 (2022)
5. Zongbin, L.: Research on 3 D digital method based on motion capture technology. Fashion Tomorrow **02**, 140–141 (2020)

Author Index

J. C. Hung et al. (Eds.): FC 2023, LNEE 1132, pp. 483–485, 2024.
https://doi.org/10.1007/978-981-99-9538-7